# THE HANDBOOK OF MANUFACTURING ENGINEERING
## Second Edition

# Parts Fabrication
## Principles and Process

T0179471

## THE HANDBOOK OF MANUFACTURING ENGINEERING
### Second Edition

# Parts Fabrication
## Principles and Process

Edited by
# Richard Crowson

## CRC Press
Taylor & Francis Group
Boca Raton  London  New York

CRC Press is an imprint of the
Taylor & Francis Group, an **informa** business

CRC Press
Taylor & Francis Group
6000 Broken Sound Parkway NW, Suite 300
Boca Raton, FL 33487-2742

First issued in paperback 2019

© 2006 by Taylor & Francis Group, LLC
CRC Press is an imprint of Taylor & Francis Group, an Informa business

No claim to original U.S. Government works

ISBN-13: 978-0-8493-5554-7 (hbk)
ISBN-13: 978-0-367-39130-0 (pbk)
ISBN-13: 978-0-8247-2341-5 (set)

**Visit the Taylor & Francis Web site at**
**http://www.taylorandfrancis.com**

**and the CRC Press Web site at**
**http://www.crcpress.com**

**Library of Congress Cataloging-in-Publication Data**

Catalog record is available from the Library of Congress

# Preface

Handbooks are generally considered to be concise references for specific subjects. Today's fast-paced manufacturing culture demands that such reference books provide the reader with "how-to" information with no frills. Some use handbooks to impart buzzwords on a particular technical subject that will allow the uninitiated to gain credibility when discussing a technical situation with more experienced practitioners.

The second edition of *Handbook of Manufacturing Engineering* was written to equip executives, manufacturing professionals, and shop personnel with enough information to function at a certain level on a variety of subjects. This level is determined by the reader.

Volume 3 of this series is a refresher of some of the most forgotten aspects of the design engineer's process. The areas deal with engineering design fundamentals, free-body diagrams, stresses, forces and strength of materials help the manufacturing engineer to understand counter-intuitive problems and help the manufacturing engineer solve these problems. This type of training assists the engineer in understanding issues that are not obvious to the senses and must be anticipated in order to develop a functioning machine resistant to stresses induced during normal operation. This book deals with the fact that, in many cases, engineers do not properly convey the design intent to the worker assembling or fabricating the parts of a machine. This lack of information unknowingly will oftentimes induce stresses that may not be overcome and may result in a machine that operates at a lower standard than desired.

After a machine is designed and the first prototype is built, a team of technicians will start the machine for the first time. When the initial startup of a machine is performed, a specialized task must occur to achieve the desired result from that machine. This specialized task is called the *debug process*. For the purposes of this book, *debug* is defined as the methodical procedure of understanding and eliminating problems encountered when initially starting up a machine.

Oftentimes, design limitations are discovered that require immediate attention from engineers and technicians to perform the machine debug task. If the debug process is not efficiently performed, the machine may not perform to the designer's expectations. An improperly performed debug process may also result in higher product cost and may cause the product to reach the market later than expected. This may negatively affect the competitiveness of the product by affording the competition an advantage of being first to market with an innovation.

The main point of this book is to suggest efficient deductive and systematic approaches to machine debugging. A second aim of this book is to introduce the concept that models of efficient communication between engineers and machine-building technicians during the machine debug process must be developed.

Both techniques may result in higher productivity and a decrease in the time required for production ramp-up. These recommendations are intended to promote lower product-development costs through improved efficiency in machine debugging.

This book recommends steps to improve technical problem solving, which may assist the design engineer and the machine builder in the successful execution of their tasks and may be learned by applying engineering design basics to the problems of fabrication and assembly. Specific guidelines for improving communication techniques are also recommended.

The problem faced by all manufacturing engineers tasked with introducing a new design to the manufacturing floor has at least two parts. The first part of the problem is the need for the design engineer to employ immediate engineering support to solve all problems encountered while guiding machine-building technicians and assemblers in the implementation of design goals through deductive debug techniques during the debug phase of product development. In other words, the manufacturing engineer must encourage the design engineer to take an active role in the assembly process, or the manufacturing engineer must fill that role in the absence of the design engineer.

The second part of the problem is that the engineer, either manufacturing or design, must develop ways to succinctly communicate the design engineer's goals for machine debug to the machine builder. The solution to both parts of this problem incorporates an understanding of deductive reasoning, systematic engineering, human interaction, psychology, and corporate cultural influences.

**Richard D. Crowson**
SET, CMfgT, CMfgE

# Editor

*Richard D. Crowson*

Richard Crowson is currently a mechanical engineer at Controlled Semiconductor, Inc., in Orlando, Florida. He has worked in the field of engineering, especially in the area of lasers and in the development of semiconductor manufacturing equipment, for over 25 years. He has experience leading multidisciplinary engineering product development groups for several Fortune 500 companies as well as small and start-up companies specializing in laser integration and semiconductor equipment manufacture.

Crowson's formal engineering training includes academic undergraduate and graduate studies at major universities including the University of Alabama at Birmingham, University of Alabama in Huntsville, and Florida Institute of Technology. He presented and published technical papers at Display Works and SemiCon in San Jose, California.

He has served on numerous SEMI task forces and committees as a voting member. His past achievements include participating in writing the SEMI S2 specification, consulting for the 9th Circuit Court as an expert in laser welding, and sitting on the ANSI Z136 main committee that regulates laser safety in the United States.

# Contributors

**Frank Altmayer**
Scientific Control Laboratories, Inc.
Chicago, Illinois

**Richard D. Crowson**
Melbourne, Florida

**John F Maguire**
Materials and Structures Division,
Southwest Research Institute
San Antonio, Texas

**Robert E. Persson**
EG&G
Cape Canaveral, Florida

**Lawrence J. Rhoades**
Extrude Hone Corporation
Irwin, Pennsylvania

**Thomas J. Rose**
Advance Processing Technology/
Applied Polymer Technology, Inc.
Norman, Oklahoma

**V. M. Torbilo**
Ben-Gurion University of the Negev
Beer-Sheva, Israel

**Jack M. Walker**[†]
Merritt Island, Florida

**Don Weed**
Southwest Research Institute
San Antonio, Texas

**Bruce Wendle**
Boeing Commercial Airplane Company
Seattle, Washington

† Deceased.

# Contents

# 1 Principles of Structural Mechanics

*Jack M. Walker*

## 1.0 INTRODUCTION TO MECHANICS

This chapter is intended for those who may not have an extensive background in structural mechanics, or who have not used some of that knowledge recently and need a refresher. The working manufacturing engineer today is involved in product design, tool design, process selection and development (both parts fabrication and assembly), materials, and many other areas that require an analytical evaluation for proper understanding. In the field of metallurgy, for example, terms such as *stiffness, stress, strength, elasticity,* and so forth are the language involved—and sometimes we tend to use some of the terms incorrectly. In the fabrication of sheet metal parts, the elongation and yield strength are key elements. In machining, shear may play an important role in understanding cutting principles and selecting the proper cutting tool for a specific material.

To understand the discussion of problems in materials and processes, it is essential to know the basic principles of forces in equilibrium. Not all of the elements of today's complex structural considerations are introduced here. Rather, the intent is to establish a set of definitions and relationships that are needed to understand terms such as *tension, shear, compression,* and *bending.* In order to use terms such as *stiffness, yield strength, deflection, stress,* and *strain* correctly, the author feels that mathematical relationships offer the only logical approach to definition and understanding. Care has been taken to avoid the use of advanced mathematics; knowledge of arithmetic and high school algebra is all that is needed. It is the desire of this author to present some simple solutions to everyday structural problems to those individuals with little or no knowledge of mechanics.

## 1.1 DEFINITIONS OF FORCE AND STRESS

### 1.1.1 Force

*Force* may be defined as that which exerts pressure, motion, or tension. We are concerned here with forces at rest or in equilibrium. If a force is at rest, it must be held so by some other force or forces. As shown in Figure 1.1, a steel column in a building structure supports a given load, which due to gravity is downward. The column

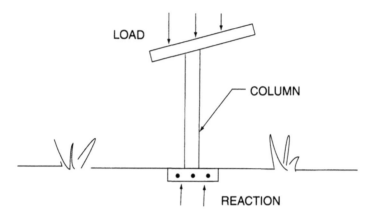

**FIGURE 1.1**    Example of forces in equilibrium. The load on the column
from the roof is reacted by the footing.

transfers the load to the footing below. The resultant upward pressure on the footing
equals the load in magnitude and is called the *reaction*. The two forces are opposite
in direction, have the same line of action, and are equal in magnitude. The system is
in equilibrium; that is, there is no motion.

The unit of force is usually pounds or kilograms. In practice, the word *kip*, meaning "a thousand pounds," is frequently used. Thus, 30 kips might also be written
30,000 lb.

## 1.1.2   Stress

Assume that a short column has a load of 100,000 lb applied to its end (see Figure 1.2a).
The load, $P$, is evenly distributed over the cross section X-X. To calculate the area of the
cross section in Figure 1.2b

$$\text{Area} = 3. \times 3.33 = 10 \text{ in.}^2$$

Or, if it was a circular cross section, as shown in Figure 1.2c, in

$$\text{Area} = \pi r^2$$

Where r (radius) = 3.568/2 = 1.784 in., or

$$A = 3.14 \times (1.784)^2 = 10 \text{ in.}^2$$

In the example, the load $P$ is evenly distributed over the cross-sectional area $A$.

We can say that 100,000 lb is distributed over 10 in.$^2$, or 100,000/10 = 10,000 lb
acting on each square inch. In this instance, the unit stress in the column is 10,000
lb/in.$^2$ (psi).

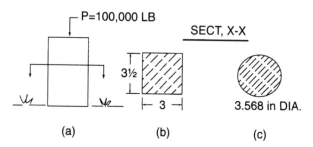

**FIGURE 1.2**  Example of the compressive stress in a short column dependent on the load applied and the column cross section.

A *stress* in a body is an internal resistance to an external force, or the intensity of the forces distributed over a given cross section.

## 1.2   TYPES OF STRESSES

The three primary stresses that we will discuss are *tension, compression,* and *shear.* Unless noted otherwise, we will assume that forces are axial and that the stresses are uniformly distributed over the cross-sectional area of the body under stress.

We normally call the load, or external force, $P$; the area of the cross section $a$; and the unit stress $f$. As discussed above, the load divided by the section area will give the unit stress. This is stated as a fundamental principle:

$$f = P/A, \text{ or } P = Af, \text{ or } A = P/f$$

### 1.2.1   Deformation

Whenever a force acts on a body, there is an accompanying change in shape or size of the body. This is called *deformation*. Regardless of the magnitude of the force, the deformation is always present, although it may be so small that it is difficult to measure even with the most delicate instruments. It is often necessary to know what the deformation of certain members will be. For example, a floor joist in the second story of the house shown in Figure 1.3 may be large enough to support a given load safely, but it may deflect (or deform) to such an extent that the floor will vibrate or bend and cause the plaster in the ceiling below to crack.

### 1.2.2   Tension

When a force acts on a body in such a manner that the body tends to lengthen or pull apart, the force is called *tensile*. The stresses in the bar due to the tensile force $P$ are called *tensile stresses* (see Figure 1.4).

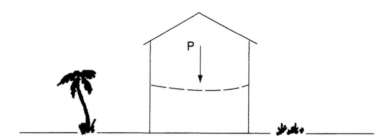

**FIGURE 1.3**    Example of the deformation of a floor joist (beam) under load (deflection).

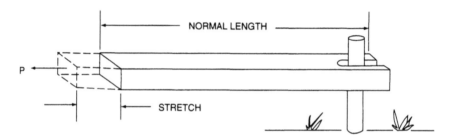

**FIGURE 1.4**    Example of the deformation in a bar under tension.

## Example

A wrought-iron bar with a diameter of 1 1/2 in. is used in a roof truss. If the tensile force supported is 20,000 lb, what is the unit stress? (See Figure 1.5.)

## Solution

To find the cross-sectional area of the bar, we square the radius and multiply by 3.1416, $A = \pi r^2$

$$A = 3.1316 \times (0.75)^2 = 1.76 \text{ in.}^2$$

The load, $P$, is 20,000 lb. These are the data; what we are looking for is the unit stress. The formula giving the relation among these three quantities is $f = P/A$

$$f = 20,000/1.76 = 11,363 \text{ psi, the unit stress in the bar.}$$

### 1.2.3   Compression

When the force acting on a body has a tendency to shorten it, the force is called *compressive*, and the stresses within the member are *compressive stresses*. (See Figure 1.6.)

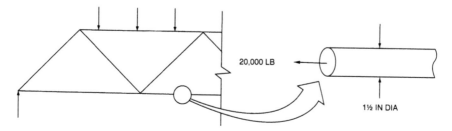

**FIGURE 1.5**   Wrought-iron bar in a roof truss under tension.

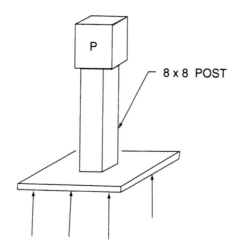

**FIGURE 1.6**   Short timber post under compression due to load P.

## Example

Suppose we have a short timber post with a cross section of 7 1/2 in. × 7 1/2 in. This cross section has an allowable stress of 1000 psi and we wish to know what load it will safely support.

## Solution

The cross-sectional area, A, is 7 1/2 × 7 1/2, or 56.25 in.² From the data, the allow-able unit stress for this timber, $f$, is 1000 psi. Substituting in the formula $P = Af$, we find:

$$P = 56.25 \times 1000, \text{ or } 56{,}250 \text{ lb}$$

Which is the maximum safe load?

*Note:* This example illustrates the definition of pure compressive strength. When the member in compression is relatively short in relation to its cross section, the above is correct. However, as the length increases or the member becomes more "slender" in relation to its cross section, it may buckle before it actually fails in compression. This long-column failure is discussed in Subchapter 1.6.

### 1.2.4   Shear

A *shearing stress* occurs when we have two forces acting on a body in opposite directions, but not on the same line. Forces acting like a pair of scissors, tending to cut a body, are an example. Figure 1.7 shows two plates held together by a rivet. The forces, P, acting on the plates tend to shear the rivet on an area equal to the cross-sectional area of the rivet at the plane of contact between the two plates.

### Example

The forces, P, on the plates shown in Figure 1.7 are each 5000 lb, and the rivet has a diameter of 3/4 in. What is the shearing unit stress?

### Solution

A 3/4 in. rivet has a cross-sectional area of 0.4418 in.$^2$. The basic formula for shearing stress is $\tau = P/a$. By substituting the known quantities,

$$\tau = 5000/0.4418 = 11,317 \text{ psi}$$

which is the average shearing stress in the rivet. The load, W, rests on a beam that is supported on walls at its ends. There is a tendency for the beam to fail by shearing at points C and D (see Figure 1.8).

### 1.2.5   Bending

Figure 1.9 illustrates a simple beam with a concentrated load P at the center of the span. This is an example of bending or flexure. The fibers in the upper part of the beam are in compression, and those in the lower part are in tension. These stresses are not equally distributed over the cross section. (A more complete discussion will be presented later in this chapter.)

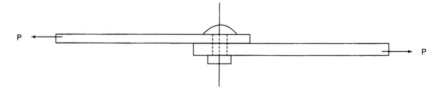

**FIGURE 1.7**   Example of shearing forces on a rivet caused by the forces in the plates.

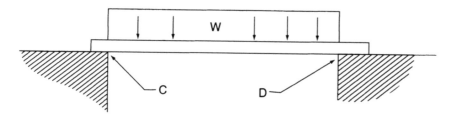

**FIGURE 1.8** Example of shearing forces on a beam at points C and D due to distributed load W.

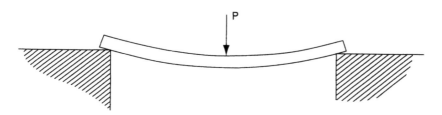

**FIGURE 1.9** Bending of a beam due to concentrated load P.

## 1.3 HOOKE'S LAW

Robert Hooke was a mathematician and physicist living in England in the seventeenth century. As a result of experiments with clock springs, he developed the theory that deformations are directly proportional to stresses. In other words, if a force produces a certain deformation, twice the force will produce twice the amount of deformation. This law of physics is of the utmost importance, though unfortunately Mr. Hooke did not carry his experiments quite far enough, for it may be readily shown that Hooke's law holds true only up to a certain limit.

### 1.3.1 Elastic Limit

Suppose that we place a bar of structural steel with a section area of 1 in. × 2 in. in a machine for making tensile tests (see Figure 1.10). We measure its length accurately and then apply a tensile force of 5,000 lb. We measure the length again and find that the bar has lengthened a definite amount, which we will call $X$ in. On applying 5,000 lb more, we note that the amount of lengthening is $2 \times X$, or twice the amount noted after the first 5,000 lb. If the test is continued, we will find that for each 5,000 lb, the length of the bar will increase the same amount noted when the first unit of 5,000 lb was applied; that is, the deformations are directly proportional to the stresses. So far Hooke's law has held true, but (and this is the significant point) after we have applied about 36,000 lb, the length increases more than $X$ in. for each additional 5,000 lb. This unit stress, which varies with different materials, is called the *elastic limit*. The

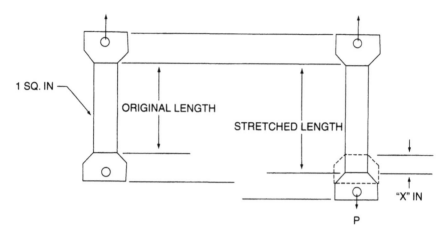

**FIGURE 1.10** Increased length of a steel bar due to applied tensile force P.

*proportional limit* is the largest value of the stress for which Hooke's law may be used. It may be defined as the unit stress beyond which the deformations increase at a faster rate than the applied loads.

Here is another phenomenon. If we make the test again, we will discover that if any applied load less than the elastic limit is removed, the bar will return to its original length. If the unit stress greater than the elastic limit is removed, we will find that the bar has permanently increased its length. This deformation is called the *permanent set,* or plastic deformation. This fact permits another way of defining the elastic limit: unit stress beyond which the material does not return to its original length when the load is removed. The property that enables a material to return to its original shape and dimensions is called *elasticity.*

Another term used in connection with these tests is the *yield point.* It is found during tests that some materials—steel, for instance—show increases in deformation without any increase in stress; the unit stress at which this deformation occurs is called the *yield point.* The yield point, although slightly higher than the elastic limit, is almost identical to the elastic limit. Nonductile materials, such as cast iron, have poorly defined elastic limits and no yield point.

## 1.3.2   Ultimate Strength

If a load of sufficient magnitude is applied to a test specimen, rupture occurs. The unit stress within the bar just before it breaks is called the *ultimate strength.* For the steel bar referred to earlier, failure occurs at a unit stress of about 70,000 psi.

## 1.3.3   Stress and Strain

We pointed out earlier that stress always implies a force per unit area and is a measure of the intensity of the force. *Strain* refers to the elongation per unit length of a member in stressed condition. *Strain* should never be used in place of the terms *elongation* and *deflection*. Most of the above terms are best shown on a stress–strain diagram (see Figure 1.11).

### 1.3.4 Modulus of Elasticity

We have seen that if a bar is subjected to a force, a deformation results. Also, if the unit stress in the bar does not exceed the elastic limit of a material, the deformations arc in direct proportion to the stresses. The key to computing the magnitude of the deformation lies in the stiffness of a material. The number that represents the degree of stiffness of a material is known as the *modulus of elasticity*. We represent this quantity by the letter $E$ and define it as the unit stress divided by the unit deformation (strain), or

$$E = f/s$$

Where

$E$ = modulus of elasticity
$f$ = unit stress
$s$ = unit deformation

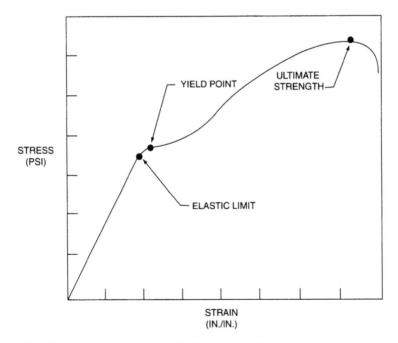

FIGURE 1.11  Typical stress–strain diagram for mild steel.

## Example

Suppose we place a steel bar with a 1 in. × 2 in. cross section in a testing machine and apply a tensile force of 1000 lb. Its length becomes greater (although we cannot see it with the naked eye). If we apply the same force to a piece of wood having the same dimensions as the steel bar, we find that the deformation is greater—probably 20 times greater. We say that the steel has a greater degree of stiffness than the wood (see Figure 1.12).

From our discussion on unit stress earlier, we saw that

$$f = P/A$$

And, from Figure 1.12, where $L$ represents the length of the member and e the total deformation, s the deformation per unit of length would equal the total deformation divided by the original length, or

$$s = e/L$$

Now, since the modulus of elasticity, $E = f/s$, this becomes

$$E = \frac{P/A}{e/L} \, or \, E = \frac{P \times L}{Ae}$$

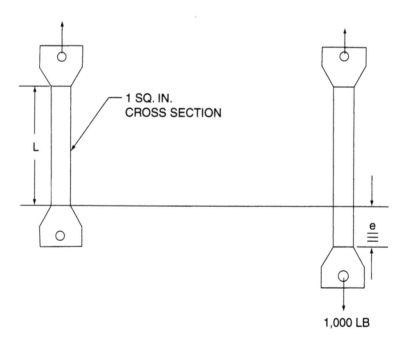

**FIGURE 1.12** Example of stress and strain for calculation of modulus of elasticity E.

which can also be written as

$$e = PL/AE$$

where

$e$ = total deformation in inches
$P$ = force in pounds (axial load)
$L$ = length in inches
$A$ = cross-sectional area in inches
$E$ = modulus of elasticity in pounds per square inch.

Remember that all of the above is valid only when the unit stress does not exceed the elastic limit of the material. Figure 1.13 gives typical values for some common materials.

## 1.4  MOMENTS

You have probably heard the term *moment* used with problems in engineering. A force of 100 lb, an area of 16 in.$^2$, or a length of 3 ft can readily be visualized. A moment, however, is quite different; it is a force multiplied by a distance. A moment is the tendency of a force to cause rotation about a certain point or axis. The moment of a force with respect to a given point is the magnitude of the force multiplied by the distance to the point.

The following examples may help explain moment.

### Example 1

Two forces are acting on the bar, which is supported at point $A$, as shown in Figure 1.14. The moment of force, $P_1$, about point $A$ is 8 ft × 100 lb, or 800 ft-lb. This force tends to produce clockwise rotation about point $A$ (the direction in which the hands of a clock revolve), called *positive moment*. The other force, $P_2$, has a lever

| MATERIAL | ELASTIC LIMIT (PSI) | | ULTIMATE STRENGTH (PSI) | | | MOD. OF ELASTICITY (PSI) | WEIGHT (LB/C) |
|---|---|---|---|---|---|---|---|
| | TENSION | COMPRESSION | TENSION | COMPRESSION | SHEAR | | |
| Structural Steel | 36,000 | 36,000 | 70,000 | 70,000 | 55,000 | 29,000,000 | 490 |
| 6061 Aluminium | 35,000 | 35,000 | 38,000 | 38,000 | 30,000 | 10,000,000 | 170 |
| Timber | 3,000 | 3,000 | 10,000 | 8,000 | 500 | 1,200,000 | 40 |
| (Perpendicular to grain) | | | | | 3,000 | | |

FIGURE 1.13  Average physical properties of some common materials.

arm of 4 ft with respect to point $A$, and its moment is 4 ft × 200 lb, or 800 ft-lb. $P_2$ tends to rotate the bar in the opposite direction (counterclockwise), and such a moment is called *negative*.

In Figure 1.14 the positive and negative moments are equal in magnitude, and equilibrium (or no motion) is the result. Sometimes it is stated: "If a system of forces is in static equilibrium, the algebraic sum of the moments is zero." In Example 1, if the system of forces is in equilibrium, the sum of the downward forces must equal the upward forces. The reaction at point $A$, therefore, will act upward and be equal to 200 lb plus 100 lb, or 300 lb. We could say: "If a system of parallel forces is in equilibrium, the algebraic sum of the forces is zero."

## Example 2

The beam shown in Figure 1.15 has two downward forces, 100 and 200 lb. The beam has a length of 8 ft between supports, and the supporting forces (called *reactions*) are 175 and 125 lb.

Check:

1. The sum of downward forces must equal the upward forces:

$$100 + 200 = 175 + 125$$

$$\text{or, } 300 = 300 \text{ (it's true)}$$

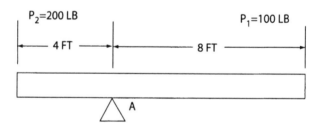

**FIGURE 1.14** Simple beam used to illustrate the moments about the support point A in Example 1.

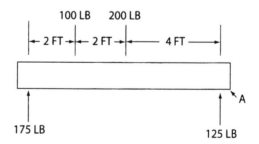

**FIGURE 1.15** Simple beam used to demonstrate calculation of moments in Example 2.

2. The sum of the moments of the forces tending to cause clockwise rotation (positive moments) must equal the sum of the moments of forces tending to produce counterclockwise rotation (negative moments), about any center of moments.

Check the moments about point A. The force tending to cause clockwise rotation about this point is 175 lb; its moment is 175 × 8, or 1400 ft-lb. The forces tending to cause counterclockwise rotation about the same point are 100 and 200 lb, and their moments are

$$(100 \times 6) + (200 \times 4) \text{ ft-lb}$$

Now we can write:

$$(175 \times 8) = (100 \times 6) + (200 \times 4)$$
$$1400 = 600 + 800$$
$$1400 \text{ ft-lb} = 1400 \text{ ft-lb (we lucked out again)}$$

*Note:* If you wonder where the force of 125 lb went in writing this equation, the 125 lb force has a lever arm of 0 ft about point A, and the moment of the force becomes $125 \times 0 = 0$. In future problems, when we write equations of moments, we can therefore omit writing the moment of the force acting through the point we have selected, because we know it can cause no rotation about the point, and its moment is zero.

## 1.5 BEAMS

A *beam* is a structural member resting on supports, usually at its ends, that supports transverse loads. The loads acting on a beam tend to bend it rather than lengthen or shorten it. In general, there are five types of beams, depending on the position and number of supports, as shown in Figure 1.16.

The two kinds of loads that commonly occur on beams are called *concentrated* (P) and *distributed* (W). A concentrated load acts at a definite point, while a distributed load acts over a considerable length. Both types are shown in Figure 1.16a. A distributed load produces the same reactions as a concentrated load of the same magnitude acting through the center of gravity of the distributed load.

### 1.5.1 Stresses in Beams

Figure 1.17a represents a simple beam. Examination of a loaded beam would probably show no effects of the load. However, there are three distinct major tendencies for the beam to fail. First, there is a tendency for the beam to fail by dropping down between the supports, as shown in Figure 1.17b. This is called *vertical shear*. Second, the beam may fail by bending, as shown in Figure 1.17c. Third, there is a

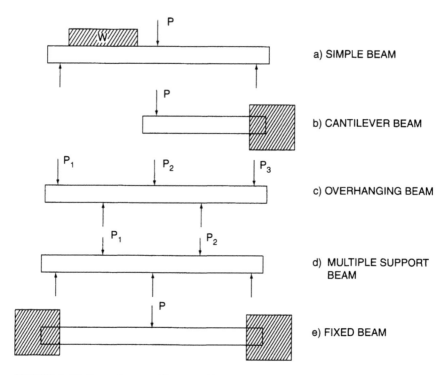

**FIGURE 1.16**  Several types of beams with concentrated (P) and distributed (W) loads.

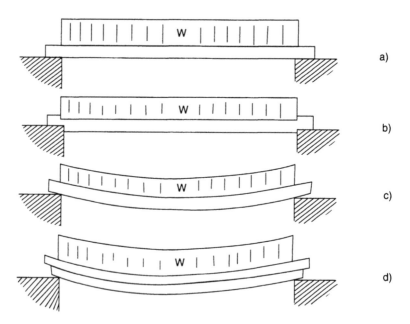

**FIGURE 1.17**  Example of a beam with a distributed load W showing vertical shear, bending, and transverse shear.

tendency for the fibers of the beam to slide past each other in a horizontal direction, as shown in Figure 1.17d. This is known as *horizontal shear,* or, in composites, *interlaminar shear.* Some of the most common beams and loads are shown in Figure 1.18. The values of $V$, the maximum shear; $M$, the maximum bending moments; and $D$, the maximum deflection, are given. If the loads are in units of pounds, the vertical shear ($V$) is also in pounds. When the loads are in pounds and the span is in feet, the bending moment will be in units of foot-pounds. Particular attention should be given to $l$ in the formulas for maximum deflection. In this case, the span length $l$ is given in inches—and the resulting deflection in inches.

## 1.5.2 Theory of Bending

Figure 1.19 shows a simple example of bending of a beam that is supported at each end with a load concentrated at the center. The load will cause the beam to deflect at the center. The deflection stretches the fibers on the lower surface and compresses those on the upper surface.

Somewhere between the compressive stresses in the upper fibers and the tensile stresses in the lower fibers is a place where there is neither. This is known as the *neutral axis* (NA), the location of which depends on the cross section of the beam. If a 2-in. × 4-in. bar, as seen in Figure 1.20, is used as the beam, we know that there will be a difference in the amount of deflection if the 2 × 4 is placed flat or if it is placed on edge. The stress due to this bending is not a uniform stress, but varies in intensity from a maximum at the extreme fiber to zero at the NA.

We discussed previously that the sum of the moments about any point on a beam must be zero—or, the positive moments (clockwise) must equal the negative moments (counterclockwise). If we cut the beam shown in Figure 1.21 at section X-X and look at the left end in Figure 1.22, we see the following.

Call the sum of the compressive stresses $C$, and all the tensile stresses $T$. The bending moment in the section of beam about point $A$ is equal to $R_1 \times X$. For our example to be in equilibrium, the *resisting moments* must be equal. The resisting moment about point $A$ is $(C \times Y) + (T \times Y)$. The bending moment tends to cause a

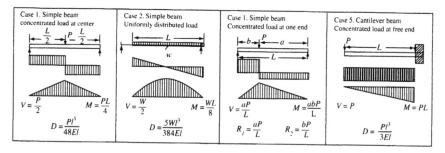

**FIGURE 1.18** Examples of calculation of moment, shear, and deflection for different types of beams.

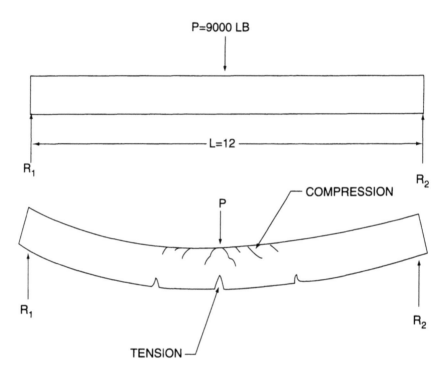

**FIGURE 1.19** Example of bending of a simply supported beam showing compression and tension of the outer surfaces.

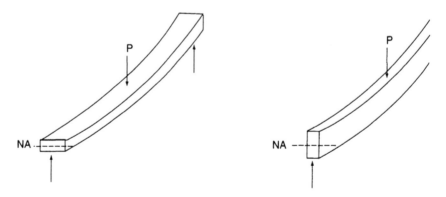

**FIGURE 1.20** Example of the deflection in a wood 2 × 4 beam turned flatwise and edgewise.

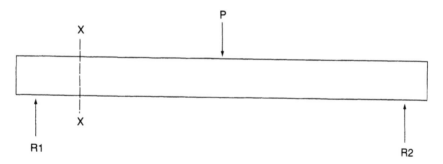

**FIGURE 1.21** Beam with a section cut X-X for analysis.

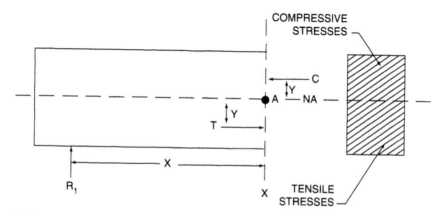

**FIGURE 1.22** Example of the compressive and tensile stresses at section X-X.

clockwise moment (+), and the resisting moment tends to cause a clockwise moment (−). We see that:

$$+ [R_1 \times X] - [(C \times Y) + (T \times Y)] = 0$$
$$\text{or } R_1 \times X = (C \times Y) + (T \times Y)$$

This is the theory of bending in beams. For any type of beam, we can compute the bending moment. If we wish to select (design) a beam to withstand this tendency to bend, we must have a member with a cross section of such shape, area, and material that the resisting moment will have a magnitude equal to the bending moment.

The maximum stress is given by the flexure formula

$$f_b = My / I$$

where

$f_b$ = bending stress
$M$ = applied bending moment
$y$  = distance from the neutral axis to the extreme fiber
$I$  = moment of inertia

This is also often expressed as

$$f_b = WS$$

where

$S = I/y$, called the section modulus.

### 1.5.3  Moment of Inertia

No attempt will be made to derive moment of inertia in this chapter, but we will discuss its use. We can say, however, that moment of inertia is defined as the sum of the products obtained by multiplying all the infinitely small areas ($a$) by the square of their distances to the neutral axis (see Figure 1.23).

For a rectangular beam, $I = bh^3\backslash12$. From our 2 × 4 beam example earlier, we can see the cross section and moment of inertia calculations, as shown in Figure 1.24 and Figure 1.25. From our flexure formula, $f_b = M/S$, we can see that the allowable bending stresses in our beam ($f_b$) due to moment M are as follows:

Flatwise 2 × 4      Edgewise 2 × 4
$f_b = M/2.7$        $f_b = M/5.35$

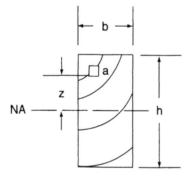

**FIGURE 1.23** Cross section of a beam used in calculation of moment of inertia, $I$.

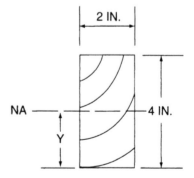

**FIGURE 1.24** Cross section of a 2 × 4 used in calculations of edgewise moment of inertia.

What all this really says is that the properties of a beam vary to a great degree with the size and shape of the cross section. In our example of the 2 × 4, the edge-wise 2 × 4 will withstand twice as much load as the same beam turned flat. Also, if we take the same cross-sectional area and convert it into a 1 × 8 beam on edge, we will see much more flexural strength than in a 1 × 8 turned flatwise.

The maximum stresses (both tension and compression) in a beam occur at the outer surfaces, which are the maximum distances from the NA. These stresses diminish toward the NA, where they are zero. The NA is at the *centroid*, or center of area, of any cross section (see Figure 1.26). Values (or formulas) for $I$, $y$, and $S$ can be obtained from a number of standard reference handbooks for standard member cross sections.

## 1.6  LONG-COLUMN FAILURE

When a column is subjected to compressive forces at its ends, one obvious but highly important result is that the distance between those ends is forcibly reduced. We discussed this in Subchapter 1.2. The action of a column under load can perhaps be most clearly understood by considering the implications of this shortening effect. One immediate result is that internal stresses are developed in each element of the column, their magnitudes depending on the amount and type of shortening involved, the elasticity of the material, and the original dimensions of the member, and their

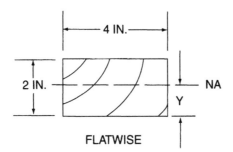

**FIGURE 1.25**  Cross section of a 2 × 4 used in calculation of flatwise moment of inertia.

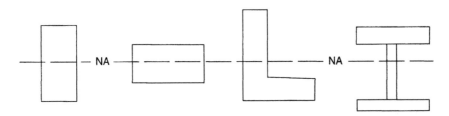

**FIGURE 1.26**  Location of the neutral axis of different cross sections used as beams.

resultant being such in any given column that end loads of definite magnitude will be held in equilibrium by these internal stresses.

With the ideal column—that is, one that is perfectly straight, perfectly homogeneous, and with the line of action of the end loads passing exactly through the centroid of each cross section—the action under load will correspond with the calculations we performed in Subchapter 1.2, on compression. The practical column never conforms completely to the characteristics of the ideal, unless it is a very stocky member, such as a cube, which cannot bend or twist appreciably under load. When this short column is subjected to loads within the elastic limit, such columns act like springs and will return to their original length after the load is removed. However, if the load is increased above this point and plastic flow occurs, the column will not return to its original length, due to the permanent deformation caused by plastic flow. Continued increase in load will eventually cause failure in a short column. In some cases of plastic flow failure of short columns, the member will be mashed out of all resemblance to its original shape but still remain in stable equilibrium. Since there is no true ultimate load in a case of this kind, failure is assumed to take place as the permanent deformation exceeds some arbitrary amount. In practice, this plastic flow failure is assumed to take place when the average compressive stress equals the column yield stress in Table 1.1. Columns, therefore, should be designed so that there will be no plastic flow (permanent deformation) under service conditions. It is convenient to call the end loads corresponding to any given shortening the "equilibrium load" for that shortening.

For a long, slender column, the imperfect column begins to exhibit different characteristics as the load is increased. Due to the column's imperfections, the degree to which its ends are held rigid or fixed, and the geometric shape of the column cross section, the long, slender column will tend to bend as well as shorten longitudinally. So long as the bending is not permitted to cause stresses beyond the proportional limit of the material at any point in the member, such columns continue to act as springs and will return to their original positions when the end loads are removed. However, if forcible movement of the ends toward each other continues beyond a certain point, the compressive strains of some of the fibers on the concave side of the column will exceed the proportional limit and there will be plastic flow of the material. At this point, as the load is increased, the long column will rapidly buckle and lose most (or all) of its structural integrity. We can visualize that the longer, or more slender, the column becomes, the less strength it will have in compression (due to buckling).

## 1.6.1   End Restraints

The load that a column can carry is influenced by the restraint imposed by the structure on the ends of the column. The tendency of the column to rotate or to move laterally would be different if the load were transmitted through frictionless pins or knife edges, as shown in Figure 1.27a, or if the ends were rigidly connected to a structure so stiff that the axis of the column was fixed in both direction and position, as shown in Figure 1.27b. In Figure 1.27a, the restraint coefficient, $c$, would be unity or 1, while

TABLE 1.1 Ultimate Allowable Stress for Various Metallic Elements

| Material | Tension Ultimate Strength | Tension Yield Strength[1] | Tension Modulus of Elasticity | Compression Block Compression | Compression Column Yield[2] | Compression Modulus of Failure | Tension Proper Limit | Tension Modulus of Rigidity | Tension Failure Limits[3] | Shear Ultimate Strength | Bending Modulus of Failure[4] | Bending Fatigue Limits[5] | Bearing Ultimate Strength[6] | Weight in Pounds Per Cu. Ft. | Weight in Pounds Per Cu. In. |
|---|---|---|---|---|---|---|---|---|---|---|---|---|---|---|---|
| Mild Steel, S.A.E. 1025 | 55 | 36 | 28,000 | 55 | 36 | 50 | 20 | 10,000 | — | 35 | 55 | 25 | 90 | 490 | 0.2833 |
| Alley Steel, not Heat-treated | 65 | 45 | 29,000 | 65 | 36 | 55 | 25 | 11,000 | — | 40 | 65 | 30 | 110 | 490 | 0.2833 |
| Chrome-Molyb. Normalized | 95 | 75 | 29,000 | 95 | 70.5 | 80 | 40 | 11,000 | — | 55 | 95 | 45 | 140 | 490 | 0.2833 |
| Chromine-Molyb. (X-4130) Welded after Heat-Treatment | 80 | 60 | 29,000 | 80 | 60 | 70 | 35 | 11,000 | — | 50 | 80 | 15 | 125 | 490 | 0.2833 |
| Alloy Steels, Heat Treated | 100 | 80 | 29,000 | 100 | 80 | 90 | 55 | 11,000 | — | 65 | 100 | 50 | 140 | 490 | 0.2833 |
| Alloy Steels, Heat Treated | 125 | 100 | 29,000 | 125 | 100 | 110 | 65 | 11,000 | — | 75 | 125 | 65 | 175 | 490 | 0.2833 |
| Alloy Steels, Heat Treated | 150 | 135 | 29,000 | 150 | 130 | 125 | 80 | 11,000 | — | 90 | 150 | 78 | 190 | 490 | 0.2833 |
| Alloy Steels, Heat Treated | 180 | 165 | 20,000 | 180 | 145 | 145 | 95 | 11,000 | — | 105 | 180 | 85 | 200 | 490 | 0.2833 |
| Alloy Steels, Heat Treated | 200 | 165 | 20,000 | 200 | 155 | 155 | 105 | 11,000 | — | 115 | 200 | 94 | 220 | 490 | 0.2833 |
| Corrosion-Resistant Steel | 125– | 65– | 26,000 | 125– | 50– | — | — | 10,000 | 30 | 90– | — | 75 | — | 490 | 0.2833 |
| Cold-Worked | 185 | 140 | 26,000 | 185 | 110 | — | — | 10,000 | 55 | 125 | — | — | — | 490 | 0.2833 |
| Annealed, nor Near Welding | 80 | 35 | 26,000 | 80 | 30 | — | — | 3,850 | — | 70 | — | — | — | 490 | — |
| 17-ST Aluminium Alloy, Street | 55 | 32 | 10,500 | 55 | 36 | 50 | 15 | 3,850 | — | 33 | 55 | 15 | 75 | 174 | 0.101 |
| Extruded Shapes | 50 | 32 | 10,500 | 50 | 32 | — | — | 3,850 | — | 30 | — | — | 75 | 174 | 0.101 |
| Unstretched Tube | 55 | 30 | 10,500 | 55 | 34.5 | 50 | 15 | 3,850 | — | 33 | 55 | 15 | 75 | 174 | 0.101 |
| Cold-Stretched Tube | 55 | 40 | 10,500 | 55 | 42.5 | 50 | 15 | 3,850 | — | 33 | — | — | 75 | 174 | 0.101 |
| 24-ST Aluminium Alloy, Sheet | 62 | 40 | 10,300 | 62 | 40 | — | — | 3,850 | — | 37 | 62 | 14 | 90 | 173 | 0.100 |
| Extruded Shapes | 57 | 42 | 10,500 | 57 | 42 | — | — | 3,850 | — | 34 | 57 | 14 | 83 | 173 | 0.100 |
| Unstretched Tube | 62 | 40 | 10,500 | 62 | 50 | — | — | 3,850 | — | 37 | 62 | 14 | 90 | 173 | 0.100 |
| Cold-Stretched Tube | 62 | 42 | 10,500 | 62 | 50 | — | — | 3,850 | — | 37 | 62 | 14 | 90 | 173 | 0.100 |
| 17-SRT Aluminum Alloy Sheet | 55 | 42 | 10,300 | 55 | 42 | — | — | 3,850 | — | 33 | — | — | 75 | 174 | 0.101 |
| 24-SRT Aluminum Alloy Sheet | 65 | 50 | 10,300 | 65 | 50 | — | — | 3,850 | — | 39 | 65 | — | 93 | 173 | 0.100 |
| 17-ST Alclad Sheet | 50 | 28 | 10,300 | 50 | 28 | — | — | — | — | 30 | — | — | 68 | 173 | 0.101 |
| 17-SRT Alclad Sheet | 50 | 37 | 10,300 | 50 | 37 | — | — | — | — | 30 | — | — | 68 | 173 | 0.101 |
| 24-ST Alclad Sheet | 56 | 37 | 10,300 | 56 | 37 | — | — | — | — | 34 | 56 | — | 82 | 173 | 0.100 |
| 24-SRT Alclad Sheet | 58 | 46 | 10,300 | 58 | 46 | — | — | — | — | 35 | 58 | — | 83 | 173 | 0.100 |

[1] Stress at which set is 0.002 in. per in.
[2] Nominal value for use in short column formulas.
[3] Maximum alternating torsional stress to withstand $20 \times 10^4$ cycles.
[4] Nominal value for shapes not subject to local buckling. See Art. 6:1 for round tubes.
[5] Maximum alternating bending stress to withstand $300 \times 10^6$ cycles for steel, $500 \times 10^5$ cycles for aluminum alloys.
[6] For connections involving no relative movement between parts.

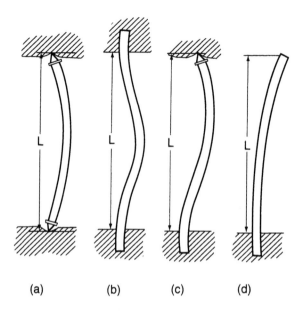

(a)            (b)          (c)            (d)

**FIGURE 1.27** Theoretical end restraint conditions for a
long column.

in Figure 1.27b, where we have a theoretically fixed-end column, the restraint coefficient of 4 might be applicable. In Figure 1.27c, the value of c would be 2.05, and in
Figure 1.27d, the value of c would be 0.25. All these conditions are purely theoretical
and apply only to ideal columns.

About two centuries ago, a Swiss mathematician, L. Euler, did some significant
work in this area. While his work is certainly not sufficient background in buckling
phenomena to use in sophisticated design problems, it does serve as a working model
to better understand the fundamentals.

Experience over a number of years has indicated that $c = 2$ may be used for the
design of tubular members where the joints are very rigid, as in welded structures,
and $c = 1$ for theoretical pin-ended columns.

If the end loads on a column are so large that they produce internal stresses in
excess of the elastic limit of the material, they will cause plastic as well as elastic
strain, and the member will not completely regain its original length when the external load is removed. Another possibility is that during part of the process of loading,
the external load may exceed the resisting force developed as a result of the elastic
strains. Thus the entire load, $P$, may be imposed on the loaded end of the column at
the start of the shortening process.

At this instant the elastic strain, and therefore the resisting stress, is zero. The
resisting force is built up as the column shortens and the elastic strain is produced,
and does not become equal to $P$ until the shortening 5′ is equal to $PL/AE$. This
makes the resulting work done greater than can be stored as strain energy, due to the

## TABLE 1.2 Column Formulas for Allowable Stress

| Material | c | Short Columns | Transitional $L/\rho$ | Long Columns |
|---|---|---|---|---|
| Spruce | 1 | $5{,}000 - 0.5\,(L/\rho)$ | 72 | $13\times10^{4}/(L/\rho)^{2}$ |
| | 2 | $5{,}000 - 0.25\,(L/\rho)$ | 102 | $26\times10^{4}/(L/\rho)^{2}$ |
| **Aluminum Alloy** | | | | |
| 17-ST Tubes Not Stretched | 1 | $34{,}500 - 245\,(L/\rho)$ | 94 | $101.6\times10^{4}/(L/\rho)^{2}$ |
| | 2 | $34{,}500 - 173\,(L/\rho)$ | 133 | $203.2\times10^{4}/(L/\rho)^{2}$ |
| 17-ST Tubes Cold-Stretched | 1 | $42{,}500 - 335\,(L/\rho)$ | 84.6 | $101.6\times10^{4}/(L/\rho)^{2}$ |
| | 2 | $42{,}500 - 237\,(L/\rho)$ | 120 | $203.2\times10^{4}/(L/\rho)^{2}$ |
| 24-ST Tubes Not Stretched | 1 | $50{,}000 - 427\,(L/\rho)$ | 78 | $101.6\times10^{6}/(L/\rho)^{2}$ |
| | 2 | $50{,}000 - 302\,(L/\rho)$ | 110 | $203.2\times10^{4}/(L/\rho)^{2}$ |
| 24-ST Tubes Cold-Stretched | 1 | $50{,}000 - 427\,(L/\rho)$ | 78 | $101.6\times10^{4}/(L/\rho)^{2}$ |
| | 2 | $50{,}000 - 302\,(L/\rho)$ | 110 | $203.2\times10^{4}/(L/\rho)^{2}$ |
| **Carbon Steels** | | | | |
| S.A.E. 1015   Y.P. = 27,000 p.s.i | 1 | $27{,}000 - 0.660\,(L/\rho)^{1}$ | 143 | $276\times10^{4}/(L/\rho)^{2}$ |
| | 2 | $27{,}000 - 0.330\,(L/\rho)^{1}$ | 202 | $552\times10^{4}/(L/\rho)^{2}$ |
| S.A.E. 1025   Y.P. = 36,000 p.s.i | 1 | $36{,}000 - 1.172\,(L/\rho)^{2}$ | 124 | $276\times10^{4}/(L/\rho)^{2}$ |
| | 2 | $36{,}000 - 0.586\,(L/\rho)^{2}$ | 175 | $552\times10^{4}/(L/\rho)^{2}$ |
| **Alloy Steels[1]** | | | | |
| S.A.E. 2330, 3120, etc.   Y.P. = 60,000 p.s.i | 1 | $60{,}000 - 3.144\,(L/\rho)^{2}$ | 98 | $286\times10^{4}/(L/\rho)^{2}$ |
| | 2 | $60{,}000 - 1.572\,(L/\rho)^{3}$ | 138 | $572\times10^{4}/(L/\rho)^{2}$ |
| Y.P. = 75,000 p.s.i | 1 | $75{,}000 - 4.913\,(L/\rho)^{3}$ | 87 | $286\times10^{4}/(L/\rho)^{2}$ |
| | 2 | $75{,}000 - 2.457\,(L/\rho)^{2}$ | 128 | $572\times10^{4}/(L/\rho)^{2}$ |
| Y.P. = 85,000 p.s.i | 1 | $85{,}000 - 6.371\,(L/\rho)^{2}$ | 83 | $286\times10^{4}/(L/\rho)^{2}$ |
| | 2 | $85{,}000 - 3.155\,(L/\rho)^{2}$ | 118 | $572\times10^{4}/(L/\rho)^{2}$ |
| Y.P. = 100,000 p.s.i | 1 | $100{,}000 - 8.735\,(L/\rho)^{2}$ | 76 | $286\times10^{4}/(L/\rho)^{2}$ |
| | 2 | $100{,}000 - 4.367\,(L/\rho)^{2}$ | 107 | $572\times10^{6}/(L/\rho)^{2}$ |
| Y.P. = 130,000 p.s.i | 1 | $130{,}000 - 14.761\,(L/\rho)^{2}$ | 66 | $286\times10^{4}/(L/\rho)^{2}$ |
| | 2 | $130{,}000 - 7.381\,(L/\rho)^{2}$ | 94 | $572\times10^{6}/(L/\rho)^{2}$ |
| Y.P. = 155,000 p.s.i | 1 | $155{,}000 - 20.98\,(L/\rho)^{2}$ | 61 | $286\times10^{4}/(L/\rho)^{2}$ |
| | 2 | $155{,}000 - 10.49\,(L/\rho)^{2}$ | 36 | $572\times10^{4}/(L/\rho)^{2}$ |
| Chromo-Molyb.X-4130 Normalised Round Tube | 1 | $79.500 - 51.78\,(L/\rho)^{2.6}$ | 91 | $286\times10^{6}/(L/\rho)^{2}$ |
| | 2 | $70.500 - 30.79\,(L/\rho)^{1.6}$ | 129 | $572\times10^{6}/(L/\rho)^{2}$ |

corresponding strain. The peculiarities of the practical column are readily discernible in the testing laboratory, but are impossible to treat mathematically (with exactness). Returning to Euler again, the following expression seems to approximate the load $P$ that should not be exceeded in safe design (usually called $P_e$):

$$P = \pi^2 EI/L^2$$

By dividing both sides of the above equation by the area, $A$, and substituting the radius of gyration, $\rho$, for $\sqrt{I/A}$, we have the following:

$$f_c = \pi^2 E/(L\rho)^2$$

The quantity $L/\rho$ appears frequently in column formulas and is usually called the *slenderness ratio*. In general, if we do not exceed $P_e$ in column design, the member probably will not fail—if we do exceed the Euler load $P_e$, it may fail. Table 1.2 shows column formulas, in terms of allowable stress, for both short and long columns, as well as the transitional $L/\rho$. The $c$ used in this table is described in the previous paragraph on end restraints.

## 1.7   STRESSES IN PRESSURE VESSELS

We are often involved in the design and fabrication of cylindrical structures used as launch tubes, rocket encasements, or rocket motor cases. The example in Figure 1.28 is for a spherical tank, or bottle. The tensile stress in the outer fibers of the sphere, due to internal pressure, is a function of the internal diameter and the wall thickness. The hoop stress, $S$, is

$$S = pD/t$$

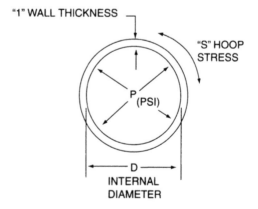

**FIGURE 1.28** Cross section of a spherical pressure vessel.

If the pressure is 1,000 psi, the wall thickness is 0.050 in. and the radius is 1 in., $S = (1{,}000 \times 2)/(0.050) = 40{,}000$ psi.

However, if we increase the inside diameter to 4 in. and keep the pressure and wall thickness the same,

$$S = (1{,}000 \times 4)/(0.050) = 80{,}000 \text{ psi}$$

In other words, the larger the diameter, the higher the hoop tensile stress in the sphere due to the same pressure. If we double the wall thickness, the hoop stress is cut in half. (This makes sense because we spread the load over twice the cross-sectional area in the wall.)

*Note:* The same formula applies to longitudinal stress in a closed tube or cylindrical tank (see Figure 1.29).

$$\text{Longitudinal stress} = S^1 = pD/t$$

The longitudinal stress in an open cylinder is only one half the hoop stress:

$$\text{Longitudinal stress} = S^2 = pD/2t$$

In the design of metal cylinders, the wall thickness must be established to withstand the hoop stress. The cylinder is then twice as strong in the longitudinal direction as is required to hold the pressure. Since metals are essentially isotropic (they have the same mechanical properties in all directions) in strength, the product of operating pressure and volume (capacity) *(PV)* per unit weight of material is less in the cylinder than in the sphere for a given peak stress.

In composite cylinders, the reinforcement can be oriented and proportioned so that the hoop strength of the cylinder wall is actually twice the longitudinal strength. Consequently, the *PV* per unit weight of material is the same in the cylinder as it is in a sphere for a given peak stress, and much higher than for the metal cylinder. This unique weight-per-unit-volume relationship, independent of size and shape, has obvious importance in rocket motor case design or any cylindrical pressure vessel.

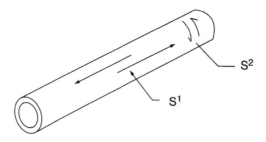

**FIGURE 1.29** Cylindrical pressure vessel (rocket launch tube).

# 2 Materials Characteristics of Metals

*Jack M. Walker*

## 2.0 INTRODUCTION TO MATERIALS CHARACTERISTICS OF METALS

Modern industry is dependent on a knowledge of metallurgy. Nearly every kind of manufacturing today is affected by the behavior of metals and alloys. Therefore, anyone who plans a career in modern industry will find a working knowledge of metallurgical processing to be a valuable asset.

Today's manufacturing engineer may not need to be a materials engineer or metallurgist, in addition to all the other skills that he or she uses in the broader role that we have discussed in several chapters of this handbook. However, to understand the forming, chip cutting, and processing principles involved in fabricating parts of metal, and in order to participate in a product design team, introductory background information is essential.

Some of us are quite familiar with the terms and properties of many of the common metals, while others are specialists in different fields and need an overview of the subject. The approach the author has taken in this chapter is to introduce the materials most commonly used, and provide an explanation of the properties that make a particular material or alloy a desirable choice for a specific application. It will make a difference in the machines selected, the design of the tooling, and the cost of the part fabrication and finishing. Different materials require different heat treatments and different surface finishes. Subchapter 2.1 discusses metallurgy, 2.2 introduces iron and steel (ferrous metals), 2.3 talks about aluminum and other nonferrous materials, and 2.4 describes the peculiarities of magnesium.

## 2.1 FUNDAMENTALS OF METALLURGY

*Metallurgy* is the art and science concerned with metals and their alloys. It deals with the recovery of metals from their ores or other sources, their refining, alloying with other metals, forming, fabricating, testing, and a study of the relation of alloy constituents and structure to mechanical properties.

The manufacturing engineer is more interested in physical metallurgy, which is concerned with the structure, properties, and associated behavior of metallic products. The properties and behavior of metals are based on their inherent crystalline structure. They do not react as amorphous (shapeless) aggregates of atoms, with a general equality of properties in all directions. They act as crystals with preferred directions of strength, flow, cleavage, or other physical characteristics, and have many limitations due to the oriented character of their particles.

## 2.1.1   Crystalline Structure

To illustrate crystalline formation, consider a metal in the fluid state, in the process of slowly cooling and solidifying. To begin with, we have a solution of free atoms. These atoms consist of a dense nucleus surrounded by several electrons. Figure 2.1 shows a diagram of the aluminum atom. With continued cooling, the atoms bond together in groups to form unit cells. A group of unit cells tends to collect as cooling continues, and forms branches, called *dendrites,* which resemble an unfinished frost pattern.

Each type of metal has its own unit cell and space lattice formation. The most common are the following four basic types, depending on the metal. Figure 2.2a shows the *body-centered unit cell,* and Figure 2.2b shows a series of these unit cells connected in a small space lattice. The *face-centered, close-packed hexagonal,* and *body-centered tetragonal space lattices* are shown in Figure 2.3. With continued cooling, the space lattices combine in groups to form crystals, and the crystals group to form grains. This process is shown in Figure 2.4. The molten state is depicted in Figure 2.4a, the formation of a unit cell in Figure 2.4b, the progressive formation of dendrites in Figure 2.4c and Figure 2.4d, and the final crystals within the grain in Figure 2.4e and Figure 2.4f. A grain is any portion of a solid that has external boundaries and an internal atomic lattice structure that is regular.

### Body-Centered Structures

The unit cell of a body-centered cubic space lattice consists of eight atoms in a square cube, plus one more in the center of the formation; see Figure 2.2a. The body-centered

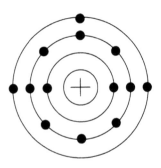

**FIGURE 2.1**    Diagram of the aluminum atom, showing 13 electrons around the nucleus.

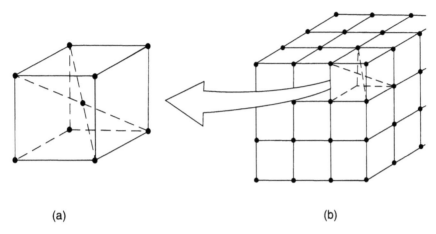

(a)                                                          (b)

**FIGURE 2.2**   The body-centered unit cell (a); and a series of these unit cells connected space lattice (b).

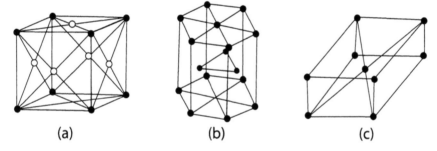

(a)                                (b)                                (c)

**FIGURE 2.3**   The face-centered (a); close-packed hexagonal (b); and body-centered tetragonal (c) unit cells.

cubic metals are at least moderately ductile. Metals included in this group are chromium, molybdenum, tantalum, tungsten, vanadium, columbium, and iron below 900°C. They always slip in planes of densest atom packing—diagonally from edge to edge through the center atom.

## Face-Centered Structures

As shown in Figure 2.3a, face-centered structures have an atom at each corner of the unit cube and an atom at the center of each face. Thus, the total number of atoms in the basic unit cell of a face-centered cubic space lattice is two. With the application of pressure, they tend to slip in the most dense atomic plane—the face diagonals. This generates 12 slip systems, consisting of the four planes and three directions. Examples of metals in this configuration are as follows:

Aluminum
Copper

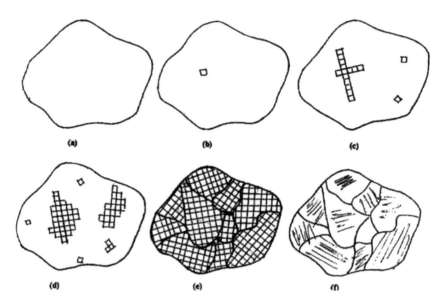

**FIGURE 2.4**    Formation of grains, starting with molten metal in (a), and the progressive cooling to achieve a solid metal grain in (f).

Gold
Iron (between 900°C and 1400°C)
Lead
Nickel
Silver
Platinum

All face-centered cubic metals are freely plastic.

## Close-Packed Hexagonal Structures

Zinc and cadmium are examples of close-packed hexagonal structures. They are freely plastic and slip horizontally. Magnesium, titanium, and zirconium have limited plasticity, while "unworkable" metals include beryllium and osmium. They are shown in Figure 2.3b.

## Body-Centered Tetragonal Space Lattice

The body-centered tetragonal space lattice is almost identical to the body-centered cube, as seen in Figure 2.3c. However, the faces of this structure are rectangular instead of square. The major example of this lattice structure is martensitic iron, which is the hardest, strongest, and most brittle type of iron. Tin is unique in that it has a tetragonal-shaped structure and is very ductile. Only iron and tin are polymorphic—they change their type of crystal at *specific temperatures*. Many other metals undergo transformations with the application of heat, which has an important effect on plasticity.

## 2.1.2   Grain Size and Characteristics

The size of the grain has a profound effect on strength, hardness, brittleness, and ductility. If metal is cooled from the molten state very, very slowly, the colonies have much more time to add on members. Therefore, if metal is cooled slowly, these colonies will have time to grow larger and larger, and very large grain size will result.

On the other hand, if metal is cooled very rapidly, many more colonies will immediately start to spring up. Then, the size of each colony is limited because so many colonies are formed. Therefore, while slow cooling produces a large grain size, rapid cooling produces a small grain size.

The larger grains are easier to tear or break or fracture. Those with small grain size have high resistance to fracture. A small crack has more difficulty moving across a series of small grains than across one large, open field. In summary, the smaller the grain size, the greater the strength; the larger the grain size, the less the strength. Since strength, hardness, and brittleness are three inseparable partners, small grain size not only will yield better strength characteristics, but will also result in a harder and more brittle material. On the other hand, if ductility is more important than strength, a larger grain size is desirable.

The ease with which metals yield to applied loads by slip processes enables them to be formed into sheets, wire, tubes, and other shapes. When deformation occurs below a certain critical temperature range, the hardness and strength properties increase, with a corresponding decrease in plasticity. This is *strain hardening*. With further deformation, plasticity may be reduced to the extent that fractures occur.

Deformation processes are used not only for shaping metals but also for improving strength properties. After cold working, plasticity (and softness) may be restored by heating to above the limiting temperature.

The term *plasticity*, as applied to metals, refers to the ability of metals to retain a change in shape brought about through deformation using pressure. This is an important metallurgical concept and is the basis for understanding formability and machinability. All metals are crystalline and owe their plasticity to the simplicity and high degree of symmetry of their crystalline structure.

## 2.2   FERROUS METALS

Historically, the principal raw materials used in steel making are ferrous scrap and pig iron, a product of the blast furnace. In the production of pig iron, iron ore containing from 50% to 60% iron, coke for fuel, and limestone for fluxing off ore impurities and coke ash are charged into the top of the blast furnace. A preheated air blast is introduced near the bottom of the furnace, burning the coke and forming carbon monoxide gas, which, in turn, reduces the iron oxides in the ore, leaving the iron in the metallic state. Molten iron dissolves considerable amounts of carbon, silicon, phosphorus, and sulfur; as a result, the pig iron contains these impurities in such quantities that it is extremely hard and brittle, making it unsuitable for applications where ductility is important. Figure 2.5 shows a typical blast furnace.

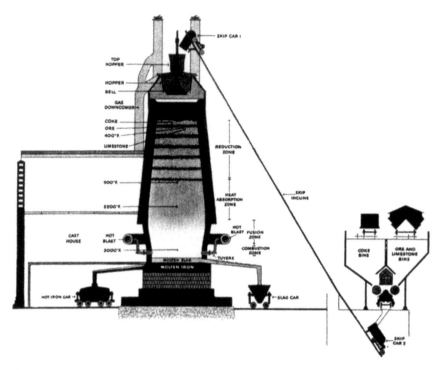

**FIGURE 2.5**   Cross section of a blast furnace in which pig iron is made. (Courtesy of Beti with permission.)

## 2.2.1   Steel Making

Steel making may be described as the process of removing impurities from pig iron and scrap and then adding certain elements in predetermined amounts so as to impart the desired properties to the finished metal. The elements added in the process may be, and in many cases are, the same ones that have been removed, but the amounts or proportions are different. Figure 2.6 shows the steel-making process. Starting with the raw materials on the left side, the farther the steel progresses to the right of Figure 2.6, the more expensive it becomes. However, the finished part or product may be of lower cost by using the more finished metal—as is the case with formed sheet metal stampings as opposed to castings.

Historically, virtually all steel was made in either the open-hearth furnace or the electric furnace, with small percentages made by the basic oxygen process and in the Bessemer converter. All processes use pig iron in varying proportions, from 60% to less than 10%. Figure 2.7 shows a cutaway diagram of a typical open-hearth furnace. The steel produced from one furnace charge, known as a *heat,* will usually weigh from 100 to 400 tons, although both larger and smaller furnaces are currently in operation. A furnace is charged with scrap, limestone, and ore. The pig iron, or hot metal, is usually added in the molten state, after the scrap is partially melted. During

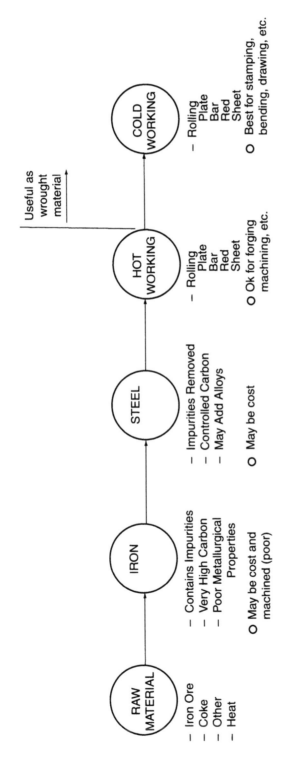

**FIGURE 2.6** The steel manufacturing process.

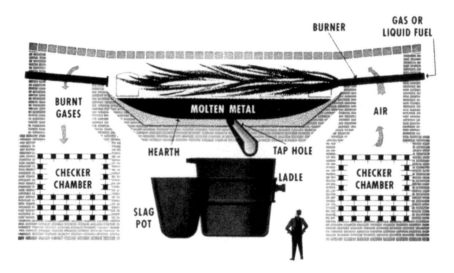

**FIGURE 2.7** Simplified cutaway diagram of a typical open-hearth furnace, viewed from the. (Courtesy of Bethlehem Steel with permission.)

subsequent refining, nearly all of the manganese, phosphorus, and silicon are oxidized and retained in the slag. The carbon is generally removed by oxidation to a percentage approximating that desired in the finished steel. At this point the heat is tapped into a ladle. To obtain the desired analysis, ferromanganese and other alloying materials are added as needed. In some cases they are added to the molten bath just prior to tapping; in others they are added to the ladle as the heat is being tapped. Aluminum or ferrosilicon also is generally added to the ladle to deoxidize the steel, as discussed later. The heat is then poured into ingot molds and solidifies into steel ingots. Recent advances in steel making allow a continuous billet to be produced by cooling the metal as it is being poured—not in an ingot but as a continuous slab of metal.

Special steels—particularly high-alloy steels, stainless steels, and tool steels, all of which have an expanding use in industry—are frequently made in electric furnaces. These vary in capacity from a few hundred pounds up to 100 tons or more. Electric arcs between the electrodes and the metal bath furnish the heat. The advantage of this type of furnace is that it is kept closed and operates with a neutral atmosphere. Oxidizing and reducing agents can therefore be applied as required, allowing close control of the chemical elements in the steel. See Subsection 2.2.2 for a discussion of the increasing utilization of "mini-mills" using a high percentage of scrap metal and direct-reduction iron.

### Definitions of Carbon and Alloy Steels

It is sometimes difficult to draw a clear dividing line between carbon and alloy steels. They have been arbitrarily defined by the American Iron and Steel Institute as follows.

*Carbon Steel*    Steel is classed as *carbon steel* when no minimum content is specified for aluminum, boron, chromium, cobalt, columbium, nickel, titanium, tungsten, vanadium, or zirconium, or any other element added to obtain a desired alloying effect; when the specified minimum for copper does not exceed 0.40%; or when the maximum content specified for any of the following elements does not exceed the percentages noted: manganese 1.65%, silicon 0.60%, copper 0.60%.

By far the most important element in steel is carbon. The following are general rules to classify steel based on its carbon content:

| | |
|---|---|
| Wrought iron | Trace to 0.08% |
| Low-carbon steel | 0.10 to 0.30% |
| Medium-carbon steel | 0.30 to 0.70% |
| High-carbon steel | 0.70 to 2.2% |
| Cast iron | 2.2 to 4.5% |

*Note:* All carbon above 2.2% is uncombined with iron and is present in the form of graphite. This presents planes of easy cleavage, which accounts for the easy breakage of cast iron.

*Alloy Steel*    Steel is classified as *alloy steel* when the maximum of the range specified for the content of alloying elements exceeds one or more of the following limits: manganese 1.65%, silicon 0.60%, copper 0.60%; or in which a definite range for a definite minimum quantity of any of the following elements is specified or required within the limits of the recognized commercial field of alloy steels: aluminum, boron, chromium up to 3.99%, cobalt, columbium, molybdenum, nickel, titanium, tungsten, vanadium, zirconium, or any other alloying element added to obtain a desired alloying effect.

In addition to differences in the steel-making processes, factors such as segregation, the type and amount of deoxidizers used, and variations in chemical analysis all profoundly affect the properties of steel.

## The Steel Ingot

Many hundreds of shapes and sizes of steel ingots have been developed over the last century. The cross section of most ingots is roughly square or rectangular, with rounded corners and corrugated sides. All ingots are tapered to facilitate removal from the molds. Depending on the type of steel, ingots may be poured big end up or big end down, as will be discussed under "Types of Steel."

All steel is subject to variation in internal characteristics as a result of natural phenomena that occur as the metal solidifies in the mold. Liquid metal, just above the freezing point, is less dense than solid metal just below it; that is, there is a shrinkage in volume during solidification. Hence a casting or an ingot is given a sink head large enough to supply the extra metal needed in the desired shape when frozen. If this extra metal can be fed in while freezing is going on, the frozen metal will be solid; otherwise it will have voids at locations where feeding has been cut off by early freezing at thin sections between those locations and the sink head. In ingots, unless the metal freezes wholly from the bottom up and not simultaneously from the sides, there will be a pipe (see Figure 2.8). The flow of metal in the sink head to fill the void must not be cut off, else secondary pipe or spongy centers will result. The extent of the piping is dependent on the type of steel involved, as well as the size and design of the ingot mold itself. Pipe is eliminated by sufficient cropping during rolling.

Another condition present in all ingots to some degree is nonuniformity of chemical composition, or segregation. Certain elements tend to concentrate slightly in the remaining molten metal as ingot solidification progresses. As a result, the top center portion of the ingot, which solidifies last, will contain appreciably greater percentages of these elements than the average composition of the ingot. Of the elements normally found in steels, carbon, phosphorus, and sulfur are most prone to segregate. The degree of segregation is influenced by the type of steel, the pouring temperature, and ingot size. It will vary with position in the ingot and according to the tendency of the individual element to segregate.

### Types of Steel

The primary reaction involved in most steel-making processes is the combination of carbon and oxygen to form a gas. If the oxygen available for this reaction is not removed prior to or during casting (by the addition of ferrosilicon or some other deoxidizer), the gaseous products continue to evolve during solidification. Proper control of the amount of gas evolved during solidification determines the type of steel. If no gas is evolved, the steel is termed *killed* because it lies quietly in the molds. Increasing degrees of gas evolution results in semikilled, capped, or rimmed steel.

*Rimmed steels* are only slightly deoxidized, so a brisk effervescence or evolution of gas occurs as the metal begins to solidify. The gas is a product of a reaction between the carbon and oxygen in the molten steel that occurs at the boundary between the solidified metal and the remaining molten metal. As a result, the outer rim of the ingot is practically free of carbon. The rimming action may be stopped mechanically after a desired period, or it may be allowed to continue until the action subsides and the ingot

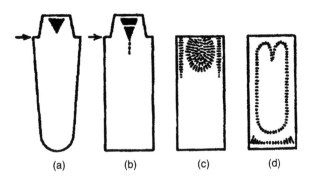

(a)          (b)          (c)          (d)

**FIGURE 2.8**   Pipe or blowholes in cast steel ingots.

top freezes over, thereby ending all gas evolution. The center portion of the ingot, which solidifies after rimming ceases, has a composition somewhat above that of the original molten metal, as a result of the segregation tendencies discussed above.

The low-carbon surface layer of rimmed steel is very ductile. Proper control of rimming action will result in a very sound surface during subsequent rolling. Consequently, rimmed grades are particularly adaptable to applications involving cold forming and where the surface is of prime importance.

The presence of appreciable percentages of carbon or manganese will serve to decrease the oxygen available for the rimming action. If the carbon content is above 0.25% and the manganese over 0.60%, the action will be very sluggish or nonexistent. If a rim is formed, it will be quite thin and porous. As a result, the cold-forming properties and surface quality will be seriously impaired. It is therefore standard practice to specify rimmed steel only for grades with lower percentages of these elements.

*Killed steels* are strongly deoxidized and are characterized by a relatively high degree of uniformity in composition and properties. The metal shrinks during solidification, thereby forming a cavity or pipe in the extreme upper portion of the ingot. Generally, these grades are poured in big-end-up molds. A hot-top brick is placed on top of the mold before pouring and is filled with metal after the ingot is poured. The pipe formed is confined to the hot-top section of the ingot, which is removed by cropping during subsequent rolling. The most severe segregation of the ingot is also eliminated by this cropping.

While killed steels are more uniform in composition and properties than any other type, they are nevertheless susceptible to some degree of segregation. As in the other grades, the top center portion of the ingot will exhibit greater segregation than the balance of the ingot. The uniformity of killed steel renders it most suitable for applications involving such operations as forging, piercing, carburizing, and heat treatment.

*Semikilled steels* are intermediate in deoxidation between rimmed and killed grades. Sufficient oxygen is retained so that its evolution counteracts the shrinkage upon solidification, but there is no rimming action. Consequently, the composition is more uniform than that of rimmed steel, but there is a greater possibility of segregation than in killed steels. Semikilled steels are used where neither the surface and cold-forming characteristics of rimmed steel nor the greater uniformity of killed steels are essential requirements.

*Capped steels* are much the same as rimmed steels except that the duration of the rimming action is curtailed. A deoxidizer is usually added during the pouring of the ingot, with the result that a sufficient amount of gas is entrapped in the solidifying steel to cause the metal to rise in the mold. With the bottle-top mold and heavy metal cap generally used, the rising metal contacts the cap, thereby stopping the action. A similar effect can be obtained by adding ferrosilicon or aluminum to the ingot top after the ingot has rimmed for the desired time. Action is stopped, and rapid freezing of the ingot top follows. Rimming times of 1 to 3 min prior to capping are most common.

Capped steels have a thin, low-carbon rim that imparts the surface and cold-forming characteristics of rimmed steel. The remainder of the cross section approaches the degree of uniformity typical of semikilled steels. This combination of properties has resulted in a great increase in the use of capped steels.

## 2.2.2   Steel Rolling

Vladimar B. Ginzburg, author of *High-Quality Steel Rolling—Theory and Practice* (Marcel Dekker, New York, 1993), has the following comments on steel rolling.

> In each stage of the development of steel rolling technology, there have been specific challenges to be met by both steel producers and designers of rolling mill equipment. In the past three decades, the most important challenges have included increasing production rates, conserving energy, increasing coil weights, and reducing the finishing gauge.
>
> These goals have been gradually achieved by the majority of steel producers, however the two main challenges that remain for steel producers today are improving quality and reducing production costs. Although these goals have always been considered in the past, they are now looked at in a completely new perspective because of the following three factors:
>
> > Excess capacity for production of flat rolled products
> > Entry of developing nations into the marketplace
> > Entry of mini-mills into flat rolled production
>
> This excess capacity for the production of flat rolled steel products has created an extremely competitive environment among the world's leading steel producers. The producers who meet the customer's high quality standards and at the same time maintain low production costs have a distinct advantage over their competitors. However, current competition is not just between the steel producers of the major industrial countries.
>
> The competition is continuously intensifying as a growing number of developing countries enter into the steel producing market. The impact of the developing nations on the world steel marketplace is not just in capacity alone, but in improved product quality and lower production costs. These developments can be directly attributed to the huge investments that the developing nations have made in modern rolling mill technology for their steel producing plants. The most recent and possibly most influential development is the transfer of the flat rolled steel production process from integrated steel mills that use iron ore and coal as the prime sources of their steelmaking process to mini-mills that utilize steel scrap and direct reduced iron. The rate of this transfer will depend on the capability of the integrated steel producers to defend their market position by further improving product quality and reducing production costs.

### Effect of the Rolling Process on the Customer Goods Manufacturer

The steel may be hot rolled to final size (when this is not too small), or the reduction in size may be completed by the addition of cold rolling or cold drawing. The internal structure and the properties of the steel may be adjusted by the temperature at which hot rolling is done, by the rate of cooling from the hot-rolling temperature, by reheating and cooling at a controlled rate (called *annealing* when the cooling is slow, *normalizing* when the cooling is done in air and is more rapid), by the amount of reduction in cold working, and by low-temperature annealing of cold-worked steel.

By such heat treatments and by the use of a trace of vanadium in unkilled steels, or by the use of vanadium, aluminum, titanium, and so forth in killed steels, the grain size can be varied from coarse to fine, with marked influence on behavior. The directional effects of rolling in one direction only can be minimized by cross-rolling, although the through-plate direction retains its differences. Cast metals have closely similar properties in different directions, but wrought metals seldom do. Rolled metals frequently show somewhat different tensile and yield strengths when the test specimen is taken longitudinally (in the direction of rolling) or transversely (at right angles to the rolling direction). The ductility is likely to be markedly lower in the transverse direction, and the properties in the through-plate direction tend to be spectacularly poorer.

Pure tensile stress in a part is rare. When stresses come in two or more directions, behavior under these biaxial or triaxial stresses cannot always be predicted from a knowledge of behavior under simple tension. Hence, combined stresses with components other than pure tension are working on material whose properties are not evaluated by the conventional tensile test. Application of the stresses to be met in service, both in magnitude and in pattern, to the particular material to be used, in full size and in its exact geometry, is often required for certainty of behavior. This testing is vital in product development today.

Handbook values are usually given only for longitudinal specimens of wrought metal; for small, fully quenched specimens of heat-treated steel; and for fully fed specimens of cast metal. It needs to be noted whether a handbook or reported test value is on a true sample, one really representing the material as it is to be used. A value determined for an as-rolled or a normalized steel in one thickness will not necessarily hold for the same steel in another thickness, nor is the strength in one direction necessarily the same as in another.

## Composition of Steel

Steel is an iron-base alloy whose strength is due primarily to its carbon content. Small amounts of manganese, and frequently silicon and a trace of aluminum, are also present in steel. The carbon in steel ready for most commercial uses is present as iron carbide ($Fe_3C$), called *cementite*. The iron matrix is called *ferrite*. The carbon may be present in plates of ferrite, in a structure called *pearlite*. Steels with a predominant ferrite matrix are called *ferritic*. If pearlite predominates, they are *pearlitic*.

At the high temperatures used for hot rolling or forging, or to prepare the steel for being hardened by quenching, the steel has a different crystal structure, called *austenite*, in which the carbon is in solid solution, interspersed in the iron, but not combined as in cementite. Upon cooling, the austenite of ordinary steels transforms to ferrite and carbide. With the presence of sufficient amounts of certain alloying elements, austenite can be retained upon cooling without transformation. Stainless steel with 18% chromium and 8% nickel, and manganese steel with about 1.25% carbon and 13% manganese, are familiar examples of austenitic steel. Figure 2.9 illustrates the heat-treating process that produces ferrite, austenite, and martensite in steel. In Figure 2.9a, ferrite is transformed to austenite and back again in stages between the lower transformation temperature and the upper transformation temperature during slow cooling. In Figure 2.9b, below the lower transformation temperature, ferrite exists without austenite. Above the upper transformation temperature, austenite exists

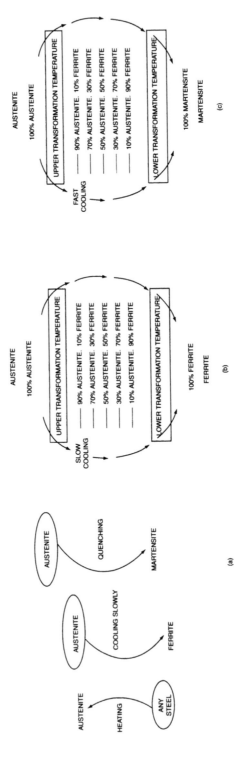

**FIGURE 2.9**  The heating and cooling processes that produce ferrite, austenite, and martensite in steel.

without ferrite. In Figure 2.9c, martensite is transformed to austenite and back again in stages between the lower transformation temperature and the upper transformation temperature during fast cooling.

## Space Lattice Structures in Iron and Steel

Iron is unusual in that it can take three different space lattice structures. As iron goes through a temperature change, its atoms realign themselves into new geometric patterns. This has a great effect on the strength, hardness, and ductility of the iron.

Ferritic iron, or ferrite, takes the body-centered cubic lattice structure formation. *Ferrite* is basic iron at room temperature that has not previously been heat-treated.

Austenitic iron, or austenite, takes the face-centered cubic lattice structure. *Austenite* is the structure that iron takes at elevated temperatures. In other words, if ferrite is heated, it gradually becomes austenitic when high temperature is reached. As ferrite is becoming austenite, the atoms are reshuffling within the crystal, realigning themselves into a new space lattice formation.

Martensitic iron, or martensite, has the body-centered tetragonal crystal lattice structure. *Martensite* is iron at room temperature that has previously been heated and suddenly quenched. The heating and quenching operation serves to produce this third geometric pattern. Heating and sudden quenching tend to harden metal. Therefore, martensite is the strongest, hardest, but most brittle of the three iron structures.

The *lower transformation temperature* is the temperature at which the body-centered cubic structure starts to change to the face-centered cubic structure. It is the temperature at which ferrite starts to change to austenite. The *upper transformation temperature* is the temperature at which the body-centered cubic lattice structure has completely changed to the face-centered cubic structure. It is the temperature at which no ferrite exists. All of the iron structure above the upper transformation temperature is austenite.

## 2.2.3   Steel Sheet Properties

Figure 2.10 is a load-versus-elongation curve for a typical steel sheet. Tensile testing is a common method of determining the mechanical properties of metal. A sample taken from a roll of steel is placed under tension and pulled until it fails. Data obtained from the test are used to plot a load-elongation (stress–strain) curve that shows the yield point, yield-point elongation, total elongation, ultimate tensile strength, and other properties. All metal forming takes place after reaching the yield point but before the ultimate tensile strength is reached. This means that a metal with a relatively low yield point and a high ultimate strength would be the easiest to form. On the other hand, if the yield point is quite high in relation to the ultimate strength, careful measures must be taken to prevent overstressing the metal, or it will fail in the forming process. The yield point, which must be exceeded to produce a permanent shape change in the metal, begins when elastic deformation ends and plastic deformation starts. Beyond this point, the steel yields discontinuously (repeated deformation followed by work hardening) up to the point where the load on the steel rises continuously. This yield-point elongation can produce strain lines if it exceeds 1.5% of total elongation.

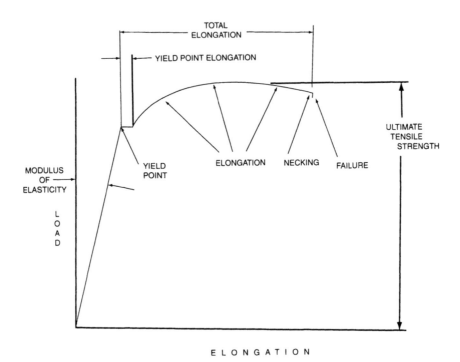

**FIGURE 2.10** Stress–strain curve for typical wrought steel sheet.

After yield-point elongation ends, the steel thins in two dimensions: through the thickness and across the width. A good way to measure ductility is by the percent elongation in the gage length of a broken test sample. In low-carbon sheet steels, for example, elongation is usually between 36 and 46% in a 2-in. gage length. The amount of springback in forming aluminum alloys is generally less than in forming low-carbon steel, and this must be considered in tool design. The amount of springback is roughly proportional to the yield strength of the metal. The slower rate of work hardening of aluminum alloys permits a greater number of successive draws than is possible with steel.

## Achieving Overall Economy

Figure 2.11 is a block diagram of the principal parties involved in producing consumer goods based on steel products. At the top of the hierarchy is the consumer, who, through the market mechanism, dictates both the quality and the price of goods. Thus, for example, if the consumer demands a better fit of car panels, the car manufacturer may consider the following approaches:

Demand tighter geometric tolerances of the coils being supplied by steel producers

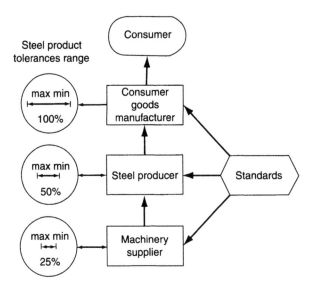

**FIGURE 2.11** Hierarchy of subordination and distribution of tolerances between consumer manufacturer, steel producer, and machinery supplier. (From V. Ginzburg, High-Quality Rolling, Marcel Dekker, New York, 1993 with permission.)

Modernize production equipment so that the quality of the assembly process can be improved without tightening the geometric tolerances of the supplied coils

Distribute improvements in quality fairly between the car manufacturers and steel producers

If the car manufacturer decides to tighten the tolerances of the purchased coils to 50% of the standard value, then the steel producer, in turn, would have to consider the following similar approaches:

Demand that the machinery supplier install mill equipment that will produce coils with geometric tolerances within 25% of the standard value

Improve maintenance, operating practices, and quality control so that the desired tolerances can be obtained without modernization of the mill equipment

Distribute improvements in quality fairly between the steel producers and machinery suppliers

Tremendous efforts have been made in the last decade by all parties to improve the quality of products produced by their own facilities. There are, however, questions one may ask. The first question is if the burden of improving quality has been fairly distributed among all parties so that the overall economy is maximized. If the answer is no, then who is going to make this fair distribution of responsibilities for improvement of quality? Obviously in the long run it will be enforced by the market system. In the short

term, however, it is very difficult to prevent the passing of a disproportionate burden by
the parties located at the higher hierarchical levels of the production process to the ones
at the lower levels.

From this point of view, standards may play an important role by stipulating technically and economically reasonable tolerances that would be required to be achieved
for each particular application. Ginzburg shows comparative analyses of the Japanese
JIS standards, the ASTM standards, the German DIN standards, and the new ISO standards for some of the parameters of flat-rolled products. An example of the thickness
tolerances of cold-rolled, high-strength sheets for automotive applications is shown in
Figure 2.12, comparing Japanese and U.S. standards. The manufacturing engineer must
carefully compare tolerances as well as price in the selection of a material for production.
Figure 2.13 plots the manufacturing cost being reduced as the quality (and cost) of the
finished steel is increased—showing the optimum quality level somewhere in between.

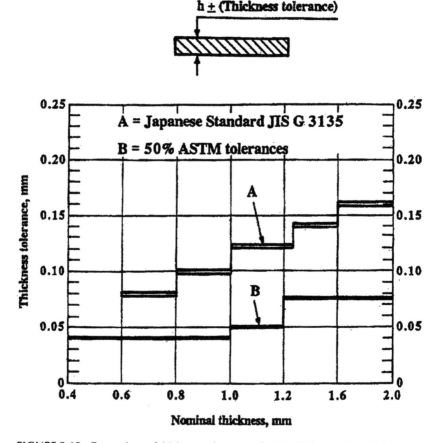

**FIGURE 2.12** Comparison of thickness tolerances of cold rolled strength steel sheets for
automobile applications. (From V. Ginzburg, High-Quality Rolling, Marcel Dekker, New
York, 1993 with permission.)

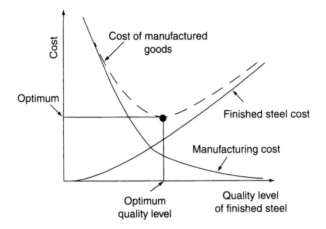

**FIGURE 2.13** Quality level/cost relationship in manufacturing plants utilizing finished products. (From V. Ginzburg, High-Quality Rolling, Marcel Dekker, New York, 1993 with permission.

TABLE 2.1  Relation of the Alloy Content in Steel and the First Two Digits of Its Name

| Steel Numerical Name | Key Alloys |
|---|---|
| 10XX | Carbon Only |
| 11XX | Carbon Only ( tree Cutting ) |
| 13XX | Manganese |
| 23XX | Nickel |
| 25XX | Nickel |
| 31XX | Nickel–Chromium |
| 33XX | Nickel–Chromium |
| 303XX | Nickel–Chromium |
| 40XX | Molybdenum |
| 41XX | Chromium–Molybdenum |
| 43XX | Nickel–Chromium–Molybdenum |
| 44XX | Nickel–Molybdenum |
| 46XX | Chromium |
| 47XX | Chromium |
| 48XX | Chromium |
| 50XX | Chromium |
| 51XX | Chromium |
| 501XX | Chromium |
| 511XX | Chromium |
| 521XX | Chromium |
| 514XX | Chromium |

*(Continued)*

TABLE 2.1  (Continued)

| Steel Numerical Name | Key Alloys |
|---|---|
| 515XX | Chromium |
| 61XX | Chromium–Vanadium |
| 81XX | Nickel–Chromium–Molybdenum |
| 86XX | Nickel–Chromium–Molybdenum |
| 87XX | Nickel–Chromium–Molybdenum |
| 88XX | Nickel–Chromium–Molybdenum |
| 92XX | Silicone–Manganese |
| 93XX | Nickel–Chromium–Molybdenum |
| 94XX | Nickel–Chromium–Molybdenum |
| 98XX | Nickel–Chromium–Molybdenum |
| XXBXX | Boron |
| XXLXX | Lead |

### 2.2.4   Steel Designation Systems

Steel is composed primarily of iron. Most steel contains more than 90% iron. Many carbon steels contain more than 99% iron. All steel contains a second element, which is carbon. Many other elements, or alloys, are contained in most steels, but iron and carbon are the only elements that are in all steels. The percent carbon in steel ranges from just above 0% to approximately 2%. Most steels have between 0.15 and 1.0% carbon.

Steels with the least carbon are more flexible and ductile (tend to deform appreciably before fracture), but they are not as strong. However, as the carbon content increases, so do strength, hardness, and brittleness.

Each type of steel has a name, usually consisting of four numbers. The first two digits refer to the alloy content. The last two digits refer to the percent carbon in the steel. In 5147 steel, for example, the "51" tells you that the steel has a lot of chromium in it. In 2517 steel, the "25" indicates that there is an unusual amount of nickel in this steel. Similarly, the "10" in 1040 steel tells you that the steel has very little alloy content except carbon. The last two digits (or three digits) indicate the percent carbon that the steel contains. In 1040 steel, for example, the "40" tells you that there is 0.40% carbon in the steel. In 1018 steel, the "18" indicates that there is only 0.18% carbon in it; thus, it is a very low carbon steel. An 8086 steel contains approximately 0.60% carbon, which makes it a medium-carbon steel. Table 2.1 relates the alloy content in steel to the first two digits of its name. Table 2.2 shows some examples of common steels with their carbon percentages, major alloying ingredients, and tensile strengths.

### 2.3  NONFERROUS METALS: ALUMINUM

Aluminum is made by the electrolysis of aluminum oxide dissolved in a bath of molten cryolite. The oxide, called *alumina,* is produced by separating aluminum hydrate from

TABLE 2.2 Alloy Content of Several Typical Steels

| STEEL | TYPE OF STEEL | TENSILE STRENGTH × 1000psi | C | Mn | P | S | Si | Ni | Cr | Mo | V |
|---|---|---|---|---|---|---|---|---|---|---|---|
| 1025 | Plain Carbon | 60–103 | 0.22–0.28 | 0.30–0.60 | 0.04 max | 0.05 max | | | | | |
| 1045 | Plain Carbon | 80–182 | 0.43–0.50 | 0.60–0.90 | 0.04 max | 0.05 max | | | | | |
| 1095 | Plain Carbon | 90–213 | 0.90–1.0 | 0.30–0.50 | 0.04 max | 0.05 max | | | | | |
| 1112 | Free Cutting Carbon | 60–100 | 013 max | 0.70–1.00 | 0.07–0.12 | 0.16–0.23 | | | | | |
| 1330 | Manganese | 90–162 | 0.28–0.33 | 1.60–1.9 | 0.035 | 0.04 | 0.20–0.35 | | | | |
| 2517 | Nickel | 88–190 | 0.15–0.20 | 0.45–0.60 | 0.025 | 0.025 | 0.20–0.35 | 4.75–5.25 | | | |
| 3310 | Nickel Chromium | 104–172 | 0.08–0.13 | 0.45–0.60 | 0.025 | 0.025 | 0.20–0.35 | 3.25–3.75 | 1.40–1.75 | | |
| 4023 | Molybdenum | 105–170 | 0.20–0.25 | 0.70–0.90 | 0.035 | 0.04 | 0.20–0.35 | | | 0.20–0.30 | |
| 52100 | Chromium | 100–240 | 0.98–1.1 | 0.25–0.45 | 0.035 | 0.04 | 0.20–0.35 | | 1.30–1.60 | | |
| 6150 | Chromium Vanadium | 96–230 | 0.48–0.53 | 0.70–0.90 | 0.035 | 0.04 | 0.20–0.35 | | 0.80–1.10 | | 0.15 min |
| 8840 | Nickel Chromium Molybdenum | 120–280 | 0.38–0.43 | 0.70–0.90 | 0.04 | 0.04 | 0.20–0.35 | 0.85–1.15 | 0.70–0.90 | 0.20–0.30 | |
| 4140 | Chromium Molybdenum | 95–125 | 0.38–0.43 | 0.75–1.00 | 0.035 | 0.04 | 0.20–0.35 | | 0.80–1.10 | 0.15–0.25 | |

the impurities associated with it in naturally occurring deposits of bauxite, and calcining to drive off the combined water.

The electrolytic process was discovered only a hundred years ago. In the short period since, the production has risen and aluminum now stands third in tonnage among the nonferrous metals, and the volume of aluminum produced is second only to that of steel.

In order to better understand the internal structural modifications that occur as a result of the various alloying ingredients and heat-treating operations, some knowledge of the physical changes taking place during solidification is necessary. When molten aluminum is cooled, its temperature drops until it reaches 1220.4°F, at which point the material gives up its latent heat of fusion and begins to solidify. Aluminum, as well as most of the other easily worked metals, crystallizes as a face-centered cubic structure, which possesses more effective slip planes than any other structure. As cooling continues, additional crystals form, building on the first ones and producing the larger units, called *grains*. The temperature will remain nearly constant at 1220.4°F until the entire mass has solidified. Then the temperature again drops as cooling continues. The solidified metal is thus composed of grains, which are in turn composed of crystals.

Figure 2.14 indicates the relation between time and temperature as a pure metal is allowed to cool from the molten state, represented by point A. As its temperature falls, it reaches a point B, where the metal begins to solidify or freeze. Note that the curve indicates that the temperature remains at this value for a period of time. This is because the change from a liquid to a solid is accompanied by the release of heat, the mechanism of the operation being such that just enough heat is released to balance that being lost, thus retaining the temperature of the metal constant during the period this solidification is taking place. Therefore, the curve is level from B to C. As soon as the metal has completely solidified, its temperature again falls gradually as it is allowed to cool, represented by the sloping line D.

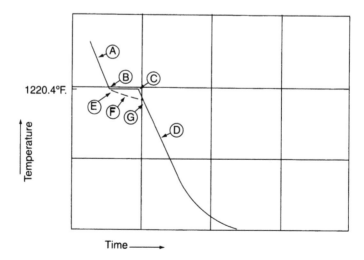

**FIGURE 2.14** Time versus temperature for a pure aluminum metal. (Courtesy Reynolds Metal Company with permission.)

It should be noted that only a pure metal follows this type of curve, and each different metal has a different solidification or freezing point; that is, the level portion of the curve, or plateau, occurs at a different temperature.

Now let us see what happens when we melt two pure metals together—say, aluminum and copper—and allow them to cool. We find that we have a curve of an entirely different shape because the combination of the two metals has a freezing range instead of a freezing point; that is, the material begins to freeze at one temperature and continues to freeze while the temperature falls to a lower value before all of it has solidified. This is shown by the dashed portion of the curve at F, where the curve slopes from E to G. The combination of aluminum and copper does not freeze completely at a single temperature because the mixture formed by the two metals behaves in an entirely different manner than a pure metal such as copper or aluminum.

At point E, the crystals forming out as the molten metal is just beginning to solidify consist of an alloy of almost pure aluminum. As the temperature falls, crystals with appreciable amounts of copper begin forming. With continued dropping temperature, the crystals forming contain more and more copper. Thus at E, the alloy particles freezing out may contain 99.9% aluminum and 0.1% copper. Just below E, the particles freezing out of solution may contain 98% aluminum and 2% copper. Similarly, particles containing 97% aluminum and 3% copper will freeze out at a lower temperature, and so on. At G, the entire mass is solidified and the temperature drops along the same type of curve as before.

Thus, as the temperature falls, the material freezing out of solution at any particular moment corresponds to the alloy of aluminum and copper that freezes at that temperature.

## 2.3.1 Alloys of Aluminum

High-purity aluminum, while it has many desirable characteristics, has a tensile strength of only about 9000 psi. Even though this strength can be doubled by cold working, the resulting strength is still not high and the alloy is not heat treatable. The small amounts of iron and silicon present in commercially pure aluminum increase the annealed condition strength by about 45%. Addition of 1.25% manganese, in addition to the impurities iron and silicon, produces a strength some 75% greater than the annealed pure metal. Addition of 2.5% magnesium increases the strength to about three times that of pure metal, and four times the pure metal strength after cold working.

Pure aluminum has inferior casting qualities. The improvement in casting qualities of aluminum alloys as compared to the pure metal is perhaps even greater than the improvement in their mechanical properties. The alloy that first gained general use contained 8% copper in addition to the impurities normally present in commercial aluminum. By reducing the copper below 5% and adding a rather large percentage of silicon, a new series of alloys came into use, followed by a group of alloys in which silicon was the only added element—or at least the major addition.

The alloys that respond to heat treatment with improvement in their physical properties all contain at least one constituent whose solid solubility is substantially greater at elevated temperatures than at room temperature. For best results, this element is added only in amounts that are completely soluble at temperatures below the melting point of the alloy, although small amounts of other alloying elements may be added. Wrought alloys are available in heat-treated tempers of 80,000 psi, and casting alloys are available that can be heat-treated to strengths of nearly 50,000 psi.

## Intermetallic Compounds

The addition of soluble elements to aluminum to produce aluminum alloys exerts a pronounced influence on the behavior of the material during the cooling period. In the molten state, certain alloying elements combine with each other and with the aluminum to form complex compounds called *intermetallic compounds.* These have characteristics entirely different from those of the elements of which they are composed.

Some of the intermetallic compounds may be dissolved in the molten aluminum alloy. Their presence lowers the solidification (or freezing) temperature of the molten metal just as the addition of alcohol to water produces a solution that freezes at a lower temperature than pure water. The freezing point of an alcohol–water solution is determined by the relative proportions. Likewise, the freezing point of molten aluminum alloy is dependent on the amount and type of dissolved constituents present.

There are many different compounds in aluminum alloy. Consequently, the material does not have one solidification point. Instead, it solidifies throughout a temperature *range.* Aluminum alloys start to freeze at a temperature just below 1220.4°F and are not completely solidified until a still lower temperature is reached. The points at which solidification starts and ends, referred to as the *liquidus* and *solidus temperatures,* are dependent on the constituent elements and the amount of each that is present in the alloy. For commercially pure aluminum, the liquidus is 1215°F and the solidus is 1190°F. Addition of alloying elements changes these figures considerably. For example, an alloy containing about 4% copper, 0.5% magnesium, and 0.5% manganese possesses a liquidus temperature of 1185°F and a solidus of 955°F.

In freezing, the first crystals to form, at approximately 1220°F, are of pure aluminum. Just below this temperature, a solution containing a major percentage of aluminum with only a minute amount of dissolved compounds will freeze. This substance solidifies separately from the primary crystals and deposits around the original pure crystals. As freezing progresses, the proportion of aluminum in the remaining molten matter becomes smaller, and the dissolved compounds form a larger and larger portion. These substances also solidify separately from the primary crystals according to their respective solidus temperatures and deposit against the core and branches of the dendrites already formed around the original pure aluminum crystals.

It is easy to see how successive layers containing increasing amounts of soluble elements solidify on the previous portion as freezing progresses. Finally, a temperature is reached at which the last portion of the melt freezes. The last metal to solidify contains a large portion of the elements added to the aluminum. Generally, this material is hard and brittle. Freezing last, these brittle components are thus concentrated at the grain boundaries (and between dendrites). The mechanical properties of an aluminum alloy are determined by the shape and size of the grains, the layer of brittle material between the dendrites, the type and amount of dissolved compounds, the shape and distribution of the insoluble constituents, and the amounts of these materials present.

In the as-cast alloy, these factors combine to produce points of weakness. This inherent weakness can be overcome in two ways: by proper heat treatment (homogenizing) or by mechanically working the material.

There is another phenomenon that occurs in casting a molten alloy into a mold. Freezing initially takes place at the surfaces of the mold, thus forming an outer solid shell with the center portions remaining liquid. Like other metals, aluminum alloys contract considerably and lose volume when passing from the liquid to the solid state. This contraction of the outer shell tends to exert hydrostatic pressure on the liquid metal inside the shell. This pressure acts to actually squeeze out the alloy-rich liquid mixtures from between the grains and dendrites. The result is that examination of a cast ingot shows few particles of the alloy-rich compounds on the outer surface. Also, a step-by-step analysis of the ingot from the center outward reveals that certain alloy percentages increase greatly as the outer skin is approached.

## Homogenizing

The resulting ingot can be made homogeneous by a special preheating technique appropriately called *homogenizing*. For this purpose, the metal is heated within the range of 900 to 1000°F and held at that temperature for a period of time—sufficient for solid diffusion to take place. *Solid diffusion* is a term used to denote the diffusion or spreading out or dissolving of one intermetallic compound into another when both are in the solid state. While it is well known that certain liquids can dissolve certain solids, as water dissolves salt, it is also true that some solids can dissolve other solids—and that is what occurs here.

While holding at temperature, not only do the alloying elements and compounds diffuse evenly throughout the ingot, but also the so-called cored structure of the grains is diffused evenly. You will recall that during solidification, the first crystals to form are almost pure aluminum and succeeding layers contain more and more of the alloying materials. Thus the inside crystals near the core of a grain are very different from the outer crystals of that grain—producing a cored structure. Homogenizing, however, allows these crystals of alloying materials to diffuse evenly throughout the structure and thus corrects the undesirable cored arrangement. Since the phenomenon of solid diffusion proceeds almost imperceptibly at room temperature and increases speed with temperature, homogenizing is done at as high a temperature as possible—just below the melting point of the compounds present in the aluminum alloy.

## Plastic Deformation

When a metal is subjected to sufficient stress, such as that produced by plastic deformation, slippage occurs along definite crystallographic planes. The number of planes on which this slippage can take place is entirely dependent on the crystal structure of the metal. Aluminum, as well as most of the other easily worked metals, crystallizes in the form of a face-centered cubic structure, which possesses more effective slip planes than any other structure. Metals that possess this crystal structure can therefore be severely plastically deformed before rupturing occurs (see Subchapter 2.2 for further explanation of crystal structure).

When aluminum is subjected to plastic deformation, slippage takes place along the slip planes that are most favorably oriented with regard to the direction of the applied stress. As slippage continues, the planes that are slipping change their positions in such a manner that they become less favorably oriented to the applied stress than other planes. Slippage then begins along these other planes. As the degree of plastic deformation progresses, the planes continue to change their orientation and the metal becomes increasingly difficult to work.

The changing of the positions of the slip planes, often referred to as *rotation* of the slip planes, produces a condition wherein a substantial number of planes have the same orientation. This condition is known as *preferred orientation,* and it is one of the reasons why some material forms "ears" when deeply drawn. The direction of preferred orientation in cold-worked aluminum depends on the thermal treatments, the degree of plastic deformation, and the direction of the applied stresses that produced the orientation.

### Fragmentation

The slipping of the planes naturally causes fragmentation of the grains. Increasing the degree of cold working increases the amount of fracturing that takes place, with the grains becoming elongated. The amount and directions of the elongation are closely associated with the reduction of cross-sectional area and the direction of working. Slippage along the slip planes is restricted in several ways. The distortion of the space lattice by atoms of other elements in a solid solution or by mechanical strain restricts slippage. The presence of insoluble or precipitated constituents can exert a keying effect that also restricts slippage. Small grain size is still another factor, due to the interference of the grain boundaries. There are many other factors, such as interatomic cohesion forces, but the above are the major ones that are closely associated with plastic deformation.

Cold-working aluminum increases the tensile strength, the yield strength, and the hardness, but decreases ductility properties such as percent elongation, the impact strength, and the formability. Excessive cold working will result in reaching a point where excessive pressures are required for further reduction, or where fracturing of the metal structure occurs. This comment applies to both the heat-treatable and the non-heat-treatable aluminum alloys. For this reason, annealing cycles are inserted at points in the fabricating cycle where cold-working stresses have been built up by large reductions during rolling or other mechanical work.

### Recrystallization

When cold-worked material is heated to a sufficiently high temperature, the fragmented particles produced by the cold-working process form new, unstrained grains, provided sufficient cold work has been performed on the material. This is *recrystallization.* The high-energy points created during the cold-working process serve as points of nucleation for the formation of the new grains. The formation of the new grains removes a substantial amount of the effects of the cold work, tending to produce properties similar to those originally possessed by the material. The degree of cold work is important. If an insufficient amount is present, recrystallization will not take place. When just enough cold work is present to cause recrystallization at the temperature used, the resulting material will possess a very coarse grain size. The presence of a substantial amount of cold work promotes the formation of fine-grained material. The fundamentals of recrystallization are as follows:

1. Increasing the degree of cold work decreases the temperature necessary for recrystallization.
2. Increasing the length of time at temperature decreases the recrystallization temperature.
3. The rate of heating to and through the recrystallization temperature affects the size of the grains formed.
4. The degree of cold work and the temperature employed affect the size of the grains formed.

**Grain Size**

It is usually desirable to have a material possessing a medium-to-fine grain size for severe drawing operations. While large-grained material actually has a greater capacity for plastic deformation than fine-grained material, such material also has a greater tendency to deform locally, or "neck down," and may produce an undesirable appearance known as *orange peel*. The final grain size of a recrystallized material is dependent on the size of the grains after recrystallization and upon grain growth. These, however, are influenced by many factors, such as:

Original grain size
Degree of cold work
Heating rate
Final temperature
Length of time at temperature
Composition

## 2.3.2   Heat Treatments

Heat treatment involves heating the aluminum alloy to a point below the melting point, where the alloying ingredients (zinc, copper, etc.) are in solution with the aluminum. The grain size is small, and the alloying ingredients are evenly distributed within the grains, along the aluminum lattices. When the metal has "soaked" for a time at this temperature to permit even distribution of the alloying ingredients, it can be *quenched* rapidly in water (or other cooling medium). The purpose of suddenly dropping the temperature in this manner is to prevent certain constituents from precipitating out, which they would do if the part were cooled slowly. Quenching from any particular temperature range tends to retain in the metal the structure present just before quenching. The result will be a slightly hard, fine-grained material. If the material is held at the elevated temperature for too long a period of time, the grains would continue to grow, and the resulting structure will be rather coarse grained and will not develop the optimum properties we are looking for.

The fast cooling to near room temperature upon quenching produces a super-saturated condition, where the material has already dissolved in it more of the constituents than it normally can carry at that temperature. The metal can be held in this condition (called the *W condition*) for some period of time by placing it in a freezer, to permit subsequent forming before the metal reaches full hardness. Such a condition is obviously unstable. The result is that certain constituents begin to separate out, or *precipitate,* from the main mass of the aluminum alloy. This precipitation occurs at room temperature with many of the alloys (for example, the copper alloys), and is known as *natural aging.* Other alloys must be heated slightly to bring this precipitation to completion within a reasonable length of time. This is called *artificial aging* and is common with the zinc alloys. In either case, this controlled reprecipitation is aimed at providing the correct size, character, and distribution of precipitated particles in the aluminum to produce maximum strength and other desired mechanical properties. The precipitation of the major alloying ingredients from the solid, homogeneous solution provides maximum keying to resist slipping or deformation.

An example of some heat treatments of a zinc-based aluminum alloy (7075) is as follows:

7075-0 Soft, annealed condition, with a minimum ultimate tensile strength of 33,000 psi and a minimum tensile yield strength of 15,000 psi

7075-W Solution heat-treated at 870°F and quenched in water. Metal is quite soft and can be held in freezer for subsequent forming operations

7075-T6 Precipitation heat treatment at 250°F (artificially aged) and allowed to cool at room temperature. Has minimum ultimate tensile strength of 78,000 psi and tensile yield strength of 69,000 psi

7075-T73 Additional precipitation heat treatment at 350°F (overaged) to improve fatigue life and reduce stress cracking. Minimum ultimate tensile strength of 67,000 psi and yield strength of 56,000 psi

Aluminum alloys hardened in this manner can be made soft and easily workable again by an annealing treatment. *Annealing* takes place when the metal is heated to about the same temperature as for solution heat-treating and then slowly cooled. This produces a precipitate in the form of large particles outside the grains along the grain boundaries and not inside the crystals. In this manner, minimum keying affects the results, and the material is soft because the crystals easily slip along their slip planes.

### 2.3.3   Wrought versus Cast Products

Commercial aluminum alloys are divided into two general types: wrought alloys and casting alloys. Wrought alloys are designed for mill products whose final physical form is obtained by mechanically working the material. This mechanical work is done by rolling, forging, drawing, or extruding the material. Wrought aluminum mill products include forgings, sheet, plate, wire, rod, bar, tube, pipe, structurals, angles, and channels, as well as rolled and extruded shapes.

---

### TABLE 2.3  Aluminum Alloy Groups

| | |
|---|---|
| Aluminum—99% minimum and greater | 1xxx |
| Aluminum alloys, grouped by major alloying element: | |
| Copper | 2xxx |
| Manganese | 3xxx |
| Silicon | 4xxx |
| Magnesium | 5xxx |
| Magnesium and silicon | 6xxx |
| Zinc | 7xxx |
| Other elements | 8xxx |

---

Casting alloys are used in the production of sand, permanent mold, or die castings—processes in which the molten alloy is allowed to solidify in a mold having the desired final size and shape.

## Aluminum Association Designation System

Wrought aluminum and wrought aluminum alloys are designated by a four-digit index system. The first digit of the designation serves to indicate the alloy group. The last two digits identify the aluminum alloy or indicate the aluminum purity. The second digit indicates modifications of the original alloy or impurity limits. Table 2.3 shows the aluminum alloy groups. Mechanical properties may be obtained from other reference sources.

## 2.3.4    Bibliography and Additional Reading

*Note:* Information on metallurgy, steel, and aluminum was largely taken from the author's lecture notes used while teaching classes at St. Louis University some years ago. It has been updated from several sources listed below. Additional source materials for reference by the reader are also included.

*The ABC's of Aluminum,* Reynolds Metals Company, Richmond, VA, 1962.
*Alcoa Aluminum and Its Alloys,* Aluminum Company of America, Pittsburgh, PA, 1960.
*Aluminum Heat Treating,* Reynolds Metals Company, Richmond, VA, 1960.
*Aluminum Standards and Data,* The Aluminum Association, New York, 1976.
Brandt, D. A., *Metallurgy Fundamentals,* Goodheart-Wilcox, South Holland, IL, 1972.
Gillett, H. W., *The Behavior of Engineering Metals,* John Wiley, New York, 1951.
Ginzburg, V. B., *High-Quality Steel Rolling,* Marcel Dekker, New York, 1993.
Jensen, J. E., (ed.), *Forging Industry Handbook,* Forging Industry Association, Cleveland, OH, 1970.
*Modern Steels and Their Properties,* Bethlehem Steel Co., Bethlehem, PA, 1964.
Tanner, John P., *Manufacturing Engineering,* Marcel Dekker, New York, 1991.

## 2.4 NONFERROUS METALS: MAGNESIUM

Robert S. Busk, International Magnesium Consultants, Inc., Hilton Head, South Carolina

### 2.4.1 Introduction to Magnesium

Magnesium, with inexhaustible raw material sources, utilizes three outstanding properties in many of its applications. It has the lowest density of any structural metal and therefore is used to produce products of light weight, such as aircraft engine parts, automotive parts, wheels, computer parts, portable tools, and materials handling equipment. Its large electronegative potential finds use in the cathodic protection of steel structures such as domestic hot-water heaters. Its high damping capacity is used to produce vibration tables and stable platforms for instruments subject to vibration damage.

Subsection 2.4.4 lists works referenced throughout this subchapter. However, two of the references are also valuable for general reading. Emley [4] contains excellent information on the chemistry, metallurgy, and practical handling of molten magnesium and its alloys. Busk [5] contains complete physical and mechanical property data, design criteria for magnesium, and detailed information on the machining, joining, forming, and finishing of magnesium and its alloys. It is well worth having on hand before starting any project involving the fabrication of magnesium.

Magnesium is produced from the magnesium ion found in ores such as seawater, the Great Salt Lake, dolomite, carnalite, or magnesite. Reduction is by electrolysis of magnesium chloride or by reaction of silicon with magnesium oxide. There are two types of electrolytic cells in use: those based on the Dow Chemical Co. design [1] and those based on the I. G. Farben design [2,3]. General descriptions of electrolytic reduction are given in Emley [4], p. 25, and Busk [5], p. 3. Two methods of reduction of magnesium oxide by silicon are also in general use. The first was developed by Pidgeon [6], the second by Pechiney [7]. General descriptions of silicon reduction are also given in Emley [4], p. 25, and Busk [5], p. 3. The pure metal is sold in various grades with total magnesium content varying from 99.80 to 99.98% [8]. Impurities that are controlled for specific effects are aluminum, copper, iron, lead, manganese, nickel, and silicon. All other impurities are found only in very small quantities.

### 2.4.2 General Characteristics of Magnesium

**Physical Properties**

For a complete listing of the physical properties of pure magnesium, see Busk [5], p. 150. Those properties that are of most common significance for manufacturing purposes are the coefficient of friction (0.36), density (1.74 g/cm$^3$ at 20°C), electrical resistivity (4.45 ohm—m $\times$ 10$^{-8}$ at 20°C), melting point (650°C), specific heat (1025 J/(kg-K) at 20°C), thermal conductivity (154 W/(m-K) at 20°C), and thermal expansion (25.2 unit strain $\times$ 10$^2$).

**Safety**

The following is largely taken from Busk [5], p. 37.

Magnesium can be ignited, upon which it will burn with a brilliant white light at a temperature of about 2800°C. However, it is only magnesium vapor that will burn. Thus, the metal must be heated to a temperature high enough to produce a sufficient quantity of vapor to support combustion. As a practical matter, this means that the metal must be melted. Because the heat conductivity of magnesium is high, all of a massive piece must be raised to the melting point for the piece to burn freely. Thus it is very difficult to start a fire with a massive piece such as a casting. Even if a torch is used to raise a part of the casting to the ignition temperature and burning starts, it will cease when the torch is removed, because of heat conduction to other parts of the casting and consequent lowering of the temperature below the ignition temperature. On the other hand, finely divided magnesium, such as some kinds of machining scrap and very thin ribbon, cannot conduct the heat away from a source, and burning can be initiated even with a match, which has a flame temperature of about 900°C.

As is true of all oxidizable materials, an air suspension of fine magnesium powder can explode, and this is the most serious hazard associated with the flammability of magnesium. Explosion will occur only if the powder is fine enough to remain suspended in the air for an appreciable period of time. As a practical matter, this means that the powder must be at least as fine as about 200 mesh, which has a nominal particle size of 74 μm.

Since magnesium will react very slowly with water, even at room temperature, to produce hydrogen, the large surface area associated with finely divided machining chips can, when wet, produce sufficient hydrogen to result in a hydrogen explosion hazard. The worst case is a large amount of damp powder, since the temperature will then rise, leading to still greater hydrogen production and even to ignition of the magnesium itself. Powder should be stored dry. If it must be wet, the amount of water should be copious enough to prevent a temperature rise, and means should be provided for hydrogen dispersal.

Those areas of manufacturing that should take precautions regarding magnesium ignition are melting (e.g., for casting), machining, welding, and heat treating. Specific precautions will be given in the sections treating each of these operations.

Once started, a magnesium fire can be extinguished by using the normal strategies of cooling, removing oxygen, or letting the magnesium be totally consumed. Machining chips that have started to burn can be cooled effectively with cast-iron chips, and this technique is used quite successfully. Since magnesium is an active chemical, it will react with oxygen preferentially to many other materials. Hence, any extinguisher that contains oxygen as part of its chemical makeup will probably support the combustion of magnesium rather than stop it. This is true of water, so one should not use water to cool magnesium without being aware of the fact that, during cooling with water, reaction will occur, producing heat and hydrogen. The hydrogen, in turn, may react explosively with the air. Water can be used, but only in large quantities in order to cool the mass of magnesium below the melting point, with recognition of the hydrogen explosion hazards that are being introduced. Water should never be used for extinguishing a fire of finely divided material, since the vigorous reaction with water will scatter the chips and spread the fire. Oxygen can be excluded by covering the magnesium with a nonreactive material such as melting fluxs, G-1 powder, Metal-X, or other proprietary materials. Any oxygen-containing material, such as sand, should be avoided. See also reference [9].

### Alloys of Magnesium

The most common alloying elements for magnesium are aluminum, zinc, and manganese. In addition, silicon, rare earth metals, and yttrium are used for higher strength, especially at elevated temperatures. If aluminum is not present as an alloying element, zirconium is added for grain refinement.

## 2.4.3 Manufacturing Processes

See Busk [5] for a detailed discussion of typical uses, effects of alloying elements, properties, and design criteria for the use of magnesium alloys. The purpose of this subsection is to point out specific effects the properties of magnesium have on manufacturing practices.

### Casting

Magnesium alloy castings are produced by all the standard techniques of sand, permanent-mold, low-pressure, high-pressure (die), and investment casting. In general, the practices used for other metals apply also to the handling of magnesium. See References [4] and [19–25] for specific details. However, there are three properties of magnesium that profoundly affect all casting processes: (1) liquid magnesium does not attack steel; (2) liquid magnesium will ignite if exposed to air; (3) magnesium has a lower density than almost any other metal or impurity.

Because the liquid metal does not attack steel, the common material for handling liquid magnesium alloys is steel. Plain carbon steel is the most common, although stainless steel is sometimes used where excessive oxidation of carbon steel is a problem. If stainless steel is chosen, a type such as 430, which is low in nickel, should be used. Steel pots are used for melting; steel pumps and steel lines for conveying liquid magnesium from one point to another; steel equipment for metering exact quantities for casting; and steel hand tools for miscellaneous handling of the molten metal. Pumps of various kinds are suitable, including electromagnetic pumps. However, the most common is a simple, rugged centrifugal pump generally built by the casting shop itself. The secret for building a good centrifugal pump for molten magnesium is to use generous, even sloppy, clearances. An air motor is best for this kind of pump since, in contrast to an electric motor, it is little affected by the heat of the molten metal.

Because molten magnesium will ignite if exposed to air, it must be protected from such exposure. The classic method is to cover the molten metal with a liquid flux consisting of a mixture of chlorides, fluorides, and magnesium oxide [4,15]. A more efficient, cleaner, and less expensive protection is by the use of small quantities of $SF_6$, or of $SF_6$ and $CO_2$, in air [16–18].

The reactivity of magnesium with other oxides, such as silicon dioxide, requires that sand used for molds be protected from the molten magnesium by mixing inhibitors with the sand, such as sulfur, boric oxide, and ammonium or potassium borofluoride [4,19].

Because of the low density of magnesium, impurities in molten metal tend to sink rather than float. Thus, sludge builds up on the bottom of the melting container and must periodically be removed by dredging. An advantage of flux over $SF_6$ is the ability of the former to trap impurities, thus making dredging easier. If scrap, for example, is melted, the high content of impurities can easily be removed with the use of flux, but can be removed when $SF_6$ is used for protection only by filtering [27].

A new casting technique has been developed for magnesium, called *thixomolding* [28–30]. Magnesium pellets are fed to a screw conveyer that advances them through a heated chamber to raise the temperature to a point between the liquidus and solidus while simultaneously shearing the metal by the action of the screw. This produces metal in a thixotropic condition [31], which is then injected by advancing the screw into a die to produce a part—the process closely resembling the injection molding of plastics. A considerable advantage for the caster is that there is no need to handle molten magnesium.

As with all molten metal, there is a severe explosion hazard if tools that are wet with water are immersed below the surface. The rapid expansion of the water as it first expands to the gaseous state and then further expands as its temperature rises will empty molten metal from the container with explosive force. Any tools must be perfectly dry before immersion.

## Wrought Semiproducts

Although magnesium alloys can be rolled, extruded, and forged in much the same way as other metals [4,5,32–34], there are certain characteristics peculiar to magnesium that affect the technology.

The crystal structure of magnesium is hexagonal close-packed. (See Subchapter 2.1 for a discussion of unit cells and lattice formations.) The major plane of deformation by slip is the basal, and all of the crystallographic directions of deformation, regardless of the plane, lie in the basal plane [33]. This crystallographic mechanism results in a preferred orientation such that the basal plane of sheet lies in the plane of the sheet; of extrusions, in the plane of extrusion; and of forgings, in the plane of major deformation. Since slip deformation does not allow any deformation out of the basal plane, deformation of polycrystalline material without cracking would be impossible if slip were the only mechanism available. Other mechanisms such as twinning, grain-boundary sliding, and kinking also exist. Twinning takes place when the direction of compressive stress is parallel to the basal plane, the twinning resulting in a reorientation of the basal plane in the twin to an angle of about 86° to the original basal plane. Slip on the twin then allows deformation out of the original basal plane at an angle of 86°.

Since grain boundary sliding and kinking are both more easily activated as the temperature increases and the grain size becomes smaller, both temperature of deformation and grain size control become important. Magnesium is usually rolled at temperatures exceeding 400°F, and both extruded and forged at temperatures exceeding 600°F. Rolling is sometimes done at room temperature, but total deformation before annealing at about 700°F is limited to about 30%, and reduction/pass to about 4%.

Because twinning occurs when compression is applied to the surface of the sheet, coiling results in twinning on the compressive side of the coil, but not on the tension side. This leads to many problems in forming and handling, so sufficient tension must be applied to the up-coiler to prevent twinning during coiling. Roller leveling, while practical from a manufacturing standpoint, results in extensive twinning in the sheet, which results in a lowered tensile yield strength. For this reason, anneal flattening is generally preferred.

## Forming

All of the forming methods used for metals, such as bending, drawing, spinning, dimpling, and joggling, are used with magnesium. Good details for these processes are given in Busk [5] and in the works cited therein. The differences between forming of magnesium and other metals are all related to the desirability of forming at elevated temperatures and to the need for controlling grain size and the effects of twinning.

## Joining

Joining of magnesium is common, using welding, brazing, adhesive bonding, and mechanical attachment such as riveting. Soldering is not recommended. Good details for all of these are given in Busk [5] and in the works cited therein. Protection of the molten metal from oxidation is required for both welding and brazing. Arc welding using shielded helium or argon gas is suitable. Fluxes are used for brazing.

## Machining

Magnesium is the easiest of all metals to machine. For example, 1 hp is required to remove 1.1 in.$^3$ of steel, but is sufficient to remove 6.7 in.$^3$ of magnesium. Busk [5] should be consulted for tool design, machining practice, and safety precautions.

## Heat Treatment

Recommended practices for heat-treating magnesium products are given in Reference [35]. Included in that reference is a recommended procedure for combatting a fire in a heat-treated furnace.

## Finishing

There are three characteristics of magnesium that influence finishing procedures:

1. Magnesium is inert in strong caustic solutions. Therefore, cleaning of surface contamination such as grease and oil is best accomplished using strong caustic solutions. If embedded material must be removed by removing some of the magnesium, acidic solutions must be used.
2. Magnesium develops hydroxide or hydroxycarbonate surface films when exposed to air. Paints such as epoxies, which are resistant to high-pH surfaces, should therefore be used.

3. Magnesium is more electronegative than all other common metals when exposed to salt solutions. It is thus the anode in galvanic couples and will corrode while protecting the cathodic material if exposed to salt water while connected electrically to the cathode. It is necessary to protect joints so that this does not happen in service [36]. Good details on finishing methods are given in Busk [5] and the works cited therein.

## 2.4.4  Bibliography

1. Hunter, R. M., *Trans. Electrochem. Soc.* 86:21 (1944).
2. Høy-Petersen, Nils, Magnesium production at the Porsgrunn Plant of Norsk Hydro, *Proc. Int. Magnesium Assoc.* (1979).
3. Strelets, Kh. L., Electrolytic production of magnesium, translated from the Russian, obtainable from the International Magnesium Association.
4. Emley, E. F., *Principles of Magnesium Technology,* Pergamon Press, Elmsford, NY, 1966.
5. Busk, R. S., *Magnesium Products Design,* Marcel Dekker, New York, 1987.
6. Pidgeon, L. M., and W. A. Alexander, *Trans. AIME,* 159:315 (1944).
7. Trocmé, F., The development of the magnetherm process, *Trans. AIME* (1974).
8. Standard specification for magnesium ingot and stick for remelting, in *Annual Book of ASTM Standards, V. 02.02,* ASTM Standard B92, ASTM, Philadelphia, PA.
9. *Storage, Handling, and Processing of Magnesium,* National Fire Protection Association Bulletin 48.
10. Standard specification for magnesium alloy sand castings, in *Annual Book of ASTM Standards, V. 02.02,* ASTM Standard B80, ASTM, Philadelphia PA.
11. Standard specification for magnesium-alloy sheet and plate, in *Annual Book of ASTM Standards, V. 02.02,* ASTM Standard B90, ASTM, Philadelphia, PA.
12. Standard specification for magnesium-alloy forgings, in *Annual Book of ASTM Standards, V. 02.02,* ASTM Standard B91, ASTM, Philadelphia, PA.
13. Standard specification for magnesium-alloy die castings, in *Annual Book of ASTM Standards, V. 02.02,* ASTM Standard B94, ASTM, Philadelphia, PA.
14. Standard specification for magnesium-alloy extruded bars, rods, shapes, tubes, and wire, in *Annual Book of ASTM Standards, V. 02.02,* ASTM Standard B107, ASTM, Philadelphia, PA.
15. *Melrasal Fluxes,* Magnesium Elektron Limited Bulletin 498.
16. Couling, S. L., F. C. Bennett, and T. E. Leontis, Melting magnesium under air/$SF_6$ protective atmosphere, *Proc. Int. Magnesium Assoc.* (1993).
17. Couling, S. L., Use of Air/$CO_2$/$SF_6$ mixtures for improved protection of molten magnesium, *Proc. Int. Magnesium Assoc.* (1979).
18. Busk, R. S., and R. B. Jackson, Use of $SF_6$ in the magnesium industry, *Proc. Int. Magnesium Assoc.* (1980).
19. *Molding and Core Practice for Magnesium Foundries,* Dow Chemical Company Bulletin 141–29, 1957.
20. Berkmortel, John, and Robert Hegel, Process improvement & machine development in magnesium cold chamber die cast technology, *Proc. Int. Magnesium Assoc.* (1991).
21. Fink, Roland, Magnesium hot chamber improvements, *Proc. Int. Magnesium Assoc.* (1993).

22. Holta, O., O. M. Hustoft, S. I. Strømhaug, and D. Albright, Two-furnace melting system for magnesium, NADCA, Cleveland, OH, October 18–21, 1993.

23. Øymo, D., O. Holta, Om M. Hustoft, and J. Henriksson, Magnesium Recycling in the Die Casting Shop, American Society for Materials, "The Recycling of Metals," Düsseldorf, May 13–15, 1992.

24. *Recommended Practice for Melting High-Purity Magnesium Alloys*, International Magnesium Association.

25. *Magnesium Die Casting Manual*, Dow Chemical Company.

26. *Permanent Mold Practice for Magnesium*, Dow Chemical Company Bulletin 141–101.

27. Petrovich, V. W., and John Waltrip, Fluxless refining of magnesium scrap, *Proc. Int. Magnesium Assoc.* (1988).

28. Erickson, Stephen C., A process for the injuction molding of thixotropic magnesium alloy parts, *Proc. Int. Magnesium Assoc.* (1987).

29. Frederick, Paul, and Norbert Bradley, Injection molding of thixotropic magnesium: update, *Proc. Int. Magnesium Assoc.* (1989).

30. Carnahan, R. D., F. Decker, D. Ghosh, C. VanSchilt, P. Frederick, and N. Bradley, The thixomolding of magnesium alloys, in *Magnesium Alloys and Their Applications*, Mordike, B. L. and Hehmann, F., Eds., DGM Informationgesellschaft, 1992.

31. Flemings, M. C., R. G. Riek, and K. P. Young, Rheocasting, *Material Sci. Eng.*, 25 (1976).

32. Ansel, G., and J. O. Betterton, The hot and cold rolling of Mg-base alloys, *Trans. AIME* 171 (1947).

33. Roberts, C. Sheldon, *Magnesium and Its Alloys*, John Wiley & Sons, New York, 1960.

34. *Magnesium Forging Practice*, Dow Chemical Company, 1955.

35. Standard practice for heat treatment of magnesium alloys, in *Annual Book of ASTM Standards, V. 02.02*, ASTM Standard B661, ASTM, Philadelphia, PA.

36. *Preventive Practice for Controlling the Galvanic Corrosion of Magnesium Alloys*, International Magnesium Association.

# 3 Conventional Fabrication Processes

*Jack M. Walker*

## 3.0 INTRODUCTION TO CONVENTIONAL FABRICATION PROCESSES

More steel is produced than any other metal, with aluminum second in volume and in use in fabricating products. The weight of aluminum is approximately one third the weight of steel. Steel is approximately 0.3 lb/in.$^3$, aluminum is 0.1 lb/in.$^3$, and magnesium is 0.06 lb/in.$^3$. Figure 3.1 is a nomogram for calculating weights of steel and aluminum stock. In this chapter we concentrate our discussions on parts fabricated from these two materials. Steel is made by heating iron ore and casting into pigs or ingots, or continuous casting into slabs. Aluminum is made from bauxite and converted into molten aluminum pigs or ingots using a complex chemical and electrolytic process. At this point, both metals are in their least expensive form, but suitable only for making rather poor castings. Gray cast iron is superior to raw aluminum for casting at this time. From this point on, the processes for steel and aluminum are quite similar. Gray cast iron is the lowest-cost ferrous material, since little work has been done to refine either its form or its material content and properties. As you continue through the process of removing impurities, adding alloying ingredients, hot rolling and cold rolling, it appears that carefully controlled sheet metal has the greatest labor content, and therefore may have the highest cost per pound. The material form is one factor that enters into the cost of a finished metal part, but it can seldom offset the fabrication expense of the total part cost.

Metal manufacturers have traditionally relied on several basic forming techniques, such as casting, forging, machining, and sheet metal stamping, to impart the desired geometric shape to their products. In recent years, because of stiffer industrial competition, the development of new alloys, shortages of certain metals, and the increase in energy costs, these traditional processing methods have been critically analyzed and reevaluated. It is becoming very desirable to produce the final product in fewer processing steps and with as little waste as possible. Several techniques for the manufacture of components to "net shape" or to "near net shape," based on the firm foundation of the traditional processes, are being developed to meet these challenges of today and the future.

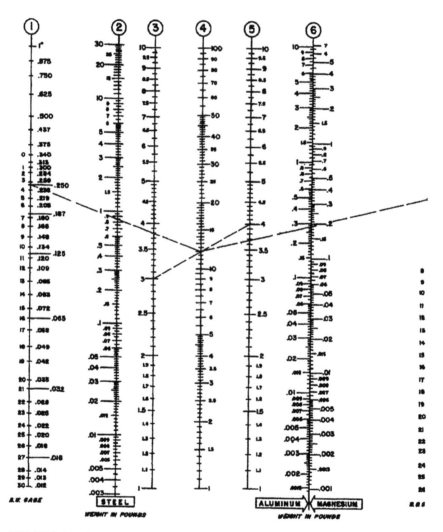

**FIGURE 3.1**  Nomogram for calculating weights of steel, magnesium, and aluminum stock.

This chapter introduces the conventional processes and equipment for sheet metal fabrication in Subchapter 3.1, machining in Subchapter 3.2, extrusion and forging in Subchapter 3.3, and casting and molding in Subchapter 3.4.

## 3.1 SHEET METAL FABRICATION PROCESSES

### 3.1.1 Introduction to Sheet Metal Fabrication Processes

Sheet metal stampings are generally the lowest-cost parts to produce. Both the machinery and the labor are relatively low in cost, and production rates can be quite high. Most of the equipment used in the forming of steel and other metals is suitable for use with aluminum alloys. Because of the generally lower yield strength of aluminum alloys, however, press tonnage requirements are usually lower than for comparable operations on steel, and higher press speeds can be used. Similarly, equipment for roll forming, spinning, stretch forming, and other fabrication operations on aluminum need not be so massive or rated for such heavy loading as for similar operations on steel.

### 3.1.2 Blanking

#### Shears

One of the most common and versatile machine tools used in sheet metal work is the vertical shear, or square shear. For small-quantity jobs, this is the most common blanking process. Figure 3.2 shows a common shear. The backstops are adjustable

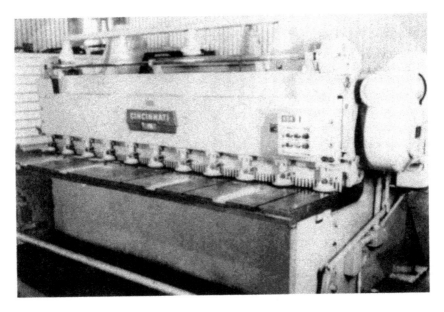

**FIGURE 3.2**    Example of vertical shear, or square shear, used for making straight cuts in sheet stock.

and can maintain cutoff lengths of 0.010 with care, and 0.030 even with old, worn equipment. A clearance of 68% of the thickness of the sheet is normally used between shear blades, although for thin sheet and foil, smaller clearances give a better edge with no burr or curvature.

## Punch Press

For larger parts runs, or for blanks requiring other than straight-sheared sides, the punch press and blanking dies are commonly used. Figure 3.3 shows a production punch press for relatively small parts that can run at 300 strokes per minute with material fed automatically from coil stock in precut widths. Die-cut blanks can hold tolerances of 0.001 with good equipment. The correct clearance between punch and die is essential to obtain a good edge with a low burr. Clearance is dependent on alloy, temper, and gauge. Recommended clearances for the more common aluminum alloys are shown in Figure 3.4. The table shows that the required clearance increases with higher mechanical properties of the metal.

## Turret Press

The turret punch has come into its own with the development of computer numerical control (CNC) fabrication equipment. The most common application is to cut the blank edges on a square shear and transfer this blank to the fabricator

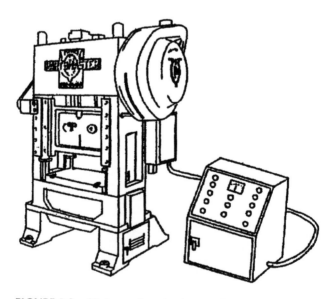

**FIGURE 3.3**   High-speed production punch press.

| t = Thickness of sheet in inches | | | | | |
|---|---|---|---|---|---|
| Alloy | Temper | Clearances O per side | Alloy | Temper | Clearances O per side |
| 1100 | O<br>H12, H14<br>H16, H19 | 0.050t<br>0.060t<br>0.070t | 5083 | O<br>H112, H323, H343 | 0.070t<br>0.075t |
| 2014 | O<br>T4, T6 | 0.065t<br>0.080t | 5086 | O, H112<br>H32, H34, H36 | 0.070t<br>0.075t |
| 2024 | O<br>T3, T361, T4 | 0.065t<br>0.080t | 5154 | O, H112<br>H32, H34, H36, H38 | 0.070t<br>0.075t |
| 3003 | O<br>H12, H14<br>H16, H18 | 0.050t<br>0.060t<br>0.070t | 5454 | O, H112<br>H32, H34 | 0.070t<br>0.075t |
| 3004 | O<br>H32, H34<br>H36, H38 | 0.065t<br>0.070t<br>0.075t | 5456 | O, H321<br>H323, H343 | 0.070t<br>0.075t |
| 5005 | O<br>H12, H14, H32, H34<br>H16, H18, H36, H38 | 0.050t<br>0.060t<br>0.070t | 6061 | O<br>T4<br>T6 | 0.055t<br>0.060t<br>0.070t |
| 5050 | O<br>H32, H34,<br>H36, H38, | 0.050t<br>0.060t<br>0.070t | 7075 | O<br>W<br>T62 | 0.050t<br>0.060t<br>0.070t |
| 5052 | O<br>H32, H34,<br>H36, H38, | 0.065t<br>0.070t<br>0.075t | 7178 | O<br>W<br>T6 2 | 0.065t<br>0.070t<br>0.075t |

Blanks over 0.080 in. thick should be sheard 1½ oversize on each side and be machined to size.

**FIGURE 3.4** Die clearances for blanking common aluminum alloys. (Courtesy Aluminum Corporation of America. With permission.)

for all the internal cutouts and special shapes. The turrets can hold as many as 42 different punchdie sets, which can be programmed XY to tolerances of 0.004 at rates up to 330 hits per minute. Station-to-station indexing time is as low as 0.5 sec. Figure 3.5 shows one of Strippit's 20-ton models, programmed on a PC. With some loss in accuracy, the latest machines allow one to rotate oversized sheets and fabricate parts up to double the machine throat capacity. Large sheets can also be repositioned, permitting fabrication of parts greater in length than the table size.

## Dimensioning Practices

If there is a single area where the manufacturing engineer can accomplish the greatest benefit in producibility and economy of manufacture, it is in ensuring the appropriate detailing practices on drawings. Following are a few basic guidelines.

First, select a meaningful datum in the body of the part passing through a hole center, if possible, rather than using an edge or corner of the part. This avoids problems

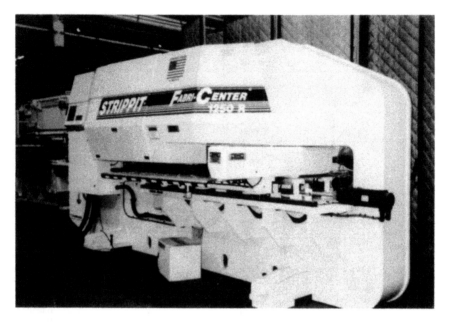

**FIGURE 3.5**   Example of a turret punch. (Courtesy Strippit, Inc. Akron, NY. With permission.)

of possible misalignment of the part, distortion from clamping, and so forth. It allows for more precise measurement by avoiding measurements from edges that may be tapered and therefore dimensionally uncertain. It facilitates accurate inspection, and it avoids unnecessary accumulation of tolerances.

Second, on related hole patterns, dimensioning and tolerances should be within this pattern, with only one dimension linking to the general datum. Better quality control and function of the product can be expected.

Third, highlight the truly significant dimensions. Critical dimensional relationships can be protected if they are known.

### Blanking Pressures

The blanking operation is usually performed on a single-action press employing a punch and die with sharp cutting edges. The dimensions of the blank correspond to the dimensions of the die.

The blanking or shearing load is calculated using the following equation:

$$P = Lts$$

where

$P$ = load

L = peripheral length of the blank
t = thickness of the material
s = shear strength of the material

As an example, to blank a circle 4.75 in. in diameter from material 0.050 in. thick with a 40,000 psi shear strength requires 3 tons of load.

P = 4.75 0.050 40,000
= 30,000 lb = 15 tons (minimum press capacity)

The shear strength of the commonly used aluminum alloys ranges from 9,000 to 49,000 psi, whereas that of low-carbon steel is from 35,000 to 67,000 psi. Because of the generally lower shear strength of aluminum alloys, lower-tonnage presses are required than for comparable operations with steel. For easy and quick determination of the load, refer to the nomogram in Figure 3.6.

Dayton Rogers Manufacturing Company uses the following formula for calculating blanking tonnage requirements:

T = P Th C

where

T = pressure required in tons
P = perimeter of blank in inches
Th = thickness of material
C = constant (see common ones below)

## Example

0.050 CR steel, half-hard; cutting edge of 12 linear inches

T = 12 0.050 32
= 20 tons required

Constants

AluminumSoft = 11
    T4/T6 = 15
SteelHR/Cold rolled = 27
    Half hard = 32
Stainless SteelAnnealed = 37
    Half-hard = 50
    4230 AQ = 40

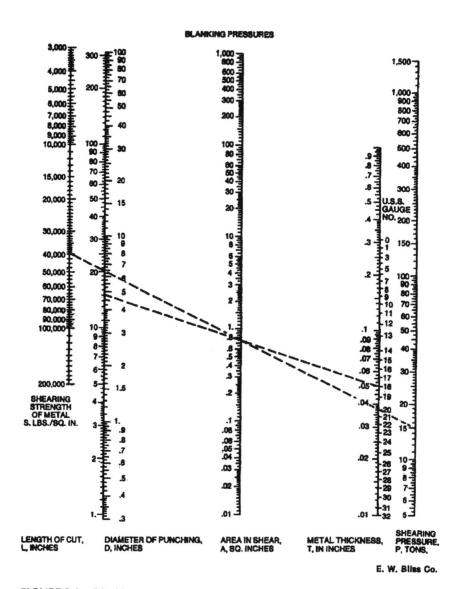

**FIGURE 3.6**    Blanking pressure nomogram.

Brasshalf-hard = 22
    hard = 25

Figure 3.7 gives the approximate pressures required for punching round holes in mild sheet metal.

To reduce the load required for blanking, the face of the die can be ground at an angle so that the cutting edge around the die opening is on a slanted plane, and a peak is formed across one of the diameters of the die face (see Figure 3.8). The maximum difference in height between the peak and the lowest point on the face should not be more than the thickness of the metal to be blanked; the minimum difference should not be less than one half the thickness of the material (referred to as incorporating shear into the die).

## Fine Blanking

There are several techniques for improving the edge finish of a blank. One method, referred to as *finish blanking,* incorporates a small radius on the cutting edge of the die. Another method, applicable to thicknesses from approximately 1/64 to 1/2 in., is called *fine blanking.* The process involves clamping the material securely throughout the entire blanking operation. A small radius is employed on the die, with almost zero die clearance, and a counterpressure is used against the blank, as shown in Figure 3.9. The clamping action, which is produced by a V knife edge on the pressure pad around the punch at a distance of about 1/16 in. from the punch, stops any lateral movement of the stock during blanking. This action, combined with the very small clearance, small-radius die, and counter-pressure, produces a blank with a smooth sheared edge, very little burr, good flatness, and close tolerances. Because fine blanking is akin to extruding, some aluminum alloys and tempers are difficult to blank with a smooth edge. Alloys 2024-T6 and 7075-T6, for example, produce a relatively rough edge because of their hardness and tendency to fracture rather than shear. Punch speeds of one half to one third that of conventional blanking are used. Therefore, special presses have been developed to give a fast punch advance followed by a decreased punch blanking speed. The production rate may vary from 10 to 100 parts/min; the actual rate depends on the thickness of the stock and the geometric shape of the blanked part. Punch and die clearance for ordinary stamping is usually about 5% of the dimension across the die. For fine blanking, this clearance is about 1% and some-times less than 0.5%. With larger presses, parts can be fine-blanked from plate as thick as 0.75 in. (19 mm).

## Laser Cutting

Current trends toward just-in-time (JIT) manufacturing, shorter parts runs, and limited product life cycles have increased the use of laser-cutting machines in production and prototype fabrication. Laser cutters are constantly evolving, as manufacturers find new and innovative ways to apply this technology.

| Gauge | 28 | 26 | 24 | 22 | 20 | 18 | 16 | 14 | 12 | 10 | 5/32 | 3/16 | 1/4 |
|---|---|---|---|---|---|---|---|---|---|---|---|---|---|
| Material thickness | .0149 | .0179 | .0239 | .0289 | .0359 | .0478 | .0598 | .0747 | .1046 | .1345 | .1562 | .1875 | .250 |
| Hole Diameter | | | | | | | PRESSURE IN TONS | | | | | | |
| .125 | .2 | .2 | .2 | .3 | .4 | .5 | .6 | .7 | 1.0 | 1.3 | 1.5 | 1.8 | 2.5 |
| .1875 | .2 | .3 | .4 | .4 | .5 | .7 | .9 | 1.1 | 1.5 | 2.0 | 2.3 | 2.8 | 3.7 |
| .250 | .3 | .4 | .5 | .6 | .7 | .9 | 1.2 | 1.5 | 2.1 | 2.6 | 3.1 | 3.7 | 4.9 |
| .3125 | .4 | .4 | .6 | .7 | .9 | 1.2 | 1.5 | 1.8 | 2.6 | 3.3 | 3.8 | 4.6 | 6.1 |
| .375 | .4 | .5 | .7 | .9 | 1.1 | 1.4 | 1.8 | 2.2 | 3.1 | 4.0 | 4.6 | 5.5 | 7.4 |
| .4375 | .5 | .6 | .8 | 1.0 | 1.2 | 1.6 | 2.1 | 2.6 | 3.6 | 4.6 | 5.4 | 6.4 | 8.6 |
| .500 | .6 | .7 | .9 | 1.2 | 1.4 | 1.9 | 2.3 | 2.9 | 4.1 | 5.3 | 6.1 | 7.4 | 9.8 |
| .5625 | .7 | .8 | 1.1 | 1.3 | 1.6 | 2.1 | 2.6 | 3.3 | 4.6 | 5.9 | 6.9 | 8.3 | 11.1 |
| .625 | .7 | .9 | 1.2 | 1.5 | 1.8 | 2.4 | 2.9 | 3.7 | 5.1 | 6.6 | 7.7 | 9.2 | 12.3 |
| .6875 | .8 | 1.0 | 1.3 | 1.6 | 1.9 | 2.6 | 3.2 | 4.0 | 5.7 | 7.3 | 8.4 | 10.1 | 13.5 |
| .750 | .9 | 1.1 | 1.4 | 1.8 | 2.1 | 2.8 | 3.5 | 4.4 | 6.2 | 7.9 | 9.2 | 11.0 | 14.7 |
| .8125 | 1.0 | 1.1 | 1.5 | 1.9 | 2.3 | 3.1 | 3.8 | 4.8 | 6.7 | 8.6 | 10.0 | 12.0 | 16.0 |
| .875 | 1.1 | 1.2 | 1.6 | 2.1 | 2.5 | 3.3 | 4.1 | 5.1 | 7.2 | 9.2 | 10.7 | 12.9 | 17.2 |
| .9375 | 1.2 | 1.3 | 1.8 | 2.2 | 2.6 | 3.5 | 4.4 | 5.5 | 7.7 | 9.9 | 11.5 | 13.8 | 18.4 |
| 1.0000 | 1.3 | 1.4 | 1.9 | 2.4 | 2.8 | 3.8 | 4.7 | 5.9 | 8.2 | 10.6 | 12.3 | 14.7 | 19.5 |

**General Information:**

Formula determining punching pressure for round holes in mild sheet steel:   D × Thickness × 25 = Pressure in Tons.

Formula for determining blanking pressure in mild sheet steel:   Shear Length × Thickness × 25 = Pressure in Tons.

Formula for stripping pressures:   Shear Length × Thickness × 3500 = Pressure in Pounds.

FIGURE 3.7  Blanking pressures for round holes. (From John Tanner, *Manufacturing Engineering*, Marcel Dekker, New York, 1982. With permission.)

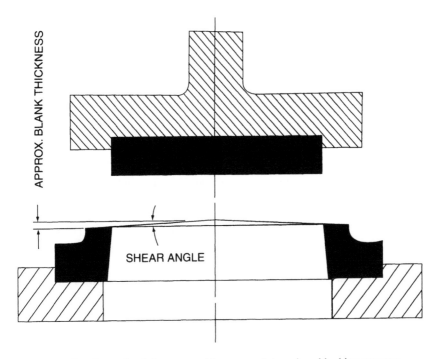

**FIGURE 3.8**   Example of shear ground into a punch to reduce blanking pressure.

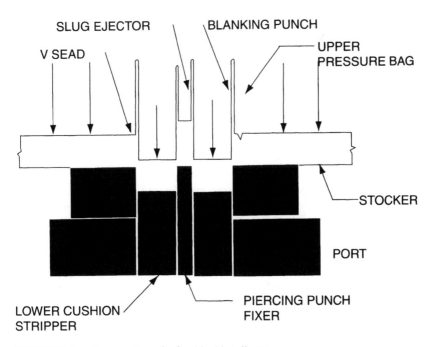

**FIGURE 3.9**   Cross section of a fine blanking die set.

Often the capabilities of lasers and turret punches can be combined. Turret presses are very fast and generate acceptable accuracy when punching many holes of the same or different diameters. Lasers are particularly accurate and economical for profiling irregular exterior contours. These capabilities can be combined to produce accurate, complex parts at acceptable production rates by using each machine to perform that part of the cutting operation for which it is best suited.

## Laser Operation

Lasers can be operated in either the continuous wave (CW) or the pulsed mode. CW operation is faster and generates a smoother edge. It is inherently less accurate because of thermal workpiece expansion due to the higher power levels reaching the work.

Where there is a need for intricate or very close-tolerance cutting, the pulsed mode generates less heat but produces a very finely serrated edge. The finest quality of the workpiece is a carefully balanced compromise between speed, workpiece cooling, and edge condition.

Lasers are most productive when applied to mild steel and stainless steel and are more difficult to employ on aluminum. Aluminum and certain other metals, such as zinc and lead, continue to reflect light when molten. This quality scatters the beam, requiring more power. In addition, aluminum and copper alloys conduct heat away from the cutting area, which again means that more power is required.

## Laser Considerations

In addition to production economics, precision, and edge condition, the knowledgeable manufacturing engineer considers these characteristics of laser-produced parts:

*Localized hardening.* Lasers cut by melting or vaporizing metal. This action can create problems when cutting heat-treatable materials, as the area around the part will become case-hardened. Laser-cut holes in stainless steel or heat-treatable steel alloys that require machining (tapping, countersinking, or reaming) can be particularly troublesome. By the same token, this characteristic can benefit a product that must be case-hardened for wear resistance.

*Edge taper.* The laser is most accurate where the coherent light beam enters the workpiece. As the beam penetrates the part, the light scatters, creating an edge-taper condition similar but opposite to breakout in a sheared or pierced part. (The hole on the side of the workpiece from which the laser beam exits is generally smaller in diameter than the entrance hole.) Thus one must carefully consider the final use of the part and, in some cases, may need to specify from which side the part should be cut.

*Minimum through-feature size.* The cutting beam is focused down to approximately 0.010 in. (0.2 mm) and is therefore capable of cutting holes and features with radii approximating 0.030 in. (0.76 mm). The limits applicable to piercing or blanking with a punch and die, such as the relationship between hole size and material thickness, or the minimum distance between features to avoid distortion, do not apply when laser cutting.

However, some limitations do exist and are also related to the material thickness. See Figure 3.10 for an illustration of the minimum through features that are possible using lasers. Laser cutting allows for through features to be one sixth to one eighth the size needed for die piercing. Also, since no mechanical force is applied, the width of material remaining between cutout features may be very narrow without distortion occurring during metal removal. A typical application is tightly spaced venting holes on a visually important surface.

## 3.1.3 Piercing

When the diameter of the punch becomes too small to hold an ejector pin in its center, the clearance between the punch and die is decreased to about 5% or less of the metal thickness to ensure that the slugs are not lifted with the punch on the return stroke. The hole in the die below the cutting edge should be tapered to permit the slug to fall freely. When working gauges of aluminum alloy up to about 0.081 in., a taper of 3/8° is ample. For greater thicknesses, the taper angle should be increased up to twice this amount. A gang punch is often used for simultaneous punching of a large number of holes. In this case, it is advisable to step the individual punches slightly to stagger their entry into the metal. If the punches are close together, stepping will also prevent crowding of the metal and the deflection of thin punches. The relative lengths of the stepped punches will depend on the gauge of the metal to be perforated. The difference in length between one punch and the next shorter one should be slightly less than the thickness of the metal. If the difference is too large, jerky operation will result. The longer punches should normally be on the outside, surrounding the smaller punch, such as drawing or blanking. Punches of large diameter, however, should always be longer than those of small diameter, regardless of position, to prevent distortion of the perforation and chipping of the smaller punches. Recommended minimum ratios of punched hole diameters to stock thickness are shown in Figure 3.11.

| minimum through-features | | | |
|---|---|---|---|
| material thickness range | | minimum hole diameter and slot width achievable | |
| in. | mm | in. | mm |
| 0-0.075 | 0-19 | 0.010 | 0.25 |
| 0.075-0.090 | 1.9-2.3 | 0.015 | 0.38 |
| 0.090-0.125 | 2.3-3.2 | 0.020 | 0.50 |
| 0.125-0.166 | 3.2-4.0 | 0.025 | 0.64 |
| 0.158-0.187 | 4.0-4.8 | 0.030 | 0.76 |

**FIGURE 3.10** Through features possible by laser cutting. (Courtesy Dayton Rogers Manufacturing Co. With permission.)

P =  Punched Hole Diameter
      (0.062 min. dia.)
T =  Stock Thickness

| Material Ultimate Tensite Strength (PSI) | Ratio P to T |
|---|---|
| 32,000 | P = 1.0T |
| 50,000 | P = 1.ST |
| 95,000 | P = 2.0T |

**FIGURE 3.11**  Minimum ratios of punch hole diameters to stock thickness. (Courtesy Dayton Rogers Manufacturing Co. With permission.)

## 3.1.4  Forming

True bending is done in a straight plane, such as in the use of a press brake and brake dies. *Forming* is a better term when the bend line is not straight, requiring stretching or shrinking of a flange. Total wear of tools used in forming aluminum is somewhat less than with steel. This results in part from the lower force levels involved, and in part from the smoother surface condition that is characteristic of aluminum alloys. Accordingly, tools can sometimes be made from less expensive materials, even for relatively long runs. However, a higher-quality surface finish is generally required on tools used with aluminum alloys, to avoid marking. The oxide film on the surface of aluminum alloys is highly abrasive, and for this reason many forming tools are made of hardened tool steels. As a rule, these tools, even if otherwise suitable, should not be used interchangeably to form steel parts, because this use could destroy the high finish on the tools.

Suggestions that will assist in successful bending are:

1. Clean bending tools thoroughly, removing particles of foreign material.
2. Remove burrs, nicks, or gouges at ends of bend lines, which can initiate fractures. (Heavy plate should be chamfered or radiused on the edges at the bend lines to reduce the possibility of cracks.)
3. Avoid nicks, scribe lines, or handling marks in the vicinity of the bend.
4. Employ rubber pads, flannel, or other intermediate materials between tools and aluminum, where high finish standards must be maintained. (Recent developments in coating tools with low-friction materials may be useful.)
5. Apply a light oil coating on tools and bend lines to minimize scoring and pickup.
6. Form metal across the direction of rolling.

Figure 3.12 shows forming characteristics for forming carbon steel strip. Minimum permissible bend radii for aluminum are shown in Figure 3.13. The minimum permissible radius varies with the nature of the forming operation, the type of forming equipment, and the design and condition of the forming tools. Minimum working radius for a given material or hardest alloy and temper for a given radius can be ascertained only by actual trial under contemplated conditions of fabrication. Figure 3.14 shows the calculation of flat pattern bend allowance development for various sheet thicknesses and bend radii. The approximate load per lineal foot required to make a 90° bend in sheet metal is shown in Figure 3.15.

## Cold Rolled Tempers

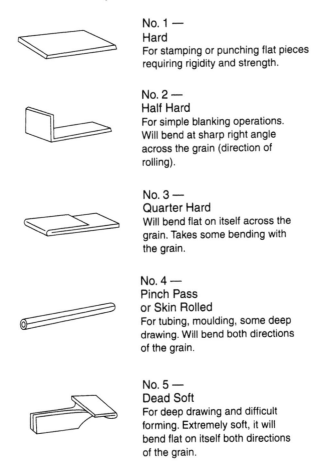

No. 1 —
Hard
For stamping or punching flat pieces requiring rigidity and strength.

No. 2 —
Half Hard
For simple blanking operations. Will bend at sharp right angle across the grain (direction of rolling).

No. 3 —
Quarter Hard
Will bend flat on itself across the grain. Takes some bending with the grain.

No. 4 —
Pinch Pass
or Skin Rolled
For tubing, moulding, some deep drawing. Will bend both directions of the grain.

No. 5 —
Dead Soft
For deep drawing and difficult forming. Extremely soft, it will bend flat on itself both directions of the grain.

FIGURE 3.12 Forming characteristics of carbon steel strip.

| Radii required for 90 bend in terms of thickness (T)<br>Approximate Thickness | | | | | | |
|---|---|---|---|---|---|---|
| Type | .016 | .032 | .064 | .125 | .187 | .250 |
| 2SO, 3SO<br>52SO | 0 | 0 | 0 | 0 | 0 | 0 |
| 2S¼H, ½H,<br>3S¼H, 24SO*,<br>61SO | 0 | 0 | 0 | 0 | 0-1T | 0-1T |
| 3S½H, 52S¼H | 0 | 0 | 0 | 0-1T | 0-1T | ½T-1½T |
| 2S½H, 52S½H | 0 | 0 | 0-1T | ½T-1½T | 1T-2T | 1½T-3T |
| 3S¼H, 61SW | 0-1T | 0-1T | ½T-1½T | 1T-2T | 1½T-3T | 2T-4T |
| 2SH, 52S¼H,<br>61ST | 0-1T | ½T-1½T | 1T-2T | 1½T-3T | 2T-4T | 2T-4T |
| 3SH, 52SH | ½T-1½T | 1T-2T | 1½T-3T | 2T-4T | 3T-5T | 4T-6T |
| 24ST* | 1½T-3T | 2T-4T | 3T-5T | 4T-6T | 4T-6T | 5T-7T |

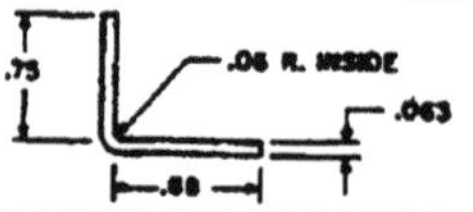

* Alclad 24S can be bent over slightly smaller radii than the corresponding tempers of the uncoated alloy.

**FIGURE 3.13** Minimum possible bend radii for aluminum. (Courtesy Dayton Rogers Manufacturing Co. With permission.)

WORKING FROM THE INSIDE, THE FIGURES SHOWN ARE THE CORRECT BEND ALLOWANCES.
SEE EXAMPLE FOR CLARITY

DEVELOPED LENGTH
.750 + .880 + .018 = 1.648

| MATERIAL THICKNESS | INSIDE RADIUS OF BEND | | | | | | | | | | | | |
|---|---|---|---|---|---|---|---|---|---|---|---|---|---|
| | .016 | .030 | .048 | .063 | .094 | .125 | .156 | .188 | .219 | .250 | .312 | .375 | .437 |
| .016 | .004 | .003 | .010 | .016 | .029 | .043 | .056 | .070 | .083 | .097 | .124 | .150 | .177 |
| .020 | .007 | .000 | .007 | .014 | .027 | .041 | .054 | .067 | .080 | .094 | .121 | .148 | .175 |
| .025 | .010 | .004 | .004 | .010 | .023 | .037 | .051 | .064 | .077 | .091 | .118 | .144 | .171 |
| .030 | .013 | .007 | .001 | .007 | .020 | .034 | .047 | .061 | .074 | .088 | .115 | .141 | .168 |
| .032 | .014 | .008 | .000 | .006 | .019 | .033 | .046 | .059 | .073 | .086 | .113 | .140 | .167 |
| .036 | .017 | .010 | .003 | .003 | .016 | .030 | .044 | .057 | .070 | .084 | .111 | .137 | .164 |
| .040 | .020 | .013 | .006 | .001 | .014 | .027 | .040 | .054 | .067 | .081 | .108 | .135 | .161 |
| .045 | .022 | .015 | .009 | .002 | .011 | .025 | .038 | .052 | .065 | .078 | .105 | .132 | .159 |
| .048 | .025 | .018 | .011 | .004 | .008 | .022 | .035 | .049 | .062 | .076 | .103 | .130 | .156 |
| .050 | .027 | .020 | .013 | .007 | .007 | .020 | .033 | .047 | .060 | .074 | .101 | .127 | .154 |
| .060 | .032 | .026 | .019 | .013 | .001 | .014 | .028 | .041 | .054 | .068 | .095 | .122 | .148 |
| .063 | .036 | .029 | .022 | .015 | .003 | .011 | .025 | .038 | .052 | .065 | .092 | .119 | .145 |
| .075 | - | .036 | .029 | .022 | .009 | .005 | .018 | .031 | .045 | .058 | .085 | .112 | .139 |
| .080 | - | .039 | .033 | .026 | .012 | .000 | .014 | .028 | .041 | .055 | .081 | .108 | .135 |
| .090 | - | - | .040 | .033 | .020 | .006 | .007 | .021 | .034 | .047 | .074 | .101 | .128 |
| .125 | - | - | - | .056 | .042 | .029 | .015 | .002 | .011 | .025 | .051 | .078 | .105 |
| .160 | - | - | - | .076 | .063 | .049 | .036 | .022 | .009 | .004 | .031 | .058 | .085 |
| .190 | - | - | - | - | .083 | .070 | .056 | .043 | .030 | .016 | .011 | .038 | .064 |

NOTE: FIGURES TO THE LEFT OF THE HEAVY LINE ARE TO BE ADDED
FIGURES TO THE RIGHT OF THE HEAVY LINE ARE TO BE SUBTRACTED

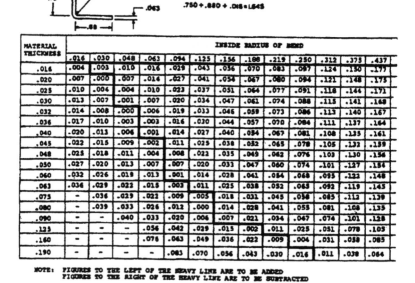

**FIGURE 3.14** Calculation of flat pattern bend allowance development. (From John Tanner, *Manufacturing Engineering*, Marcel Dekker, New York, 1991. With permission.)

**APPROXIMATE LOAD PER LINEAL FOOT REQUIRED TO MAKE RIGHT ANGLE BEND IN 3003-H14, TONS**

| Thickness of metal, | Width of female die opening, w, in. | | | | | | | | | | | | | | | | | | | | | | | | | | |
|---|---|---|---|---|---|---|---|---|---|---|---|---|---|---|---|---|---|---|---|---|---|---|---|---|---|---|---|
| | ¼ | ⅜ | ½ | ⅝ | ¾ | ⅞ | 1 | 1⅛ | 1¼ | 1⅜ | 1½ | 1¾ | 2 | 2¼ | 2½ | 2¾ | 3 | 3½ | 4 | 4½ | 5 | 5½ | 6 | 7 | 8 | 10 | 12 |
| 1/16 | | 1.7 | 1.2 | 0.91 | 0.73 | 0.58 | | | | | | | | | | | | | | | | | | | | | |
| 3/32 | | | | 2.3 | 1.8 | 1.5 | 1.3 | 1.1 | | | | | | | | | | | | | | | | | | | |
| 1/8 | | | | | 3.4 | 2.8 | 2.4 | 2.1 | 1.8 | 1.6 | 1.4 | | | | | | | | | | | | | | | | |
| 5/32 | | | | | | | 3.9 | 3.3 | 2.9 | 2.6 | 2.4 | 2.0 | 1.7 | | | | | | | | | | | | | | |
| 3/16 | | | | | | | | 5.0 | 4.5 | 3.9 | 3.5 | 3.0 | 2.5 | 2.0 | | | | | | | | | | | | | |
| 1/4 | | | | | | | | | | | 6.8 | 5.5 | 4.7 | 4.1 | 3.6 | 3.2 | 2.9 | | | | | | | | | | |
| 5/16 | | | | | | | | | | | | | 7.7 | 6.6 | 5.9 | 5.3 | 4.7 | 3.9 | 3.3 | | | | | | | | |
| 3/8 | | | | | | | | | | | | | | 10 | 9.0 | 7.8 | 7.1 | 5.9 | 5.0 | 4.4 | | | | | | | |
| 7/16 | | | | | | | | | | | | | | | | 11 | 9.6 | 8.3 | 6.8 | 5.9 | 5.2 | 4.8 | | | | | |
| 1/2 | | | | | | | | | | | | | | | | | 14 | 11 | 9.4 | 8.2 | 7.3 | 6.4 | 5.7 | | | | |
| 5/8 | | | | | | | | | | | | | | | | | | | 15 | 13 | 12 | 11 | 9.5 | 7.8 | 6.6 | | |
| 3/4 | | | | | | | | | | | | | | | | | | | | 20 | 18 | 16 | 14 | 12 | 10 | 7.5 | |
| 7/8 | | | | | | | | | | | | | | | | | | | | | | 22 | 19 | 17 | 14 | 10 | 8.2 |
| 1 | | | | | | | | | | | | | | | | | | | | | | | 27 | 22 | 19 | 17 | 11 |
| 1-1/4 | | | | | | | | | | | | | | | | | | | | | | | 44 | 36 | 31 | 23 | 19 |
| 1-1/2 | | | | | | | | | | | | | | | | | | | | | | | | | 46 | 36 | 28 |

**Conversion factor for various alloys**

| Alloy | 3003-H14 | 3004-H32 | 3004-H34 | 5052-H32 | 5052-H34 | 5083-H113 | 5154-H32 | 6061-T4 | 6061-T6 | 2024-O | 2024-T3 | 5456-H321 |
|---|---|---|---|---|---|---|---|---|---|---|---|---|
| factor | 1.00 | 1.4 | 1.6 | 1.5 | 1.7 | 2.1 | 1.8 | 1.6 | 2.0 | 1.2 | 3.2 | 2.3 |

NOTES: 1. Given values are for a male die having a radius, *r*, equal to the thickness of the metal, *t*. For the alloys for which a larger radius is used because of minimum bend requirements (see Table 3-1), the loads will be greater. These loads can be approximated by using an effective width of opening, determined by subtracting the value of (*r*-*t*) from the actual die opening in the above table. For example, if it were desired to bend a 5-ft long, 1/8-in. thick 6061-T6 sheet to a radius of 2*t* in a die with a 1-3/8-in. opening, the effective opening would be 1-3/8—1/4—1/8 = 1-1/4. Therefore, the required load would be 5 x 1.8 x 2.0 = 18 tons.

2. The factor for any other alloy can be determined by dividing the typical tensile strength by 22 ksi.

FIGURE 3.15 Approximate force per lineal foot required to make a 90° bend. (Courtesy Aluminum Corporation of America. With permission.)

## Dimensioning Practices for the Press Brake

Practical experience has shown that dimensioning and measuring practices must be understood and agreed on by all parties to achieve a mutual, workable standard. Formed sheet metal parts present a unique problem in that angular tolerances as well as flatness conditions interact with single plane dimensions because of the flexibility of sheet metal, especially the thinner gauges. To achieve consistent results when measuring formed parts, a standard must be established on where and how dimensions are to be taken.

Form dimensions should be measured immediately adjacent to the bend radius in order not to include any angular and flatness discrepancy. See Figure 3.16a for a simple illustration. Figure 3.16b is a typical stress–strain curve, showing the elastic springback in the straight portion of the curve (within the elastic limit) and the plastic strain, or permanent deformation, beyond this point, shown as a curved line. The deformation of a formed flange of a sheet metal part varies with the type of forming, the material, the height of the flange, and so forth.

Feature-to-feature dimensions on formed legs of any length on flexible parts will be assumed to be measured in constrained condition, holding the part fixtured to the print's angularity specification. This standard is appropriate for the majority of thin sheet metal parts and results in a functional product. This is always true when the assembled part is, by design, held in the constrained condition. For the most economical production, dimension the part in a single direction whenever possible. Because of the sequential nature of the forming process, and the fact that dimensional variation is introduced at each bend, dimensioning in a single direction parallels the process and helps to control tolerance accumulation. It is generally recommended that dimensioning be done from a feature to an edge. Feature-to-feature dimensions may require special fixtures or gauging.

## Elements of Forming

Figure 3.17a shows some of the elements of formed stampings. A stretched flange is easier to form than a flange that needs to be shrunk. It may tend to thin out, but it will not buckle as a shrink flange will. The continuous corner is especially difficult and is usually limited in flange height to prevent buckling. Figure 3.17b shows the drawing of the part as it would be seen by the manufacturing engineer, the tool designer, the toolmaker, and the inspector. The preferred dimensioning system is shown.

## 3.1.5   Deep Drawing

The force exerted by the punch varies with the percent of reduction, the rate of strain hardening, and the depth of the draw. Figure 3.18 shows the drawing force to take about 43% reduction in 5052–O aluminum alloy. It can be seen that maximum load occurs at about 50% depth of draw. For most alloys, peak loads occur at one-half to two-thirds depth of draw. Aluminum, like other metals, strain-hardens during draw operations and is changed to a harder temper with a corresponding increase in tensile

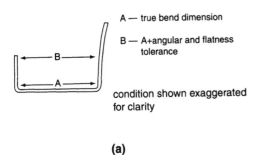

A — true bend dimension

B — A+angular and flatness
    tolerance

condition shown exaggerated
for clarity

**(a)**

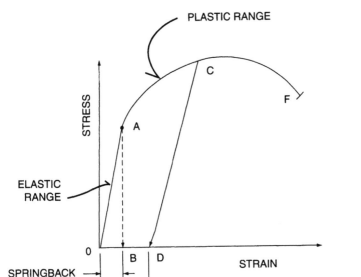

**(b)**

**FIGURE 3.16** (a) Measurement of form dimension adjacent to bend radius, (b) Typical stress–strain curve showing elastic springback.

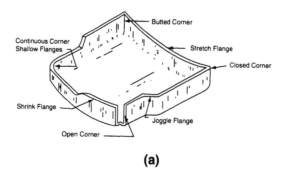

**(a)**

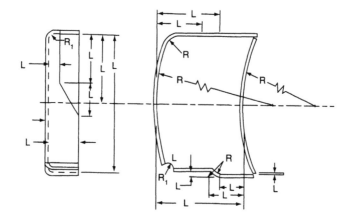

Preferred dimensioning and points to measure:

L   = Linear dimensions; corner radius

R   = Radil

R1  = Typical inside bend or

R2  = Radius in flat blank

T   = Material thickness

**(b)**

**FIGURE 3.17** (a) Elements of formed stamping, (b) Drawing of the part as designed. (Courtesy Dayton Rogers Manufacturing Co. With permission.)

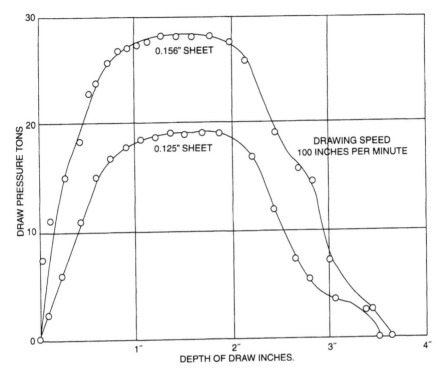

**FIGURE 3.18** Effect of depth of draw on drawing force (43% reduction on 10.0-in.-diameter blank, 5052-O alloy). (Courtesy Aluminum Corporation of America. With permission.)

and yield strength. Figure 3.19 shows the effect of drawing on the mechanical properties of 3003–0 and 5052–O alloys.

If a part is to be drawn successfully, the force exerted by the face of the punch must always be greater than the total loads imposed on that portion of the blank between the blankholder and the die. Also, the metal between the edges of the punch and die must be strong enough to transmit the maximum load without fracturing. This set of conditions establishes a relationship between blank size and punch size for each alloy-temper combination and dictates the minimum punch size that can be employed in both draw and redraw operations. For circular parts, this relationship, stated in terms of percent reduction of the blank, allows the punch diameter to be 40% smaller than the blank diameter and still produce good parts consistently in the first draw operation, and 20% and 15% smaller for second and third operations, respectively, when drawing annealed tempers of most alloys (Figure 3.20). The full hard tempers of low-strength alloys can often withstand up to 50% reduction, while high-strength alloys, which work-harden rapidly, are limited to 30–35% reduction for first draw operations. In the latter case, redrawing may involve intermediate annealing operations. The edge condition of drawn parts is shown in Figure 3.21. The top view shows "earring," which is a function of material properties.

| Alloy | Number of draws | Tensile strength psi | Yield strength psi | Elongation, in 2 in. percent |
|---|---|---|---|---|
| 3003 | 0 | 16,000 | 6,000 | 30 |
| | 1 | 19,000 | 17,000 | 11 |
| | 2 | 22.000 | 21,000 | 9 |
| | 3 | 23,500 | 22,000 | 8 |
| | 4 | 24,500 | 22,500 | 8 |
| 5052 | 0 | 28,000 | 13,000 | 25 |
| | 1 | 34,500 | 32,000 | 6 |
| | 2 | 39,500 | 36,000 | 6 |
| | 3 | 43,000 | 37,000 | 6 |
| | 4 | 44,000 | 38,000 | 6 |

Specimens taken from sidewall at top of shell.

**FIGURE 3.19** Effect of deep drawing on mechanical properties of aluminum. (Courtesy Aluminum Corporation of America. With permission.)

## DIE DIMENSIONS FOR DRAWING CYLINDRICAL SHAPES

First draw . . . . . . . . . . . . . . . . . . . . Punch diameter plus 2.2 times thickness of blank
Second draw . . . . . . . . . . . . . . . . . . Punch diameter plus 2.3 times thickness of blank
Third and succeeding draw . . . . . . . . . Punch diameter plus 2.4 times thickness of blank
Final draw of tapered shells . . . . . . . . . Punch diameter plus 2.0 times thickness of blank

## REDUCTIONS IN DIAMETER FOR DEEP SHELLS[1]

| Operation | 1100, 3003, 3004, 3005, 5005, 5050, 5052, 5457, 6061 | 2014, 2024, 5083, 5086, 5154, 5456 |
|---|---|---|
| Blank (D) | — | — |
| First draw ($D_1$) | 0.40D | 0.30D |
| Second draw ($D_2$) | $0.20D_1$ | $0.15D_1$ |
| Third draw ($D_2$) | $0.15D_2$ | $0.10D_2$ |
| Fourth draw ($D_1$) | $0.15D_2$ | — |

[1] Based on annealed blanks

**FIGURE 3.20** Relationship of blank size to punch size, in terms of percent reduction. (Courtesy Aluminum Corporation of America. With permission.)

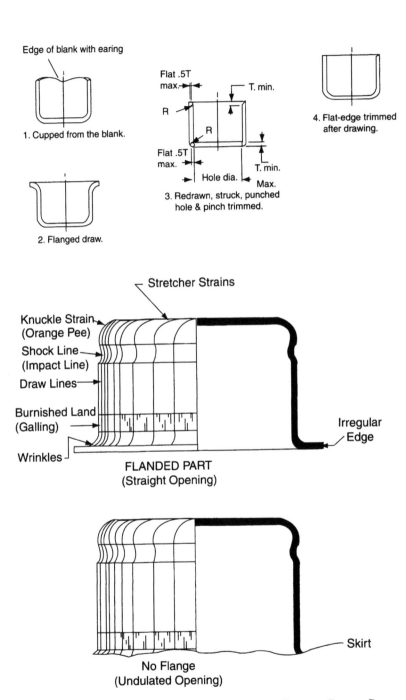

**FIGURE 3.21** Edge condition of deep-drawn parts. (Courtesy Dayton Rogers Manufacturing Co. With permission.)

## 3.2  MACHINING

### 3.2.1   Introduction to Machining

The basic machine tools were introduced in Chapter 2, to help explain the meaning of some of the dos and don'ts of design for machining. This subchapter goes a little deeper into the machine functions and operations. Chip formation, cutters, feeds and speeds, power requirements for machining, and so forth are defined and discussed. We discuss the software systems that are so critical in converting a product design (although it may be in digital form on CAD) into machine codes, and actually cutting a part in the shop. Due to page limitations, there is not sufficient space to introduce all of the new machines and machining centers on the market today. It is to the point now where one cannot always distinguish between a milling machine and a lathe, since many machines are able to do both functions, as well as drilling, reaming, broaching, and so forth. References to some of the many books available are supplied. The author's recommendation is first to understand the product requirements, then to define the machining requirements, and finally to contact machine suppliers to obtain the latest machine-tool information for a particular project.

In this subchapter, the basic principles of the machining processes are described and fundamental definitions given, after which chip formation and the process conditions are discussed. Material removal can be based on four fundamental removal methods, which illustrate the relationship between the imprinting of the information and the energy supply. Figure 3.22 shows the classification of mass-reducing processes in terms of the process and methods of material removal. The mechanical processes of turning, milling, and drilling are the main subjects of this subchapter. Blanking, punching, and shearing were covered in Subchapter 3.1.

Mass-reducing processes are used extensively in manufacturing. They are characterized by the fact that the size of the original workpiece is sufficiently large that the final geometry can be circumscribed by it, and the unwanted material is removed as chips, particles, and so on (i.e., as scrap). The chips or scrap are a necessary means to obtain the desired geometry, tolerances, and surfaces. The amount of scrap may vary from a few percent to 70–80% of the volume of the original work material. Most metal components have been subjected to a material-removal process at one stage or another.

Owing to the rather poor material utilization of the mass-reducing processes, the anticipated scarcity of materials and energy, and increasing costs, development in the last decade has been directed toward an increasing application of mass-conserving processes. These include casting, forging, powder metal, and deforming processes resulting in a near-net-shape product, without extensive metal removal. However, die costs and the capital cost of machines remain rather high; consequently, the mass-reducing processes are in many cases the most economical, in spite of the high material waste. Therefore, it must be expected that the material-removal processes will maintain their important position in manufacturing for the next several years. Furthermore, the development of automated production

| Category of basic process | Fundamental removal method | Example of processes |
|---|---|---|
| Mechanical | I 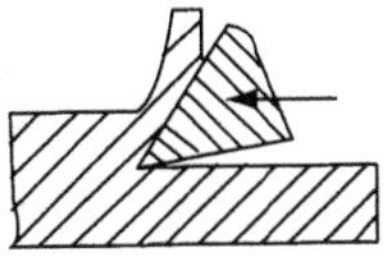 | Cutting:<br> Turning<br> Milling<br> Drilling<br> Grinding, etc. |
| | II 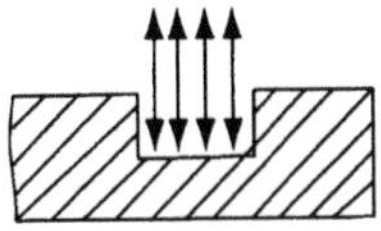 | Water jet cutting<br>Abrasive jet matching<br>Sand blasting, etc. |
| | III 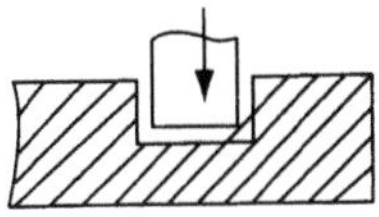 | Ultrasonic machining |
| | IV 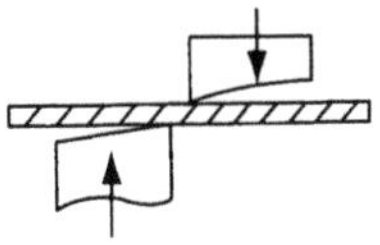 | Blanking<br>Punching<br>Shearing |
| Thermal | II 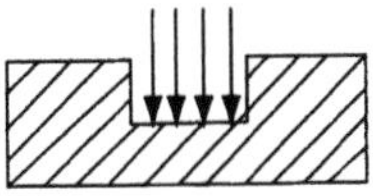 | Thermal cutting (melting)<br>Electron beam machining<br>Laser machining |
| | III 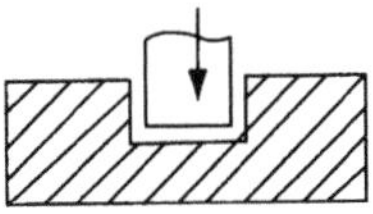 | Electrodischarge machining |
| Chemical | II 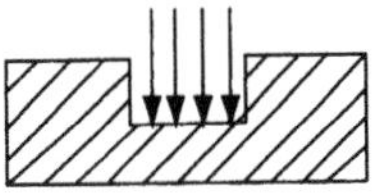 | Etching<br>Thermal cutting (combustion) |
| | III 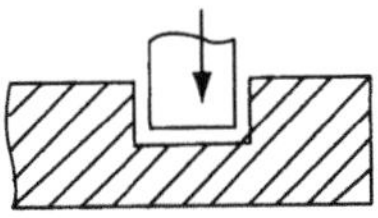 | Electrochemical machining |

**FIGURE 3.22** Classification of mass-reducing and mass-conserving processes. (From Leo Alting, *Manufacturing Engineering Processes,* Marcel Dekker, New York, 1982. With permission.)

systems has progressed more rapidly for mass-reducing processes than for mass-conserving processes.

## 3.2.2  Machining Fundamentals

The unwanted material in mass-reducing processes, based on mechanical removal method I in Figure 3.22, is removed by a rigid cutting tool, so that the desired geometry, tolerances, and surface finish are obtained. Most of the cutting or machining processes are based on a two-dimensional surface creation, which means that two relative motions are necessary between the cutting tool and the work material. These motions are defined as the *primary motion,* which mainly determines the cutting speed, and the *feed motion,* which provides the cutting zone with new material. In turning, the primary motion is provided by the rotation of the workpiece, and the feed motion is a continuous translation of the cutting tool (see Figure 3.23). In milling, the primary motion is provided by the rotating cutter, and the feed motion by moving the workpiece.

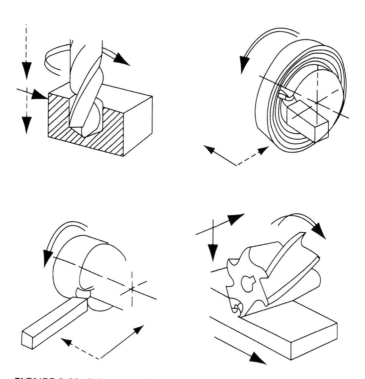

**FIGURE 3.23**  Primary motions and feed motions in machining. (From Leo Alting, *Manufacturing Engineering Processes,* Marcel Dekker, New York, 1982. With permission.)

## Cutting Speed

The cutting speed $v$ is the instantaneous velocity of the primary motion of the tool relative to the workpiece (at a selected point of the cutting edge). The cutting speed for these processes can be expressed as

$$v = dn \text{ (in m/min)}$$

where v is the cutting speed in m/min, d is the diameter of the workpiece to be cut in meters, and n is the workpiece or spindle rotation in rev/min. Thus $v$, $d$, and $n$ may relate to the work material or the tool, depending on the specific kinematic pattern of the machine.

## Feed

The feed motion $f$ is provided to the tool or the workpiece, and when added to the primary motion leads to a repeated or continuous chip removal and the creation of the desired machined surface. The motion may proceed by steps or continuously. The feed speed $v_f$ is defined as the instantaneous velocity of the feed motion relative to the workpiece (at a selected point on the cutting edge).

## Depth of Cut (Engagement)

In turning, the *depth of cut* (sometimes called "back engagement") is the distance that the cutting edge engages or projects below the original surface of the work-piece. The depth of cut determines the final dimensions of the workpiece. In turning with an axial feed, the depth of cut is a direct measure of the decrease in radius of the workpiece; and with a radial feed, the depth of cut is equal to the decrease in the length of the workpiece. For milling, the depth of cut is defined as the working engagement $a_e$ and is the radial engagement of the cutter. The axial engagement (back engagement) of the cutter is called $a_p$. In drilling, the depth of cut is equal to the diameter of the drill.

## Chip Formation

The cutting process is a controlled interaction among the workpiece, the tool, and the machine. This interaction is influenced by the selected cutting conditions (cutting speed, feed, and depth of cut), cutting fluids, the clamping of the tool and the workpiece, and the rigidity of the machine. Figure 3.24 illustrates this interaction. The clamping of the tool and the workpiece is not discussed here, and it is assumed that the machine possesses the necessary rigidity and power to carry out the process.

The chip formation mechanism is shown in Figure 3.25a. It can be seen that the shear deformation in the model is confined to the shear plane AB, extending from the tool cutting edge to the intersection of the free surfaces of the workpiece and

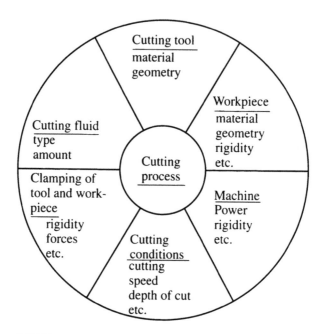

**FIGURE 3.24** Interaction among the workpiece, the tool, and the machine in chip cutting. (From Leo Alting, *Manufacturing Engineering Processes*, Marcel Dekker, New York, 1982. With permission.)

chip. In practice, shearing is not confined to the plane AB but to a narrow shear zone. At low cutting speeds, the thickness of the zone is large; at practical speeds, the thickness is comparable to that shown in Figure 3.25b and can be approximated to a plane. The angle that the shear plane forms with the machined surface is called the *shear angle*.

The chip can be considered as built up of thin layers, which slide relative to each other, as in Figure 3.25c. These layers can be compared to a stack of cards pushed toward the tool face. High normal pressures exist between the chip and the tool, causing high frictional forces and resulting in a chip with a smooth rear surface. The influence of friction is not shown in Figure 3.25c.

In the cutting process, the properties of the tool, the work material, and the cutting conditions ($h_1$, $\gamma$, and $v$) can be controlled, but the chip thickness $h_2$ ($< h_1$) is not directly controllable. This means that the cutting geometry is not completely described by the chosen parameters. The cutting ratio or chip thickness ratio, which is defined by

$$r = h_1 / h_2 \, (< 1)$$

can be measured and used as an indicator of the quality of the cutting process.

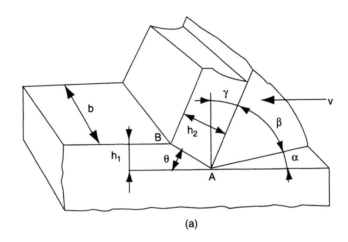

(a)

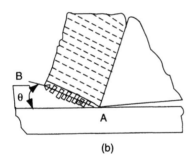

(b)

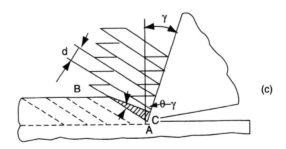

(c)

**FIGURE 3.25** Chip formation mechanism. (From Leo Alting, *Manufacturing Engineering Processes*, Marcel Dekker, New York, 1982. With permission.)

The shear angle $\phi$ can be expressed by the rake angle $\gamma$ and the inverse cutting ratio $\lambda_h$, as seen in Figure 3.25c:

$$\tan \phi = \cos \gamma/(\lambda_h - \sin \gamma)$$

The inverse cutting ratio (also called the chip compression) and the rake angle determine the shear angle $\phi$. The smaller $\phi$ is, the larger $h_2$ is, which means that the shear zone increases in length (i.e., the force and power requirements increase). Consequently, a large shear angle gives the best utilization of the supplied power. The chip compression must thus be kept as small as practically possible, since this increases the shear angle and, consequently, decreases the power consumption.

Hard work materials give lower chip compression values than do soft materials but require higher cutting forces. Friction increases the chip compression and can be reduced by introducing suitable cutting fluids. Chip compression can be further reduced by increasing the cutting speed or the feed. These increases in cutting speed and feed have an upper limit, however, because the tool life decreases, which might have a greater economic effect than the resulting increases in material-removal rate. The actual shear angle can be determined experimentally by measuring $h_2$.

## Types of Chips

Much valuable information about the actual cutting process can be gained from the appearance of the chip, as some types of chips indicate more efficient cutting than others. The type of chip is determined mainly by the properties of the work material, the geometry of the cutting tool, and the cutting conditions. It is generally possible to differentiate three types of chips: (1) discontinuous (segmental) chips, (2) continuous chips, and (3) continuous chips with built-up edges.

Discontinuous chips, shown in Figure 3.26a, represent the cutting of most brittle materials, such as cast iron and cast brass, with the stresses ahead of the cutting edge causing the fracture. Fairly good surface finish, in general, is produced in these brittle materials, as the cutting edge tends to smooth the irregularities. Discontinuous chips can also be produced with more-ductile materials such as steel, causing a rough surface. These conditions may be low cutting speeds or low rake angles in the range of 0–10° for feeds greater than 0.2 mm. Increasing the rake angle or the cutting speed normally eliminates the production of discontinuous chips.

Continuous chips, shown in Figure 3.26b and Figure 3.26c, represent the cutting of most ductile materials that permit shearing to take place without fracture. These are produced by relatively high cutting speeds, large rake angles ($\gamma = 10$–30°), and low friction between the chip and the tool face. Continuous and long chips may be difficult to handle; consequently, the tool must be provided with a chip breaker, which curls and breaks the chip into short lengths. The chip breaker can be formed by grinding a stop or a recess in the tool, or brazing a chip breaker onto the tool face.

Continuous chips with built-up edges represent the cutting of ductile materials at low speeds, where high friction exists on the tool face. This high friction causes

**FIGURE 3.26** Types of chips. (From Leo Alting, *Manufacturing Engineering Processes*, Marcel Dekker, New York, 1982. With permission.)

a thin layer of the underside of the chip to shear off and adhere to the tool face. This chip is similar to the continuous chip, but it is produced by a tool having a nose of built-up metal welded to the tool face. Periodically, portions of the built-up edge separate and escape onto the chip undersurface and the material surface, resulting in a rough machined surface, as shown in Figure 3.27a. The built-up edge effectively increases the rake angle and decreases the clearance angle, as shown in Figure 3.27b. At sufficiently high cutting speeds, the built-up edge normally disappears, and this upper limit is called the *free machining cutting speed*. A hard material will generally have a lower free machining speed than a softer material. At increasing feed, the curve in Figure 3.27c will shift to the left. In most processes, cutting speeds above the free machining speed are chosen, but for broaching, for example, it is sometimes necessary to approach the minimum.

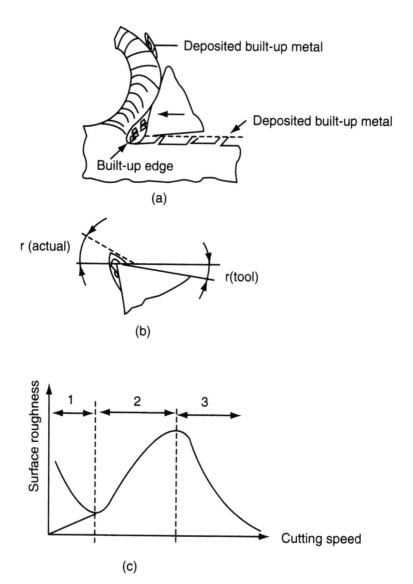

**FIGURE 3.27** Chip cutting creating a built-up edge on the tool face. (From Leo Alting, *Manufacturing Engineering Processes,* Marcel Dekker, New York, 1982. With permission.)

## Tool Material

Chip formation involves high local stresses, friction, wear, and high temperatures; consequently, the tool material must combine the properties of high strength, high ductility, and high hardness or wear resistance at high temperatures. The most important tool materials are carbon tool steels (CTS), high-speed steels (HSS), cemented or sintered carbides (CC), ceramics (C), and diamond (D).

*Carbon Tool Steel*
Plain carbon steels of about 0.5–2.0% C, when hardened and tempered, have a high hardness and strength, and can be used as hand tools for cutting softer materials at low speeds. The wear resistance is relatively low, and cutting-edge temperatures must not exceed about 300°C. This material is now used only for special purposes and has generally been replaced by the materials below.

*High-Speed Steel*
High-speed steels are alloyed steels that permit cutting-edge temperatures in the range of 500 to 600°C. The typical alloying elements are tungsten, chromium, vanadium, and cobalt. The higher cutting-edge temperatures make it possible to increase the cutting speed by about 100 over carbon tool steels—hence the name "high-speed steels." This steel is used quite extensively in twist drills, milling cutters, and special-purpose tools and is, in fact, the most common tool material.

*Sintered Carbide*
Sintered (or cemented) carbides are produced by powder metallurgical processes. Sintered carbides of tungsten carbide with cobalt as a binder are hard and brittle and are used in cutting cast iron and bronze. If titanium carbide is added or used as the main constituent, the strength and toughness can be increased, and these materials can be used in cutting hard materials. A large variety of sintered carbides exist, and each is generally developed to fulfill the requirements of effective cutting of different material groups.

Carbide cutters are very hard, and they permit an increase in cutting speeds of about 200 to 500% compared to high-speed steel tools. But it must be remembered that they have a relatively low ductility and, consequently, care must be taken to avoid high-speed impacts such as those that occur during interrupted cutting operations. Sintered carbides are, in general, used as throwaway inserts supported in special holders or shanks. The inserts may have from three to eight cutting edges, and when one edge becomes dull, the insert is indexed to a new cutting edge. This procedure continues until all edges are used, at which time a new insert is substituted.

In recent years, coated sintered carbide tools have been developed, which allow both higher cutting speeds and higher temperatures. Production rate increases of about 200% are obtainable compared to conventional sintered carbides. Titanium carbide, titanium nitride, aluminum oxide, and so on can be used as coating materials to prolong the life of the tool.

*Ceramics*
Ceramic tool materials have been developed within the last couple of decades. The material most frequently used is aluminum oxide, which is pressed and sintered. For light finishing cuts, the cutting speeds obtainable are two or three times greater than the cutting speeds for sintered carbides. Ceramics are used mainly where close

tolerances and high surface finish are required. The ceramics are produced as throw-away inserts or tips.

*Diamond*

Diamond is the hardest of all tool materials and is used mainly where a very high surface finish is required as well as close tolerances.

## The Work Material

When an economical machining operation is to be established, the interaction among the geometry, the material, and the process must be appreciated. It is not sufficient to choose a material for a product that merely fulfills the required functional properties; the suitability of the material for a particular process must also be considered. It must have properties that permit machining to take place in a reasonable way; these properties are collectively called its *machinability*.

The term *machinability* describes how the material performs when cutting is taking place. This performance can be measured by the wear on the tool, the surface quality of the product, the cutting forces, and the type of chip produced. In most cases, tool wear is considered the most important factor, which means that a machinability index can be defined as the cutting speed giving a specified tool life. Machinability tests are carried out under standardized conditions (i.e., specified quality of tool material, tool geometry, feed, and depth of cut).

The machinability of a material greatly influences the production costs for a given component. In Table 3.1, the machinability for the different material groups is expressed as the removal rate per millimeter depth of cut when turning with carbides. The table can be used only as a general comparative guide; in actual situations, accurate values must be obtained for the particular material.

The machinability of a particular material is affected primarily by its hardness, composition, and heat treatment. For most steel materials, the hardness has a major influence on machinability. A hardness range of HB from 170 to 200 is generally optimal. Low

TABLE 3.1 Removal Rate per Millimeter Depth of Cut for Different Groups of Materials When Turning with Carbides

| Material | Removal rate/mm depth of cut (mm² /min) |
| --- | --- |
| Construction steel | 47.000–63,000 |
| Tool steel (annealed) | 15.000–37,000 |
| Stainless steel | 17.000–43,000 |
| Cast steel | 20.000–27,000 |
| Cast iron | 13.000–23,000 |
| Copper alloys | 50.000–63,000 |
| Brasses | 60.000–70,000 |

hardness tends to lead to built-up edge formation at low speeds. High hardness, above HB = 200, leads to increased tool wear, as seen in Figure 3.28, which gives the machinability as the cutting speed for a tool life of 30 min ($T_{30}$) for hardened and tempered alloy steel. Sometimes it is preferable to accept a lower tool life when machining hard materials (HB from 250 to 330) instead of annealing and rehardening the material.

The heat treatment of the work material can have a significant influence on its machinability. A coarse-grained structure generally has better machinability than a fine-grained structure. The distribution of pearlite and cementite has a definite influence too. It should be mentioned, however, that hardened plain carbon steels (< 0.35% C) with a martensitic structure are very difficult to machine. Inclusions, hard constituents, scale, oxides, and so on have a deteriorating effect on the machinability as the abrasive wear on the tool is increased.

Figure 3.28 shows machinability as a function of hardness for different material groups. The machinability is again defined as the cutting speed giving a tool life of 30 min. In the figure it can be seen that hardened and tempered materials, in spite of their higher hardness, have machinabilities approximately as high as the softer materials in turning and milling. In drilling, increased hardness results in poorer machinability.

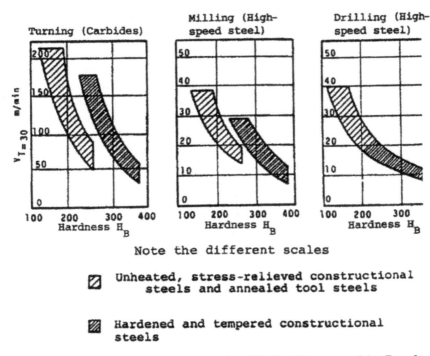

FIGURE 3.28 Machinability as a function of tool life for different materials. (From Leo Alting, *Manufacturing Engineering Processes*, Marcel Dekker, New York, 1982. With permission.)

**Surface Quality (Roughness)**

In a machining process, a specific geometry is produced, which also implies that a surface of satisfactory quality must be achieved. A machined surface always deviates from the theoretical surface. The real surface looks like a mountain landscape. Figure 3.29a shows definitions of roughness height and arithmetic average; Figure 3.29b shows that the corner radius of the cutting tool and the feed determine the surface roughness in turning. Roughness height in face milling is shown in Figure 3.29c. It can be concluded that the roughness decreases (i.e., the surface quality improves) when the feed is decreased, the nose radius is increased, and both the major cutting-edge angle and the minor cutting-edge angle are reduced. Furthermore, increasing cutting speeds and effective cutting lubricants can improve the surface quality. Typical roughness values are shown in Table 3.2 for the various machining processes.

## 3.2.3   The Lathe

The turning process is characterized by solid work material, two-dimensional forming, and a shear state of stress. The work-piece ($W$) is supported [e.g., clamped in a chuck ($C$) and supported by a center] and rotated (the primary motion, $R$). Through the primary motion ($R$) and the translatory feed ($T_a$ = axial feed for turning and $T_r$ = radial feed for facing) of the tool ($V$), the workpiece is shaped. (See Figure 3.30.)

Turning is used primarily in the production of various cylindrical components, with a nearly unlimited number of external and internal axial cross-sectional shapes (including tapers, threads, etc.). Facing is used for both regular and irregular shapes. Turning is the most extensively used industrial process. The material should not be too hard (HB <300) and should possess a minimum of ductility to confine deformation mainly to the shear zone. Turning provides close tolerances, often less than ±0.01 mm. Tighter tolerances may be obtained. The surface quality is good, normally in the range of 3 < $R_a$ < 12 μm. A wide variety of lathes are on the market: the engine lathe, the turret lathe, single and multispindle screw machines, automatic lathes, and CNC lathes.

The engine lathe shown in Figure 3.31 forms the basis of our introduction to lathe work. The tool post shown probably contains a single cutting tool, fixed to the carriage and therefore capable of movement fore and aft, and left or right. The old machine would be fitted with cranks to permit the operator to change the travel of the cutting tool. He or she would also be capable of changing the rotational speed of the spindle and the feed of the carriage by changing the lead screw speed. This was sometimes done by changing the drive belt to different steps of the drive pulleys. Later, this was done by changing gears, manually at first, then by shifting a lever to a transmission set of gears. This required an experienced machinist to make accurate parts in a reasonable amount of time. It is used today for toolroom work or for making one-of-a-kind parts.

The lathes used in production shops today are progressively improved versions of this machine. First, by mounting a turret containing several types of cutting tools on the carriage, the operator could rotate the turret to a new position and perform different operations more quickly. A second turret could be added opposite the one on the cross-slide, or a turret could be mounted on the tail stock. See Figure 3.32 for examples of various lathe cutting tools. The addition of cams and gears allows automatic change in

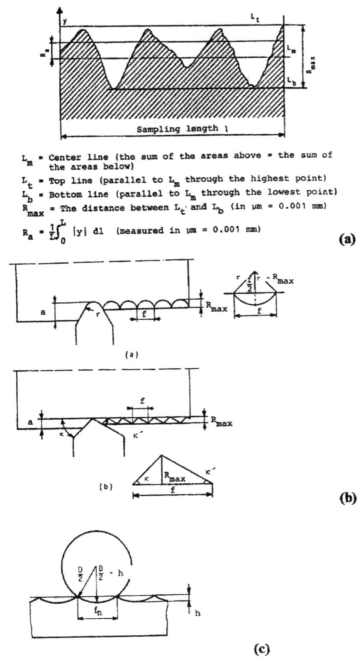

**FIGURE 3.29** Roughness of a machined surface (surface finish). (From Leo Alting, *Manufacturing Engineering Processes*, Marcel Dekker, New York, 1982. With permission.)

TABLE 3.2  Typical Roughness Values (Arithmetic
Mean Value $R_a$) for Different Processes

| Process | Roughness, $R_a$ ($+\mu$m) |
|---------|---------|
| Turning | 3–12 |
| Planing | 3–12 |
| Drilling | 3–25 |
| Milling | 1–10 |
| Grinding | 0.25–3 |

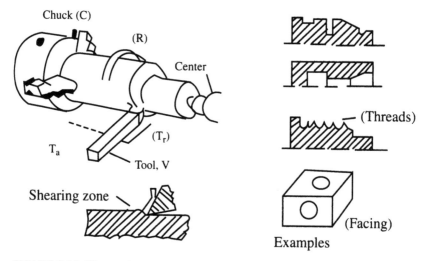

**FIGURE 3.30**  The turning process. (From Leo Alting, *Manufacturing Engineering Processes*, Marcel Dekker, New York, 1982. With permission.)

feed or speed, and indexing the turrets to different positions. After proper set-up for a particular part to be machined, a less skilled operator can now make repeatable parts.

By providing power to the turret tool holders, we can use drills, reamers, milling cutters, and so forth, and perform operations with the spindle not rotating. (For example, we can drill and tap a hole through the diameter of the cylindrical part that had been turned.) Now, however, the problem of control is much more complex. Electrical limit switches replace cams, gears, and hydraulic valves. The next logical step is to replace limit switches with controllers that can simply be programmed to sequence the desired functions. With today's microprocessors, PCs are being used on the more complex programs and machines, for both basic housekeeping functions as well as tool path programs.

A lathe today looks like any other machine in the shop. It probably has a nicely painted housing around it; it may also have two turrets that can perform

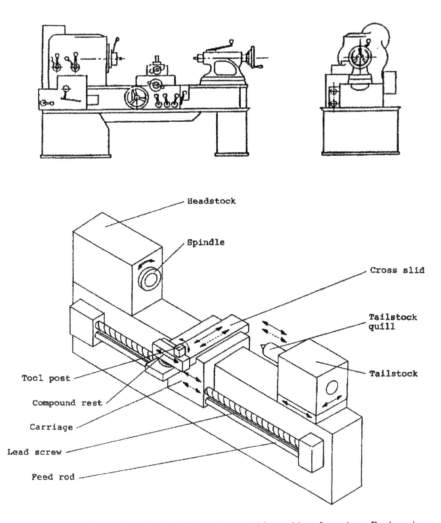

**FIGURE 3.31** The engine lathe. (From Leo Alting, *Manufacturing Engineering Processes*, Marcel Dekker, New York, 1982. Courtesy Cincinnati Machine Tool Co. With permission.)

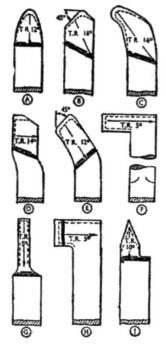

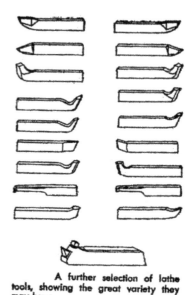

A selection of lathe tools. A, round-nose bore; B, a better shape; C, a roughing tool; D, knife tool; E, tool for sliding and surfacing motions; F, square-nose boring tool; G, parting tool; H, recessing tool; J, screw-cutting tool.

A further selection of lathe tools, showing the great variety they may have.

FIGURE 3.32 Examples of lathe cutting tools. (From Leo Alting, *Manufacturing Engineering Processes*, Marcel Dekker, New York, 1982. With permission.)

operations simultaneously, both of them with power drives, powerful electrical motors with controllable speeds, and numerical controls (NC) with computers or advanced controllers to act as the director of this multipurpose machine. Rather than just cutting cylinders or tapers, the electronic controls permit cutting an infinite number of contours and shapes. Bar stock can be fed automatically through the headstock, or a programmable robot can load and unload piece parts to the machine and to a moving conveyor. Milling and drilling can be performed with the piece part either rotating or stopped. Measurements of power consumption of the various operations can alert the operator that a cutting tool is getting dull or has broken. Measurements of various features of the part can be taken automatically, and the machine will adjust that parameter to correct the out-of-tolerance trend before the parts are actually made wrong. This can feed a statistical process control (SPC) program in parallel for verification of the piece-part accuracy. One of the latest lathes that the author purchased is even capable of performing broaching from the powered tool turret while the spindle is stopped. Figure 3.33 is an

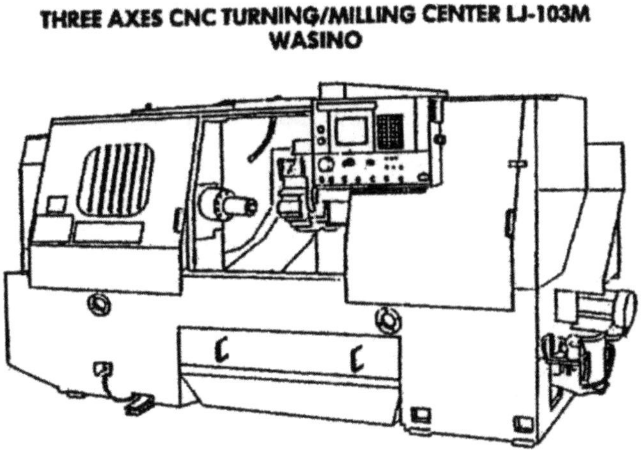

## THREE AXES CNC TURNING/MILLING CENTER LJ-103M
### WASINO

**FIGURE 3.33** Wasino Turn-Mill Center (lathe). (Courtesy McDonnell Douglas Corp. With permission.)

example of such a lathe. It can automatically feed bar stock through the spindle, or act as a "chucker" by changing the collet to a faceplate of sorts to mount castings or forgings. This particular machine is outfitted with a five-axis Fanuc robot to load and unload parts.

With all the advancements made in the lathe, there are several other important actions that have been taken. Material selection is now a very precise process, and cutting tools have been characterized and improved. Without these advances, the higher feeds and speeds in use today would be impossible. The machines themselves have been designed to be much more rigid, and can hold tighter tolerances consistently. In general, the higher the cutting speed, the lower the forces required, and the better the finish.

The computer programs that control today's NC machines can be generated on the floor by an operator using CNC, or they can be developed in an office environment and sent to the floor controls through a coax or fiber optics cable by distributed numerical control (DNC). When the product design is in digitized format and uses one of the many computer-aided design (CAD) programs available, many of the piece-part machining programs can almost be prepared automatically using computer-aided manufacturing (CAM) software. Although CAD/CAM has a long way to go to become more universally used, it is technically available today. The machine manufacturers and software companies have not yet agreed on a standard that would make all machines capable of running all program instruction sets. The difficulty of actually converting the different product design CAD programs to programs that direct the machine to make a part should not be minimized. The wide variety of hardware and software available today is rapidly changing along with the computer industry as a whole. Conversion can be a difficult problem.

Figure 3.34 shows the definitions of turning, and mathematical relationships of the different elements of lathe work.

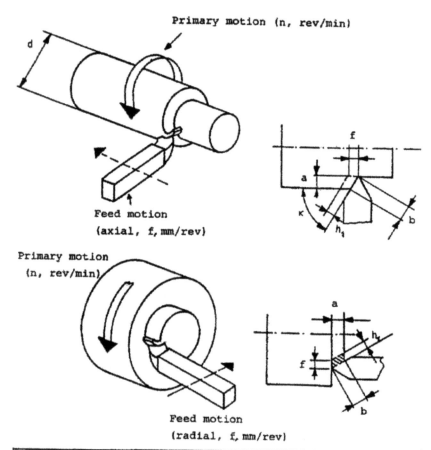

| Feed | (mm/rev) | $f$ |
|---|---|---|
| Depth of cut (Back engagement) | (mm) | $a$ |
| Cutting speed | (m/min) | $v = \pi dn$ |
| Area of cut | (mm²) | $A = bh_1 = fa$ |

**FIGURE 3.34** Mathematical relationship of the elements of lathe work. (From Leo Alting, *Manufacturing Engineering Processes,* Marcel Dekker, New York, 1982. With permission.)

### 3.2.4   The Milling Machine

The milling process is characterized by solid work material, two-dimensional forming (one-dimensional forming may be used in a few cases), and a shear state of stress. The workpiece ($W$) is clamped on the table ($B$), which is given a translatory feed ($T$) that together with the primary motion ($R$) of the cutter ($V$) provides the many geometric possibilities. The milling process through the various types of cutters and the wide variety of machines is a versatile, high-production process. Through various accessories (dividing head, attachments, etc.), many different special shapes can be produced. The milling process comes close to turning in extensive industrial use, since the geometric possibilities are enormous and the removal rate high. See Figure 3.35 for examples of a horizontal mill and a vertical mill.

The hardness of the material should not be too high (HB < 250–300), and a minimum of ductility is advisable. The obtained tolerances are normally good ($\pm0.05$ mm), and the surface quality high ($1 < R < 10$ $\mu$m). A wide variety of milling machines are available: the plain column-and-knee type (general purpose), the universal column-and-knee type, the bed type, and the planer type.

Much of the previous discussion on lathes applies to milling machine progress. In a sense, the milling machine is the opposite of the lathe in that it provides cutting action by rotating the tool while the sequence of cuts is achieved by reciprocating the workpiece. The sequence of consecutive cuts is produced by moving the workpiece in a straight line, and the surface produced by a milling machine will normally be straight in at least one direction. A milling machine, however, uses a multiple-edged tool, and the surface produced by such a tool conforms to the contour of the cutting edges. If the milling cutter has a straight cutting edge, a flat surface can be produced in both directions. The workpiece is usually held securely on the table of the machine, or in a fixture clamped to the table. It is fed to the cutter or cutters by the motion of the table. Multiple cutters can be arranged on a spindle, separated by precision spacers, permitting several parallel cuts to be made simultaneously. Figure 3.36 shows some of the typical cutters in use.

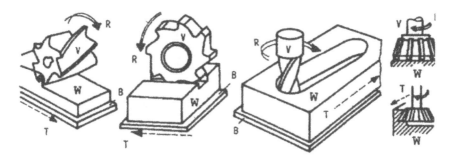

**FIGURE 3.35** Examples of horizontal and vertical milling machines. (From Leo Alting, *Manufacturing Engineering Processes,* Marcel Dekker, New York, 1982. With permission.)

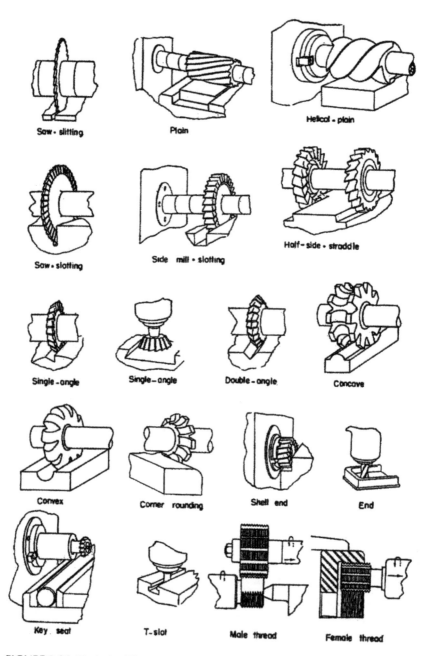

**FIGURE 3.36** Typical milling cutters in use. (Drawings courtesy of Illinois Tool Works. With permission.)

The material is moved in an $x$–$y$ direction for each pass, and can be moved toward or away from the cutting tool to change the depth of cut. The plain column-and-knee-type horizontal milling machine shown in Figure 3.37 has been a standard for many years. The universal milling machine resembles the plain-type mill shown in Figure 3.37. The chief difference lies in the fact that the table is supported on and carried by the housing, which swivels on top of the saddle. Thus, the table can also be rotated in the horizontal plane. This arrangement permits cutting helices, for milling flutes in twist drills or milling cutters. With an indexing arrangement, it is useful for cutting gear teeth.

The vertical milling machine derives its name from the position of the spindle, which is located vertically and at right angles to the top surface of the table. The vertical milling machine is especially adapted to operations with end mills and face mills, for profiling interior and exterior surfaces, for milling dies and metal molds, and for locating and boring holes. Figure 3.38 shows one of the "old standard" vertical mills.

There are special-purpose production milling machines of various types. A duplex mill, for example, has two horizontal spindles mounted on independently adjustable spindle carriers, which slide on two headstocks placed on opposite sides of the bed. Two identical or two different milling operations can be performed simultaneously on one or more workpieces.

The improvement history of milling machines is much like that of the lathe, and perhaps even more features and capabilities are now available. Figure 3.39 shows a modern vertical milling machine. It is truly difficult to identify today's machining centers as either a lathe or a milling machine. The question becomes one of the capability of a particular machine to make a particular part, versus the cost of producing the part on another type of machine. Of course, the lathe still primarily rotates the material to be cut, and the mill primarily rotates the cutters.

For years the milling machine has been considered a three-axis machine, with the cutter fixed (either horizontal or vertical), the table moving past the cutter in

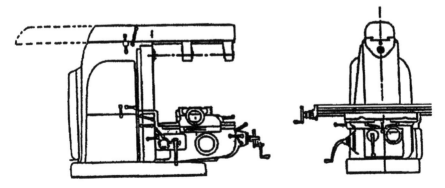

**FIGURE 3.37** Column-and-knee-type horizontal milling machine. (Courtesy Cincinnati Machine Tool Co. With permission.)

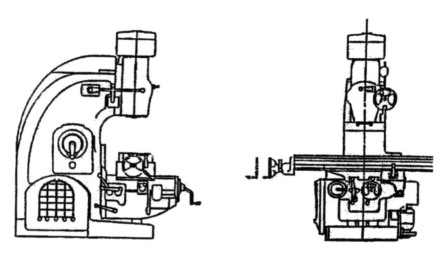

**FIGURE 3.38**  Example of an "old standard" vertical milling machine. (Courtesy of Kerney & Trecker Machine Tool Co. With permission.)

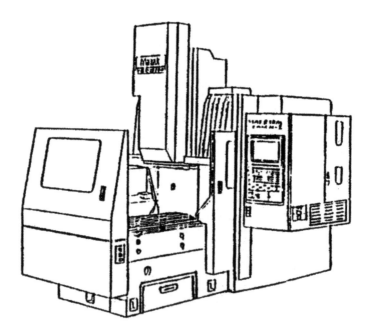

**FIGURE 3.39**  Example of a modern vertical milling machine.

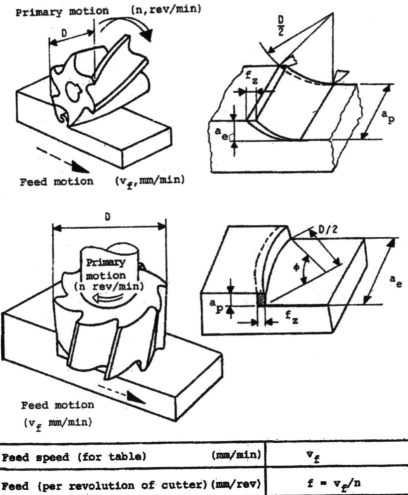

| Feed speed (for table) | (mm/min) | $v_f$ |
|---|---|---|
| Feed (per revolution of cutter) | (mm/rev) | $f = v_f/n$ |
| Feed (per tooth) | (mm/tooth) | $f_z = v_f/nz$ |
| Cutting speed | (m/min) | $v = \pi Dn$ |
| Removal rate V | (cm$^3$/min or mm$^3$/min) | $V = a_e a_p v_f$ |

**FIGURE 3.40** Mathematical relationships of milling work. (From Leo Alting, *Manufacturing Engineering Processes,* Marcel Dekker, New York, 1982. With permission.)

the $x$ axis, toward the cutter (90° to the table motion) as the $y$ axis, and up and down movement of the table called the $z$ axis. With today's machines, it is possible to tilt or rotate the table, or tilt the cutter spindle or both, creating five- and six-axis machines. Some grinders have six axes of motion. The use of multiple axes simplifies the number of cutters required and permits smooth transitions, but greatly increases the complexity of the programs and the training of the programmers. Figure 3.40 shows the definitions of milling, and mathematical relationships of the different elements of mill work.

## 3.2.5 Drilling

The drilling process is also characterized by solid work material, two-dimensional forming, and a shear state of stress. The workpiece ($W$) is clamped on a table ($B$) and the tool ($V$) is given a rotation (the primary motion, $R$) and a translatory feed ($T$). In drilling on lathes, the workpiece is rotated and the feed is applied to the tool. See Figure 3.41. The drilling process is used primarily to produce interior circular, cylindrical holes. Through various tools (twist drills, combination drills, spade drills, gun drills, etc.), different hole shapes can be produced (cylindrical holes, drilled and counterbored, drilled and countersunk, multiple-diameter holes, etc.). Drilling is an important industrial process.

The hardness of the material should normally not exceed HB = 250. For diameters less than 15 mm, the normal tolerance is around ±0.3 mm. Finer tolerances may be obtained, but finishing is often carried out by a special reaming process. The surface roughness is typically $3 < R_a < 25$ μm.

Many types of drilling machines are available: bench, upright, radial, deep-hole, and multispindle. Figure 3.42 shows the definitions of drilling, and mathematical relationships of the different elements of drill work.

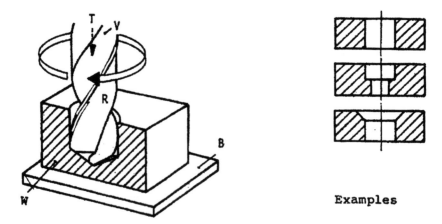

Examples

FIGURE 3.41 The drilling process. (From Leo Alting, *Manufacturing Engineering Processes*, Marcel Dekker, New York, 1982. With permission.)

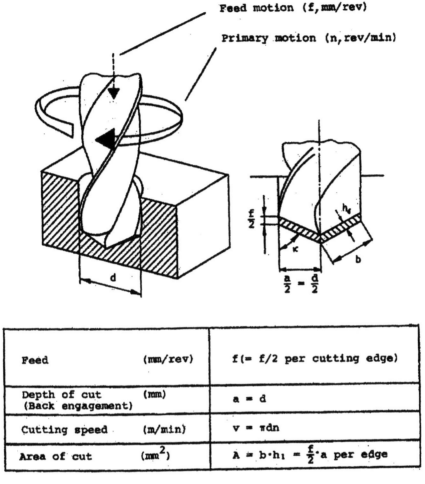

| Feed | (mm/rev) | f(= f/2 per cutting edge) |
|---|---|---|
| Depth of cut (Back engagement) | (mm) | $a = d$ |
| Cutting speed | (m/min) | $v = \pi dn$ |
| Area of cut | (mm$^2$) | $A = b \cdot h_1 = \frac{f}{2} \cdot a$ per edge |

**FIGURE 3.42** Mathematical relationships of the elements of drilling. (From Leo Alting, *Manufacturing Engineering Processes*, Marcel Dekker, New York, 1982. With permission.)

## 3.3  EXTRUSION AND FORGING PROCESSES

### 3.3.1  Introduction to Extrusion and Forging Processes

In the previous two subchapters, we discussed sheet metal fabrication (3.1) and conventional machining processes (3.2). While these processes are important to the manufacturing industry, there are other methods of making parts; and with the correct consideration of some of these alternates, distinct advantages are possible.

In Chapter 2 we introduced many of the conventional processes of forging and casting, in order for the manufacturing engineer and product designer to understand the basic characteristics of each process. The advantages and disadvantages of the processes include material, fabrication cost, mechanical properties, and tolerances.

In this subchapter we look in more detail at some of the mass-conserving processes as defined by Leo Alting in *Manufacturing Engineering Processes* (Marcel Dekker, New York, 1982). There are advantages to fabricating a product in a minimum number of process steps. If a part can be made in one operation, by the closing of a mold, for example, and obtain the desired net shape or even its "near net shape" and still possess the desired mechanical properties, tolerances, and so forth, we can reduce or eliminate costly machining finishing operations. In other words, there is more than one way to skin a rabbit once you have decided you want rabbit fur.

The author was once faced with the problem of buying small precision aluminum investment castings, made to close tolerances at a cost of $2.40 each. This purchase was for an assembly that had a very aggressive low-price target and a quantity of 100,000 pieces. One option was to buy a low-cost plaster molding of the same material for $0.70 and finish machine it to obtain the required final shape and tolerances. Our final decision was to buy the low-cost shell molding and "straighten" it by pressing it in a warm mold exceeding the yield strength in compression. This process corrected any dimensional errors of thickness, warpage, surface finish, and dimensions between planes. Total cost was less than $1.00 for material and labor, including the steel-forming die amortization. Side benefits were the increased strength and stiffness of the part. Later production of the same part by a different supplier was by machining from aluminum plate on an NC milling machine, which was done at a cost of over $5.00.

The message of this subchapter is to look at some of the near-net-shape processes in the general field of pressing and casting. While they may not be applicable in all cases, they are certainly worth looking at. An example is the progress made in cold forging, warm forging, and pressing in general. On the other side, squeeze casting, oversize investment castings, and the like offer some of the mechanical properties originally thought to be obtainable only from wrought alloys. Replacing the variations in processes due to an operator with the precision process control of pressures, time, temperature, and so forth by some type of electronics permits us to accomplish operations on a production basis that were previously not cost effective. The electronics may range from a low-cost programmable logic controller to CNC or DNC computer controls.

Newer, more accurate cost accounting systems now let us see the real cost of production. Things such as energy costs, work-in-process inventory, cycle time, worker's compensation insurance costs, parts cleaning, and so forth are now visible in the modern factory as well as direct labor and material costs.

While Subchapter 3.2 covered the mass-reducing process of machining, this subchapter is more concerned with the mass-conserving processes, which come closer to a "near net shape" achieved with the primary operation.

## 3.3.2   Form and Structure of Fabrication Processes

Figure 3.43 shows the morphological structure of material processes. This figure shows the material flow, the energy flow, and the information flow. The material flow deals with the state of the material for which the geometry or the properties are changed,

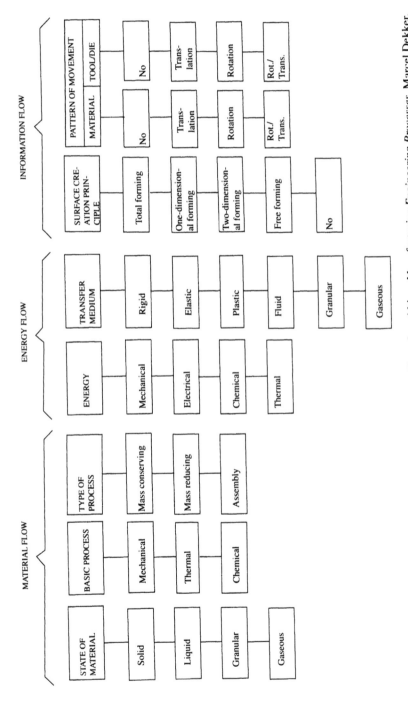

**FIGURE 3.43** Morphological structure of material processes. (From Leo Alting, *Manufacturing Engineering Processes*, Marcel Dekker, New York, 1982. With permission.)

the basic processes that can be used to create the desired change in geometry or properties, and the type of flow system characterizing the process. The material may be solid, liquid, granular, or gaseous. To carry out the basic processes described under material flow, energy must be provided to the work material through a transmission medium. The second part of Figure 3.43 shows the energy flow required. This consists of the tool-and-die system and the equipment system. The tool-and-die system describes how the energy is supplied to the material and the transfer medium used. The equipment system describes the characteristics of the energy supplied from the equipment and the type of energy used to generate this. The right side of Figure 3.43 shows the information flow, which is the impressing of shape information on the work material. The principles on which information impressing is based can be analyzed in relation to the type of process (material flow), the material, and the basic process.

Figure 3.44 shows examples of information impressing by mass-conserving processes with solid materials. The basic principles of surface creation include four possibilities:

*Free forming.* Here the medium of transfer does not contain the desired geometry (i.e., the surface/geometry is created by stress fields).

*Two-dimensional forming.* Here the medium of transfer contains a point or a surface element of the desired geometry, which means that two relative motions are required to produce the surface.

*One-dimensional forming.* Here the medium of transfer contains a producer (a line or a surface area along the line) of the desired surface, which means that one relative motion is required to produce the surface.

*Total forming.* Here the medium of transfer contains (in one or more parts) the whole surface of the desired geometry, which means that no relative motion is necessary.

Figure 3.45 shows examples of shape impressing on liquid materials. Figure 3.45a shows examples of processes where shaping and stabilizing are separate, and Figure 3.45b shows examples of processes where shaping and stabilizing are integrated.

### 3.3.3   Engineering Materials Properties

Previously, we discussed some of the basics of metallurgy and the properties of metals that we would probably be working with in the manufacturing industry. As we consider the selection of materials and processes for a particular product, we need to look at the basic information a little differently. Figure 3.46 shows typical tensile test diagrams obtained at room temperature and slow test speed. The vertical axis is the applied stress, and the horizontal axis is the strain, or deformation. The straight portion of each curve for the different materials defines the proportional limit, with the elastic limit being the point where increases in stress create a strain (deformation) that does not recover completely when the load is released. Within this proportional limit (where strain is truly

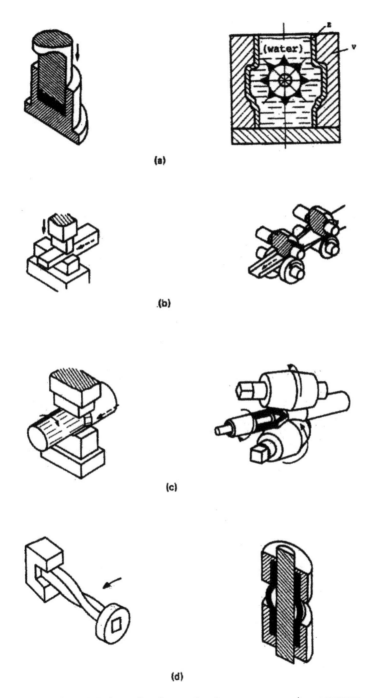

**FIGURE 3.44** Information impressing by mass-conserving processes with solid materials. (From Leo Alting, *Manufacturing Engineering Processes,* Marcel Dekker, New York, 1982. With permission.)

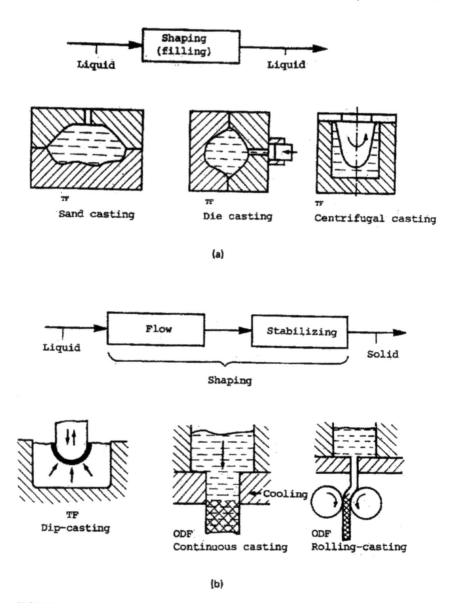

**FIGURE 3.45** Shape impressing on liquid materials.

proportional to the amount of applied stress), the ratio of stress divided by strain is called the *modulus of elasticity* (E) and defines the stiffness of the material. We could say that alloyed steel is stiffer than aluminum. (E for steel is approximately 30,000,000 psi, and E for aluminum is about 10,000,000 psi.)

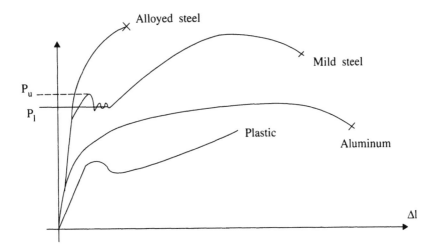

**FIGURE 3.46** Typical tensile test diagrams obtained at room temperature and slow speed. (From Leo Alting, *Manufacturing Engineering Processes,* Marcel Dekker, New York, 1982. With permission.)

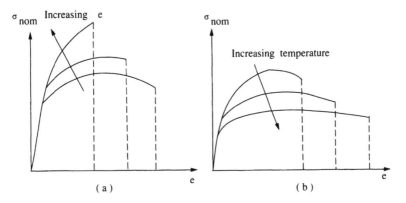

**FIGURE 3.47** Results of the rate of increasing (a) strain, and (b) temperature on strength and ductility. (From Leo Alting, *Manufacturing Engineering Processes,* Marcel Dekker, New York, 1982. With permission.)

We also know that plastic deformation is required for any permanent change in the shape of a metal, and we must exceed the proportional limit in our forming processes in order to form a part. Figure 3.47 is an important characteristic of metals in our use of mass-conserving processes. In Figure 3.47a we can see that the rate of applying strain increases the strength of the material in tension (or compression) and reduces the amount of deformation possible. In Figure 3.47b, we can see that increasing the temperature (at the same rate of strain as in Figure 3.47a) reduces the strength of the material and increases the ductility, or amount of deformation allowed before failure. Figure 3.48 shows the properties of some aluminum alloys

| Alloy and Temper | Property | 75 °F | 300 °F | 400 °F | 500 °F | 600 °F | 700 °F |
|---|---|---|---|---|---|---|---|
| 3S-H14 | Ultimate Strength (psi) | 21,500 | 17,500 | 14,000 | 10,000 | 5,000 | 3,000 |
|  | Yield Strength (psi) | 19,000 | 12,500 | 8,000 | 4,000 | 2,500 | 2,000 |
|  | Elongation % in 2 in. | 16 | 17 | 22 | 25 | 40 | 60 |
| 14S-T6 | Ultimate Strength (psi) | 70,000 | 47,000 | 18,000 | 11,000 | 6,500 | 4,500 |
|  | Yield Strength (psi) | 60,000 | 40,000 | 12,000 | 8,500 | 5,000 | 3,500 |
|  | Elongation % in 2 in. | 13 | 15 | 35 | 45 | 65 | 70 |
| 17S-T4 | Ultimate Strength (psi) | 62,000 | 40,000 | 22,000 | 12,000 | 6,500 | 4,500 |
|  | Yield Strength (psi) | 40,000 | 30,000 | 17,000 | 9,500 | 5,000 | 3,500 |
|  | Elongation % in 2 in. | 22 | 16 | 28 | 45 | 95 | 100 |
| 24S-T4 | Ultimate Strength (psi) | 68,000 | 43,000 | 26,000 | 14,000 | 7,000 | 5,000 |
|  | Yield Strength (psi) | 48,000 | 37,000 | 22,000 | 10,000 | 5,000 | 3,500 |
|  | Elongation % in 2 in. | 19 | 17 | 22 | 45 | 75 | 100 |
| 61S-T6 | Ultimate Strength (psi) | 45,000 | 32,000 | 19,000 | 7,000 | 4,000 | 3,000 |
|  | Yield Strength (psi) | 40,000 | 30,000 | 16,000 | 5,000 | 2,500 | 2,000 |
|  | Elongation % in 2 in. | 17 | 18 | 25 | 65 | 90 | 105 |

FIGURE 3.48 Elevated temperature effect on mechanical properties of aluminum. (Courtesy Aluminum Corporation of America. With permission.)

at elevated temperatures. Typical deformation velocities for various processes are shown in Figure 3.49.

Figure 3.50 shows the melting points of various metals. This is the point when all the alloying ingredients in the metal (or alloy) are in solution. This temperature is a good reference point for casting. Figure 3.51 shows the lowest possible recrystallization point for four different metals. We must exceed this point in order to obtain some

| Process | Tool/die velocity (deformation velocity) (m/s) |
|---|---|
| Tension test | $10^{-6}$–$10^{-2}$ |
| Hydraulic press | $2 \times 10^{-2}$–$3 \times 10^{-1}$ |
| Tube drawing | $5 \times 10^{-2}$–$5 \times 10^{-1}$ |
| Sheet rolling | $5 \times 10^{-1}$–25 |
| Forging | 2–10 |
| Wire drawing | 5–40 |
| High-velocity forging | 20–50 |
| Explosive forming | 30–200 |

**FIGURE 3.49**  Deformation velocities of various processes.

a. Pure metals (°C)

| | | | |
|---|---|---|---|
| Iron | 1535 | Lead | 327 |
| Copper | 1083 | Tin | 232 |
| Aluminum | 660 | Magnesium | 650 |
| Nickel | 1455 | Chromium | 1850 |
| Zinc | 419 | | |

b. Alloys (°C)

| | |
|---|---|
| Stainless steel (18% Cr, 9% Nil) | 1400–1420 |
| Brass (35% Zn, 65% Cu) | 905–930 |
| Bronze (90% Cu, 10% Sn) | 1020–1040 |
| Aluminum –bronze | 1050–1060 |
| Aluminum (1% Si, 0.2% Cu) | 643–657 |

**FIGURE 3.50**  Melting points of various materials.

| Metal | Lowest recrystallization temperature (°C) | Melting point (°C) | Upper limit for hot working (°C) |
|---|---|---|---|
| Mild steel | 600 | 1520 | 1350 |
| Copper | 150 | 1083 | 1000 |
| Brass (60/40) | 300 | 900 | 850 |
| Aluminum | 100 | 660 | 600 |

**FIGURE 3.51** Lowest possible recrystallization temperatures for various metals.

of the more desirable characteristics of the metal during the annealing and hardening processes. Also shown is the upper limit for hot working (e.g., forging). Figure 3.52 shows physical and mechanical properties of some structural materials.

### 3.3.4  Aluminum Extrusions

Extruded aluminum alloy shapes are produced by slowly forcing cast or wrought cylindrical billets, heated to plastic condition that is approximately 600 to 800°F, under hydraulic pressure through a steel die opening of a desired cross section. Billet diameters range from 4 to 16 in. and larger. Pressures up to 5500 tons are required to push the hot metal through the die opening. A schematic illustration of extrusion equipment and tools is shown in Figure 3.53. In subsequent operations, the aluminum shapes, which emerge from the extrusion press in lengths up to 80 ft, are heat-treated, straightened, and cut to the desired length.

Since extrusion dies are relatively inexpensive (often costing less than $1000), it will generally pay to design special extruded sections to meet the specific requirements of the structure to be built, rather than to sacrifice efficiency by using available standard sections nearest the desired shapes. Extruded sections can be economically tailored to meet the needs of the design from the standpoint of strength and stiffness. Metal can be placed where it will do the most good, and wall thicknesses of sections in aluminum extrusions may vary within broad limits. Typical mechanical properties of aluminum extrusion alloys are given in Figure 3.54.

### Manufacturing Possibilities

Extruded aluminum shapes offer many interesting possibilities as to sizes, weights, strengths, and types of shapes that can be produced. The maximum cross-sectional dimensions generally considered commercial are governed by a circumscribing circle approximately 12 in. in diameter; a limited number of 5500 ton presses can handle sections circumscribed by a 17-in. circle. The circumscribing circle is the smallest circle that completely encloses a shape. Section thicknesses may vary from about 0.050 in. to several inches, depending on requirements.

| Material | Tensile Strength, psi | Yield Strength, psi | Elongation, Percent In 2 Inches | Shear Strength, psi | Modular of Elasticity, psi | Spacific Gravity |
|---|---|---|---|---|---|---|
| BRASS (35% Zinc) | | | | | | |
| Hard | 76,000 | 45,000 | 7 | 43,000 | 15,000,000 | 8.46 |
| Annealed | 45,000 | 12,500 | 50 | 33,000 | 15,000,000 | 8.46 |
| BRONZE (5% Tin) | | | | | | |
| Hard | 81,000 | 75,000 | 10 | ............ | 16,000,000 | 8.86 |
| Annealed | 47,000 | 19,000 | 64 | ............ | 16,000,000 | 8.86 |
| COPPER | | | | | | |
| Hard | 50,000 | 45,000 | 6 | 28,000 | 17,000,000 | 8.90 |
| Hot Rolled | 34,000 | 10,000 | 45 | 23,000 | 17,000,000 | 8.90 |
| IRON | | | | | | |
| Gray Cast | 30,000 | 25,000 | 0.5 | 44,000 | 14,000,000 | 7.10 |
| Wrought Plate | 51,000 | 31,000 | 21 | 42,000 | 28,000,000 | 7.65 |
| MAGNESIUM | | | | | | |
| Wrought, JT Alloy | 44,000 | 32,000* | 14 | 20,500 | 6,500,000 | 1.80 |
| Sand Cast, C Alloy | 39,000 | 14,000 | 10 | 20,000 | 6,500,000 | 1.82 |
| MONEY METAL (67% NE, 30% Cu) | | | | | | |
| Hard | 110,000 | 100,000 | 8 | 87,000 | 26,000,000 | 8.80 |
| Annealed Sheet | 80,000 | 35,000 | 40 | 46,000 | 26,000,000 | 8.80 |
| PHENOLIC SHEET | | | | | | |
| Laminated Fabric-Base | 9,500 | ............ | 2 | 10,000 | 1,000,000(3) | 1.33 |
| STEEL | | | | | | |
| Carbon Cast | 75,000 | 42,000 | 24 | 60,000 | 30,000,000 | 7.86 |
| Structural, Hot Rolled | 60,000 | 38,000 | 30 | 45,000 | 28,000,000 | 7.85 |
| Stainless 18-8, Annealed | 90,000 | 40,000 | 55 | 67,000 | 29,000,000 | 7.90 |
| Stainless 18-8, Cold Rolled | 150,000 | 125,000 | 15 | 112,000 | 29,000,000 | 7.90 |
| WOOD | | | | | | |
| Hard Maple | 10,000(s) | ............ | 1.5 | 1,500 | 1,600,000 | 0.67 |
| ZINC | | | | | | |
| Die Cast | 40,000 | 26,000 | 5 | 31,000 | ................ | 6.64 |
| ALUMINUM | | | | | | |
| 3S-O | 16,000 | 6,000 | 30 | 11,000 | 10,000,000 | 2.73 |
| 3S-H18 | 29,000 | 27,000 | 4 | 16,000 | 10,000,000 | 2.73 |
| 24S-O | 26,000 | 11,000 | 19 | 18,000 | 10,600,000 | 2.77 |
| 24S-T4 | 64,000 | 42,000 | 19 | 40,000 | 10,600,000 | 2.77 |
| 52S-O | 28,000 | 13,000 | 25 | 18,000 | 10,200,000 | 2.68 |
| 52S-H38 | 41,000 | 36,000 | 7 | 24,000 | 10,200,000 | 2.68 |
| 61S-O | 18,000 | 8,000 | 22 | 12,500 | 10,000,000 | 2.70 |
| 61S-T6 | 45,000 | 40,000 | 12 | 30,000 | 10,000,000 | 2.70 |

NOTES:
(1) Values shown are typical approximate values.
(2) Per Cent of International Annealed Copper Standard.
(3) Approximate average value. Initial modules values of reinforced plastics vary over wide range.

**FIGURE 3.52** Physical and mechanical properties of some structural materials.

| Weight, Lb./Cu.in. | Melting Range (F. °) | Electrical Conductivity, Percent of Copper [2] | Thermal Conductivity, (at 212°F.) C. G. S. Units[2] | Coefficient of Thermal Expansion (68°— 212°F.), °F. x $10^{-5}$ | Specific Heat (68°— 212°F.), Cal/s/°C. | Specific Heat (68°— 212°F.), Cal/Cm$^3$/°C. |
|---|---|---|---|---|---|---|
| 0.306 | 1660-1715 | 26 | 0.29 | 10.2 | 0.091 | 0.769 |
| 0.306 | 1660-1715 | 26 | 0.29 | 10.2 | 0.091 | 0.769 |
| 0.320 | 1750-1920 | 18 | 0.19 | 9.9 | 0.09 | 0.797 |
| 0.320 | 1750-1920 | 18 | 0.19 | 9.9 | 0.09 | 0.797 |
| 0.322 | 1949-1981 | 100 | 0.93 | 9.3 | 0.092 | 0.819 |
| 0.322 | 1949-1981 | 100 | 0.93 | 9.3 | 0.092 | 0.819 |
| 0.257 | 2000-2400 | 2 | 0.12 | 5.6 | 0.13 | 0.923 |
| 0.277 | approx. 2800 | 16 | 0.17 | 6.5 | 0.114 | 0.872 |
| 0.065 | 950-1150 | 13 | 0.19 | 14.4 | 0.249 | 0.448 |
| 0.066 | 760-1110 | 12 | 0.17 | 14.8 | 0.249 | 0.453 |
| 0.318 | 2370-2460 | 3.6 | 0.06 | 7.8 | 0.128 | 1.126 |
| 0.318 | 2370-2460 | 3.6 | 0.06 | 7.8 | 0.128 | 1.126 |
| 0.042 | 320[4] | ..... | 0.0007 | 14.0 | 0.35 | 0.465 |
| 0.284 | 2670-2750 | 11 | 0.13 | 6.5 | 0.177 | 1.392 |
| 0.283 | 2765 | 12 | 0.14 | 6.5 | 0.114 | 0.895 |
| 0.288 | 2600-2680 | 2.4 | 0.04 | 9.6 | 0.12 | 0.948 |
| 0.288 | 2600-2680 | 2.1 | 0.04 | 9.6 | 0.12 | 0.948 |
| 0.024 | ................ | ..... | 0.0004 | 3.5 | 0.55 | 0.368 |
| 0.240 | 717 | 27 | 0.27 | 15.2 | 0.010 | 0.066 |
| 0.099 | 1190-1210 | 50 | 0.46 | 12.9 | 0.23 | 0.628 |
| 0.099 | 1190-1210 | 40 | 0.37 | 12.9 | 0.23 | 0.628 |
| 0.100 | 935-1180 | 50 | 0.45 | 12.9 | 0.23 | 0.637 |
| 0.100 | 935-1180 | 30 | 0.29 | 12.9 | 0.23 | 0.637 |
| 0.097 | 1100-1200 | 35 | 0.33 | 13.2 | 0.23 | 0.616 |
| 0.097 | 1100-1200 | 35 | 0.33 | 13.2 | 0.23 | 0.616 |
| 0.098 | 1080-1205 | 45 | 0.41 | 13.1 | 0.23 | 0.621 |
| 0.098 | 1080-1205 | 40 | 0.37 | 13.1 | 0.23 | 0.621 |

(4) Beginning distortion.
(5) 3-ply Plywood, parallel to grain faces.
(6) For aluminum alloys, data given at 25°C.

FIGURE 3.52 (Continued)

The permissible ratio of thickness to width of a section depends on the alloy. The softer alloys can be extruded to thinner sections than can medium- and high-strength alloys. For example, the minimum thickness for a shape about 8 in. wide is 3/32 in. for alloys 3S and 63S, but 1/8 in. for alloy 14S. The wider the shape, the more the minimum thickness may be varied.

Minimum and maximum weights per foot of aluminum extrusions are governed by the limits of the extrusion ratio, which normally should not be smaller than 16:1 nor greater than 45:1. The *extrusion ratio* is the ratio of cross-sectional area of the cast extrusion ingot to the cross-sectional area of the extruded shape. Within these limitations the weight per foot of aluminum extrusions can normally range from an ounce or less to about 20 lb, which corresponds to a cross-sectional area of over 15 in.$^2$.

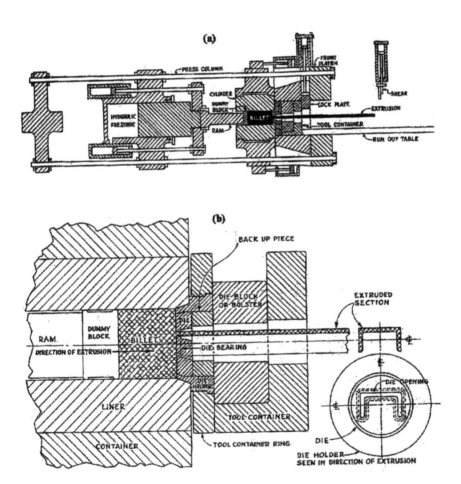

**FIGURE 3.53** Schematic of extrusion press used for producing aluminum shapes (a) A hydraulically operated ram pushes hot aluminum through the die opening, (b) Steel tools for producing aluminum shapes consist of die, backup block, die holder, die block, and tool container. (Courtesy Reynolds Metals Company. With permission.)

| Alloy and Temper | Specific Gravity | Weight, lb. per cu. in. | Approximate Melting Range (°F.) | Electrical Conductivity(1) | Theramal Conductivity at 25°C. C.G.S. Units(2) | Coefficiant of Thermal Expansion per °F. x 10$^{-6}$ | |
|---|---|---|---|---|---|---|---|
| | | | | | | 68°–212 °F. | 68°–572 °F. |
| 3S-F | 2.73 | .099 | 1190-1210 | 41 | .38 | 12.9 | 13.9 |
| 14S-O | 2.80 | .101 | 950-1180 | 50 | .46 | 12.9 | 13.6 |
| 14S-T4 | 2.80 | .101 | 950-1180 | -- | -- | 12.8 | 13.6 |
| 14S-T6 | 2.80 | .101 | 950-1180 | 50 | .37 | 12.8 | 13.6 |
| 24S-O | 2.77 | 100 | 935-1180 | 30 | .45 | 12.9 | 13.7 |
| 24S-T4 | 2.77 | .100 | 935-1180 | 30 | .29 | 12.9 | 13.7 |
| 61S-O | 2.70 | .098 | 1080-1205 | 45 | .41 | 13.1 | 14.1 |
| 61S-T4 | 2.70 | .098 | 1080-1205 | 40 | .37 | 13.1 | 14.1 |
| 61S-T6 | 2.70 | .098 | 1080-1205 | 40 | .37 | 13.1 | 14.1 |
| 63S-T42 | 2.70 | .098 | 1140-1205 | 50 | .46 | 13.0 | 14.0 |
| 63S-T5 | 2.70 | .098 | 1140-1205 | 55 | .50 | 13.0 | 14.0 |
| 63S-T6 | 2.70 | .098 | 1140-1205 | 55 | .50 | 13.0 | 14.0 |
| 75S-O | 2.80 | .101 | 890-1180 | -- | -- | 13.1 | 14.4 |
| 75S-T6 | 2.80 | .101 | 890-1180 | 30 | .29 | 13.1 | 14.4 |

**Notes:** (1) Percent of International Annealed Copper Standard.

(2) C.G.S. units = calories per second, per square centimeter, per centimeter of thickness, per degree Centigrade.

Specific heat of commercially pure aluminum is 0.226 cal/g/°C for the temperature range of 68° to 212° F. Values for the commercial alloys differ slightly.

| | | | | | | |
|---|---|---|---|---|---|---|
| 3S-F | 16,000 | 6,000 | 40 | 11,000 | 7,000 | 28 |
| 14S-O | 27,000 | 14,000 | 18 | 18,000 | 13,000 | 45 |
| 14S-T4 | 62,000 | 42,000 | 22 | 38,000 | 18,000 | 105 |
| 14S-T6 | 70,000 | 60,000 | 13 | 42,000 | 18,000 | 135 |
| 24S-O | 27,000 | 11,000 | 22 | 18,000 | 13,000 | 47 |
| 24S-T4 | 68,000 | 48,000 | 19 | 41,000 | 18,000 | 120 |
| 61S-O | 18,000 | 8,000 | 30 | 12,000 | 9,000 | 30 |
| 61S-T4 | 35,000 | 21,000 | 25 | 24,000 | 13,500 | 65 |
| 61S-T6 | 45,000 | 41,000 | 17 | 30,000 | 13,500 | 95 |
| 63S-T42 | 22,000 | 13,000 | 20 | 14,000 | ---- | 42 |
| 63S-T5 | 30,000 | 25,000 | 12 | 18,000 | ---- | 65 |
| 63S-T6 | 35,000 | 30,000 | 12 | 22,000 | ---- | 73 |
| 75S-O | 33,000 | 15,000 | 17 | 22,000 | ---- | 60 |
| 75S-T6 | 82,000 | 72,000 | 11 | 49,000 | 21,000 | 150 |

**Notes:** (1) The modulus of elasticity of all aluminum alloys is approximately 10,300,000 psi. Poisson's Ratio is about 0.33.

(2) The endurance limit values are based on 500,000,000 cycles of reversed stress using the R. R. Moore type of machine and specimen.

FIGURE 3.54 (a) Physical and (b) mechanical properties of aluminum extrusion alloys. (Courtesy Reynolds Metals Company. With permission.)

Small castings or forgings, such as brackets, clamps, or hinges, can often be changed to aluminum extrusions with considerable cost advantage if their dimensions are symmetrical about one plane. For instance, 2-in.-wide coupling clamps for aluminum irrigation pipe are cut from 63S-T6 alloy extrusion, as shown in Figure 3.55. A cut

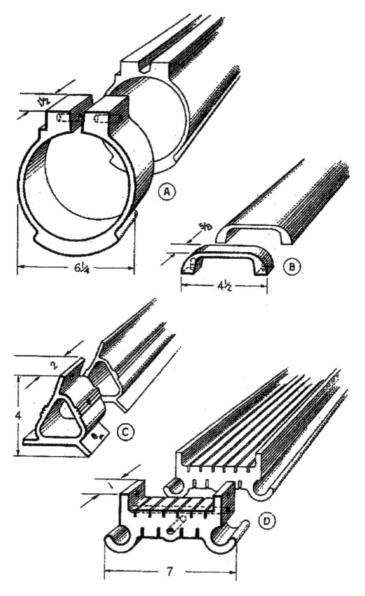

**FIGURE 3.55** Short parts cut from extruded shapes are often cheaper than castings, forgings, or parts machined from bar stock. (a) Clamp for coupling aluminum irrigation tubing. (b) Drawer pull. (c) Tripod clamp. (d) Loom part. (Courtesy Reynolds Metals Company. With permission.)

between the two heavy bosses changes the solid ring into a tightening clamp. Forging stock for aluminum die forgings is often extruded in cross sections designed to produce the desired metal flow during the forging operation and to reduce forging and trimming costs. Figure 3.56 shows a typical example.

In the author's experience, with the high-rate production of pyrotechnics, rocket motors, and so forth, the use of extrusions for the machining of rocket motors is a requirement for safety and reliability. Normal rolled aluminum stock, which is lower in cost than an extrusion, occasionally has inclusions that might permit a weak sidewall in the motor, causing an occasional failure upon rocket-motor ignition. A large gas generator used 7075 aluminum extruded stock, cut into 6-in. lengths, as the billet for forging. Again, the consistency of the extrusion aided in the reliability of the generator case.

One of the problems that an extruder has in maintaining tolerances in an extruded cross section is in the straightening process that follows extrusion. The extrusion die, of course, can be machined very accurately, which permits the metal coming through the orifice to be quite accurate. However, the length of the extrusion (40 to 80 ft) may exhibit some waviness down its length. This is clamped at each end, and pulled to slightly exceed the tensile yield strength. Upon release, the extrusions are quite straight, and when they are cut into lengths for shipment, the waviness is within the advertised limits. A more serious problem is in the cross-section tolerances. For instance, a narrow U-shaped extrusion with a relatively thick base and thin, long legs will not look the same after straightening. Normally, the legs tend to close up at the top (or unsupported) end of the U. The straightening operation may occur immediately after extrusion, as in the case of non-heat-treatable aluminum. However, it is always required after heat treatment and quenching, before the aging process. The best policy here is to talk with the extruder, to make certain that you both understand the risks and that you are able to use the product. Many times the extruder will refuse to make a quotation for an extruded shape because of this problem. If your finished part is quite short in length, neither waviness nor angle of the legs may cause a problem in your particular product.

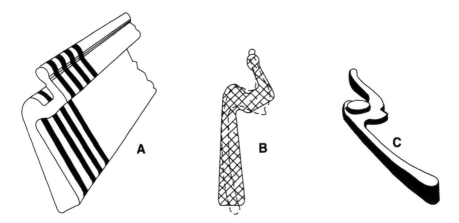

**FIGURE 3.56** Extruded stock for forging reduces cost of forging and trimming operation. (a), (b) Extruded blanks, (c) Finished forging. (Courtesy Reynolds Metals Company. With permission.)

### Dimensional Tolerances

Although extruded aluminum shapes minimize and often eliminate the need for machining, they do not possess the dimensional accuracy of machined parts, and the dimensional tolerances to which extrusions are commercially furnished must be taken into account. These tolerances, shown in Figure 3.57 and Figure 3.58, generally cover straightness, flatness, twist, and cross-sectional dimensions such as section thickness, angles, contours, and corner and fillet radii.

The tolerances on any given dimension vary somewhat depending on the size and type of the shape, relative location of the dimension involved, and other factors. Figure 3.59 illustrates many tolerances as applied to an arbitrary section used as an example.

## 3.3.5  Precision Aluminum Forging

All forgings fall into two general classes: hand forgings and die forgings. Hand forgings are sometimes called open die forgings. As the name suggests, the metal is not laterally confined when being forged to the desired shape. The forger manipulates the stock between repeated squeezes of the hydraulic press or ring roller, or blows of a hammer, in progressively shaping the forging to the desired form. These forgings have some of the desired grain-flow characteristics and require less machining than when making a finished part out of a billet or large bar stock. However, they do require significant machining to achieve a finished part.

The next step in producing a more complete part is blocker-type forgings. These are generously designed, with large fillet and corner radii and with thick webs and ribs, so that they can be produced in a single set of finishing dies only. Producing such forgings may typically require a unit pressure of 10 to 15 tons per square inch of projected plan area, depending on the alloy and the complexity of the design. This is less pressure than is necessary to make more intricate forgings. The projected plan area of the forging is used to arrive at the estimated total tonnage required.

A blocker-type forging generally requires machining on all surfaces. Economics may dictate such a design if quantity requirements are limited or if the finished-part tolerances necessitate complete finishing. A blocker-type forging is an end product and should not be confused with a blocker forging, which is a preliminary shape requiring a subsequent finishing die operation to attain its final shape. See Figure 3.60a for an example of a blocker-type forging, and Figure 3.60b for an example of a conventional forging.

Conventional forgings are the most common of all die forging types. A conventional forging is more intricate in configuration than a blocker-type forging, having proportionately lighter sections, sharper details, and closer tolerances, and thus is more difficult to forge. The design differences between the two types are graphically illustrated in Figure 3.61a and Figure 3.61b. A conventional forging requires only partial final machining. A typical unit pressure of 15 to 25 tons per square inch of plan area is required, and usually a blocking operation is required prior to the finishing operation.

The manufacturing engineer and the designer must evaluate the cost difference: A blocker-type forging has a lower die cost but will be heavier, requiring more extensive machining; a conventional forging has a higher die cost but will be lighter, requiring

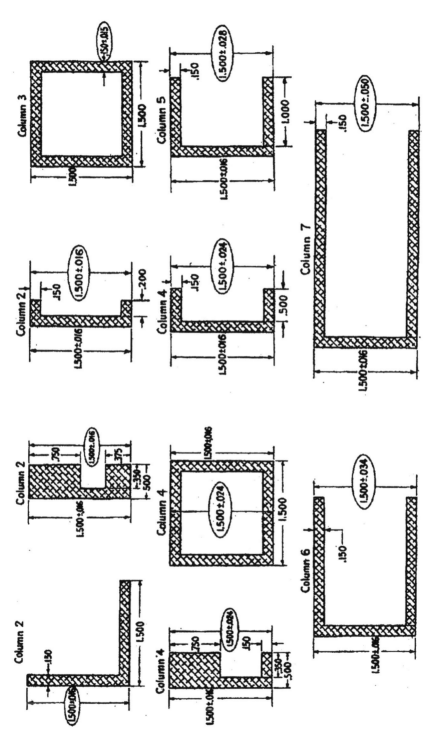

**FIGURE 3.57** Standard tolerances for aluminum extrusions (cross-sectional dimensions). (Courtesy Reynolds Metals Company.)

NOTE: See examples on opposite page (Fig. 122)

**TOLERANCES (1)—Inches Plus or Minus**

| Specified Dimension, Inches | For dimensions taken at a point where 75 percent or more of the dimension is metal | | Allowable deviation from specified dimension, where more than 25 percent of the dimension is space (2, 3) | | | |
|---|---|---|---|---|---|---|
| | All except those covered by column 3 | Wall thickness(4) completely enclosing space 0.11 sq. in. and over | .250-.624 inch from base of leg | .625-1.249 inches from base of leg | 1.250-2.499 inches from base of leg | 2.500 inches or more from base of leg |
| Column 1 | Column 2 | Column 3 | Column 4 | Column 5 | Column 6 | Column 7 |
| Up thru .124 | .006 | | .010 | .012 | .014 | .016 |
| .125– .249 | .007 | | .012 | .014 | .016 | .020 |
| .250– .499 | .008 | | .014 | .016 | .018 | .022 |
| .500– .749 | .009 | Plus or minus 10% | .016 | .018 | .020 | .026 |
| .750– .999 | .010 | max. ±.060 | .018 | .020 | .022 | .030 |
| 1.000– 1.499 | .012 | min. ±.010 | .020 | .022 | .026 | .034 |
| 1.500– 1.999 | .016 | | .024 | .028 | .034 | .050 |
| 2.000– 3.999 | .024 | | .032 | .036 | .048 | .044 |
| 4.000– 5.999 | .034 | | .042 | .050 | .064 | .088 |
| 6.000– 7.999 | .044 | | .054 | .062 | .082 | .112 |
| 8.000– 9.999 | .054 | | .064 | .074 | .100 | .136 |
| 10.000–11.999 | .064 | | .074 | .088 | .116 | .160 |
| 12.000–13.999 | .074 | | .084 | .100 | .134 | .184 |
| 14.000–14.999 | .080 | | .090 | .106 | .142 | .196 |

NOTE: The tolerances applicable to a dimension composed of two or more component dimensions is the sum of the tolerances of the component dimensions if all of the component dimensions are indicated.

NOTES:
(1) The tolerance applicable to a dimension composed of two or more component dimensions is the sum of the tolerances of the component dimensions if all of the component dimensions are indicated.

(2) At points less than 1/4 inch from base of leg, the tolerances shown in Column 2 are applicable.

(3) Where the space is completely enclosed (hollow shapes), the tolerances in Column 4 are applicable.

(4) Where the dimensions specified are outside and inside, rather than the wall thickness itself, tolerance on wall thickness shall be plus or minus 10 percent of mean wall thickness, max. ±0.060, min. ±0.010.

FIGURE 3.57 (Continued)

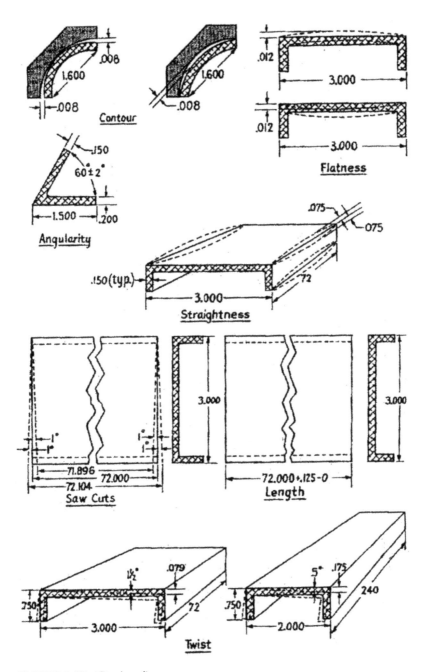

**FIGURE 3.57** (Continued)

| Type of Tolerance | Dimension to Which Tolerance Applies (1) | Tolerance |
|---|---|---|
| STRAIGHTNESS (2) | Circumscribing Circle Diameter (3):<br>Up through 1.499"<br><br>Up through 1.499"<br><br>1.500" and up | ± .0500" per foot (4) for minimum thickness up through .094"<br>± .0125" per foot for minimum thickness .095" and up<br>± .0125" per foot |
| TWIST (2) | Circumscribing Circle Diameter (3):<br>Up through 1.499"<br>1.500"—2.999"<br>3.000" and up | ± 1° per foot<br>± ½° per foot; 5° total<br>± ¼° per foot, 3° total |
| CONTOUR | Deviation from specified contour | ± .005" per inch of chord width (± .005" minimum) (5) |
| CORNER AND FILLET RADII | Sharp Corners<br>Specified Radius up through .197"<br>Specified Radius .188 and up | + 1/64"<br>± 1/64"<br>± 10 percent |
| ANGLES | Minimum Specified Leg Thickness<br>Under .188"<br>.188" to .750"<br>.750 to solid | ± 2°<br>± 1½°<br>± 1° |
| FLATNESS | | ± .004" per inch of width (± .004" minimum) |
| SURFACE ROUGH-NESS (6) | Section Thickness:<br>Up through .063"<br>.064"—.125"<br>.126"—.188"<br>.189"—.250"<br>.251"— and up | .0015" maximum depth of defect<br>.002" maximum depth of defect<br>.0025" maximum depth of defect<br>.003" maximum depth of defect<br>.004" maximum depth of defect |
| SQUARENESS OF SAWCUTS | | ± 1° |
| LENGTH | Specified Length up to 10'<br>Specified Length 10' to 30'<br>Specified Length 30' and up | + 1/8"<br>+ 1/4"<br>+ 1/2" |

NOTES:
  (1) These tolerances are applicable to the average shape. Wider tolerances may be required for some shapes, and closer tolerances may be possible for others.
  (2) Not applicable to annealed (0 temper) material.
  (3) The smallest circle that will completely enclose the shape.
  (4) When weight of shape on flat surface minimizes deviation.
  (5) Applicable to not more than 90° of any arc.
  (6) Includes die marks, handling marks, polishing marks.
  (7) Tangent values used in calculation of twist and angularity tolerances:

| Tan | 30' | .0087 | 2° 30' | .0437 |
|---|---|---|---|---|
| | 1° 0' | .0175 | 3° 0' | .0524 |
| | 1° 0' | .0262 | 4° 0' | .0699 |
| | 2° 0' | .0349 | 5° 0' | .0875 |

**FIGURE 3.58** Standard tolerances for aluminum extrusions (straightness, flatness, twist, contours, radii, angles, roughness). (Courtesy Reynolds Metals Company. With permission.)

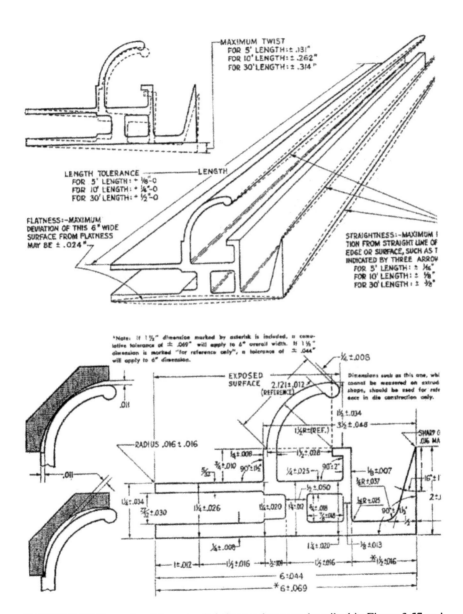

**FIGURE 3.59** Example of the extruded shape tolerances described in Figure 3.57 and Figure 3.58. (Courtesy Reynolds Metals Company. With permission.)

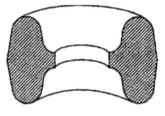 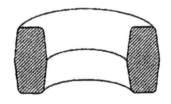

BLOCKER-TYPE FORGING – REQUIRES ONLY ONE DIE OPERATION AND A TRIMMING AND PIERCING OPERATION TO REMOVE SURPLUS METAL.

CONVENTIONAL FORGING – REQUIRES TWO DIE OPERATIONS AND A TRIMMING AND PIERCING OPERATION TO REMOVE SURPLUS METAL.

**FIGURE 3.60** Examples of (a) a blocker-type forging, and (b) a conventional forging. (From *Aluminum Forging Design Manual,* The Aluminum Association, New York, 1975. With permission.)

much less machining. Only a cost comparison by the customer can determine which type of forging will give lowest total cost.

A precision forging denotes closer-than-normal tolerances. It may also involve a more intricate forging design than a conventional type, and may include smaller fillet radii, corner radii, and draft angles, and thinner webs and ribs. The higher cost of a precision forging, including increased cost of dies, must be justified in the reduced machining required for its end use. This type of forging typically requires forming pressures of 25 to 50 tons per square inch of plan area.

The Forging Industry Association (FIA) defines a precision forging as "any forging which, by reason of tolerance, draft angle, web-to-rib ratio, or other specific requirement, falls outside the design suggestions applicable to conventional forgings" (from *Forging Industry Handbook,* Forging Industry Association, Cleveland, Ohio, 1970). While this definition is all-inclusive, it suits the wide capabilities of the precision forging process. Only recently adopted, this definition eliminates some persistent confusions. As long ago as World War II, parts were produced that today would be classified as precision forgings. As the use of such parts increased, names such as "net forgings," "no-draft forgings," "pressing," and others came into use. The FIA definition standardizes terminology and includes all these inexact, and sometimes inaccurate, labels. The "standard" tolerances were used to introduce forging design in Chapter 2 of this handbook.

Other types of forgings are can-and-tube forgings and no-draft forgings. These are special forms, and really a continuation of the chain of increasingly precision forgings introduced above. They will not be covered in any detail in this subchapter, since the peculiarities must be discussed with the forger as the product design evolves. However, the general field of impacts, as discussed in the next subchapter of this handbook, encompasses most of the more widely used special forgings, pressings, and so forth that are generally grouped as impacts because of the higher punch velocities used in the forming process.

## Precision Forgings or Conventional Forgings?

Precision forgings are used in assemblies where the basic characteristics of forgings are desirable or required, but where a conventionally forged part would require

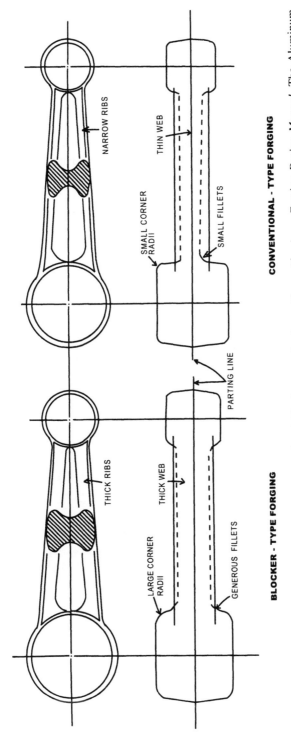

NARROW RIBS

THIN WEB

SMALL CORNER RADII

SMALL FILLETS

**CONVENTIONAL - TYPE FORGING**

PARTING LINE

THICK RIBS

THICK WEB

LARGE CORNER RADII

GENEROUS FILLETS

**BLOCKER - TYPE FORGING**

**FIGURE 3.61** Design differences between conventional and blocker-type forgings. (From *Aluminum Forging Design Manual*. The Aluminum Association, New York, 1975. With permission.)

extensive machining. Although most precision forgings are more expensive than conventional forgings, they are less expensive than the same part machined from a conventional forging, or from a hand forging, bar, or plate. The appropriate comparison is between the full costs of the parts ready for use. Most precision forgings are designed to eliminate the need for machining, aside from drilling attachment holes for installation. But for extremely complex designs, it may be more economical to precision forge only those sections of the part that are expensive to machine, and allow for machining the remaining sections. The intent is always to achieve the lowest cost yet highest quality for the finished part. The precision-forged part shown for cost comparison in Figure 3.62 does not show an unusually favorable cost in comparison to other methods of production. It represents an average finished part. For actual cost comparisons, Alcoa and other reputable forging houses will quote parts as precision forgings and also as other types of forgings, if requested.

## Mechanical Properties

Aluminum precision forgings are ordered to the same specifications, quality assurance provisions, and mechanical property levels that apply to conventional forgings.

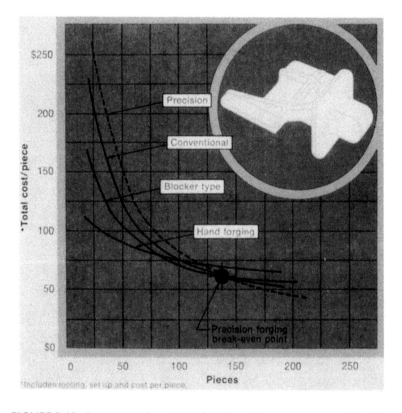

**FIGURE 3.62** Cost comparison example of precision versus conventional forgings. (From *Alcoa Aluminum Precision Forgings*, Aluminum Company of America, Pittsburgh, PA, 1950. With permission.)

Nevertheless, many users feel that precision forgings used without machining have better mechanical properties, fatigue characteristics, and resistance to stress corrosion cracking. This superiority is attributed to the high degree of work during forging, the grain orientation, parting plane location, and metallurgical advantages retained when the as-forged surfaces are not removed. See Figure 3.63 for minimum mechanical properties of popular aluminum precision forging alloys per specification QQ-A-367g.

## Size

The usual method of determining overall precision forging size limitation is to compare its plan view area (PVA) and pressure required against the rated capacity of the producer's largest forging equipment. For further discussion, PVA can be used as the unit of measure. The PVA is the length multiplied by the width (L W = PVA) when viewing the forging in the same direction as forging pressure will be applied. As a rule of thumb, 30 tons of pressure per square inch of PVA is required in precision forging. Very few precision forgings are in the over-200-in.$^2$ PVA range, due to press tonnage limitations. Very few parts are limited by bed size, press stroke, and "daylight" in the press. However, since new, larger presses are placed in service periodically, it is best to contact your forging supplier.

## Tolerances

Terminology and definitions shown in Figure 3.64 are a good reference for some of the terms that follow.

> Rib: Thin members, normal to web, confined within other forging members
> Web: Thin panel members essentially parallel to plan view of forging or dimensioned by length or width

| Alloy and temper | Longitudinal[b] | | | Transverse | | | Maximum cross section when heat treated (in.) |
|---|---|---|---|---|---|---|---|
| | Tensile strength (KSI) | Yield strength (KSI) | Elongation (%) | Tensile strength (KSI) | Yield strength (KSI) | Elongation (%) | |
| 2014-T6 | 65 | 55 | 6 | 64 | 54 | 3 | 4 |
| 2219-T6 | 58 | 38 | 8 | 56 | 36 | 4 | 4 |
| 6061-T6 | 38 | 35 | 7 | 38 | 35 | 5 | 4 |
| 6151-T6 | 44 | 37 | 10 | 44 | 37 | 6 | 4 |
| 7075-T6 | 75 | 65 | 7 | 71 | 62 | 3 | 3 |
| 7075-T73 | 66 | 56 | 7 | 62 | 53 | 3 | 3 |
| 7175-T66 | 86 | 76 | 7 | 77 | 66 | 4 | 3 |
| 7175-T736 | 76 | 66 | 7 | 71 | 62 | 4 | 3 |
| 7079-T6 | 72 | 62 | 7 | 70 | 60 | 3 | 6 |

FIGURE 3.63 Minimum mechanical properties of aluminum precision forgings. (From *Alcoa Aluminum Precision Forgings,* Aluminum Company of America, Pittsburgh, PA 1950. With permission.)

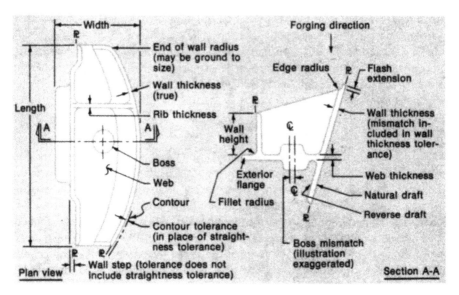

**FIGURE 3.64** Example of a part showing terminology and definitions of tolerances. (From *Alcoa Aluminum Precision Forgings,* Aluminum Company of America, Pittsburgh, PA 1950. With permission.)

Wall: Outside members essentially normal to webs

Length: Usually considered the longest dimension of the forging

Width: Usually considered the next-longest dimension of the forging and approximately at right angles to the length dimension

Recommended design proportions and tolerances for precision aluminum forgings are as follows:

Draft: $0° + 0°$.

Edge radii: 0.06 in. + 0.03 in. 0.06 in.

Fillet radii: 0.13 in. + 0.03 in. 0.06 in. (for parts heights from web up to 1.500 in.) 0.25 in. + 0.03 in. 0.13 in. (for parts heights from web exceeding 1.500 in.).

Web thickness: See Figure 3.65.

Wall or rib thickness: See Figure 3.65.

Mismatch tolerance: In general, mismatch tolerance is the misalignment of one die with another. In the production of precision forgings, the tooling may consist of several pieces or sections fitted together. Thus, mismatch may occur within a precision forging. Wall thickness tolerance is normally specified to include mismatch tolerance. However, other members of the forging, such as the boss in section AA of Figure 3.64, must also be recognized as requiring a separate mismatch tolerance of 0.00 to 0.015 in.

Thickness tolerance (die closure): See Figure 3.65.

Length and width tolerance (does not apply to wall thickness): See Figure 3.65.

**WALLTRIB HEIGHT THICKNESS GUINDELINE**

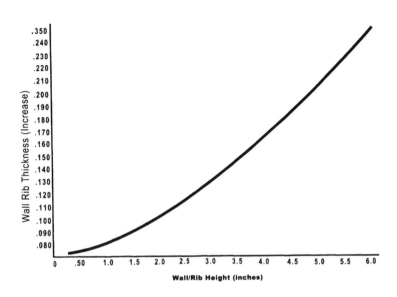

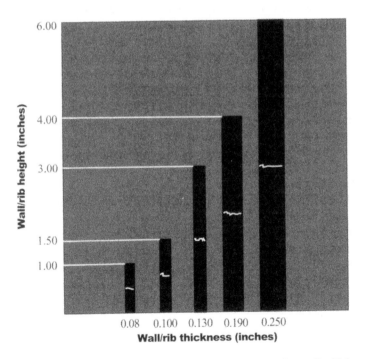

**FIGURE 3.65** Proportions and tolerances of web, wall, or rib thicknesses for precision aluminum forgings. (From *Alcoa Aluminum Precision Forgings*, Aluminum Company of America, Pittsburgh, PA, 1950. With permission.)

Normal straightness tolerance in any plane of the forging: 0.016 in. per 10 in.

Step tolerance (joggle) of ±0.010 in. does not include straightness.

Contour tolerance: ±0.015 of basic contour over the entire length.

Flash extension: 0 to 0.03 in. in the plane of the metal where interference is not normally encountered (see Figure 3.64).

Angular tolerance: ±0° 30'.

Surface finish: 125 rms (except flash extension).

Grain direction: parallel to length dimension, unless specified otherwise.

Parting plane: location optional with forger, unless specified otherwise.

## 3.3.6   Impacts

An *impact* is "a part formed in a confining die from a metal slab, usually cold, by rapid single-stroke application of force through a punch causing the metal to flow around the punch and/or through an opening in the punch or die" (The Aluminum Association definition).

### Basic Process

Impacting combines extrusion and forging and has therefore variously been called impact extruding, cold pressing, extrusion forging, cold forging, extrusion pressing, impact extruding, and the like. Today the process is simply called impacting and the parts so formed, impacts.

Although the design and fabrication of the punches and dies used is a highly skilled area, the process itself is relatively simple. The aluminum slug to be impacted, its volume carefully predetermined with an accuracy of anywhere from 1 to 10%, is placed in a die. A single-stroke punch, sometimes called a ram or a mandrel, comes down at high speed. The developed pressure extrudes the aluminum through designed openings. These orifices may be in the bottom of the die, between the die and the punch, or sometimes in the sides of the die. The aluminum that is not extruded is held between the bottom of the punch and the inside of the die. This portion of the metal is forged.

Plastic flow begins when the yield point of the aluminum is exceeded, and extrusion ensues when the pressure has increased to approximately 7 to 15 times initial yield pressure. The pressure necessary for impacting depends on the alloy as well as the complexity of the shape to be formed: the harder the alloy, the greater the pressure. Alloy 7075, for example, requires two to three times as much pressure as alloy 1100 when cold impacted. Pressures required for aluminum alloys vary between 25 and 110 tons/in.$^2$ of punch area, and punch-deforming velocities in the 20- to 50-m/sec range.

### Impact Types

There are three types of impacts: reverse, forward, and combination, named after the principal direction in which the aluminum flows under pressure developed by the punch.

Reverse impacting is used to make shells with a forged base and extruded sidewalls. A blank of material (slug) to be extruded is placed in a die cavity and struck by a punch, forcing the metal to flow upward around the punch, through the opening between the punch and the die, to form a simple shell (see Figure 3.66). Outside diameters can be stepped, but the inside diameter should be straight. Short steps, however, if necessary, can be incorporated on the inside, near the bottom of the part.

The clearance between the punch and the die determines the wall thickness of the impact. The base thickness is determined by adjustment of the bottom position of the press ram and should be a minimum of 15% greater than sidewall thickness as impacted. In general, the sidewalls should be perpendicular to the base. Multiwall shells, internal or external ribs, and circular, oval, rectangular, square, or other cross sections can be produced.

Advantages of reverse impacts include:

1.  Single operation, resulting in low cost
2.  Simplified tooling
3.  Ease of removal
4.  Inside and outside bottom contours easily achieved
5.  Irregular symmetrical shapes and broader range of shapes possible

In the reversing process there is a tendency for the punch to "wander" in producing the longer pieces, making consistency in wall thickness more difficult to maintain. See Figure 3.67 for specific design recommendations.

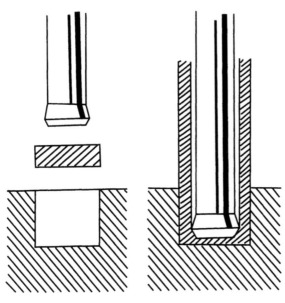

**FIGURE 3.66** The reverse impact process. (From *Aluminum Impacts Design Manual,* The Aluminum Association, New York, 1982. With permission.)

| Outside Diameter (inches) | Alloy and Wall Thicknesses (inch) | | | |
|---|---|---|---|---|
| | 1100 | 6061 | 2014 | 7075 |
| 1 | 0.010 | 0.015 | 0.035 | 0.040 |
| 2 | 0.020 | 0.030 | 0.070 | 0.080 |
| 3 | 0.030 | 0.045 | 0.105 | 0.120 |
| 4 | 0.040 | 0.060 | 0.140 | 0.160 |
| 5 | 0.050 | 0.075 | 0.175 | 0.200 |
| 6 | 0.060 | 0.090 | 0.210 | 0.240 |
| 7 | 0.075 | 0.110 | 0.245 | 0.280 |
| 8 | 0.100 | 0.130 | 0.280 | 0.320 |
| 9 | 0.110 | 0.145 | 0.315 | 0.360 |
| 10 | 0.125 | 0.165 | 0.350 | 0.400 |

Based on length-to-diameter ratio of 8:1 or less. Wall thicknesses less than those shown may be obtained by secondary operation.

**FIGURE 3.67** Design recommendations for reverse impacts. (From *Aluminum Impacts Design Manual,* The Aluminum Association, New York, 1982. With permission.)

Forward impacting, sometimes called the Hooker process, somewhat resembles conventional extruding in that the metal is forced through the orifice of a die by the action of a punch, causing the metal to flow in the direction of punch travel (see Figure 3.68). The punch fits the walls of the die so closely that no metal escapes backward. The method is used for forming round, nonround, straight, and ribbed rods and thin-walled tubing with one or both ends open, and with parallel or tapered sidewalls. Some large parts, such as transmission shafts, may be made by forward impacting.

Hollow or semihollow parts with a heavy flange and multiple diameters formed on the inside and outside are often made by forward impacting. Some of the advantages of forward impacting are:

1. Improved wall tolerances
2. Greater length–diameter ratios
3. Improved concentricities
4. Ease of producing thinner sections

However, shapes are more limited than in reverse impacting, and bases must be plain, as no forging action takes place against the bottom of the die.

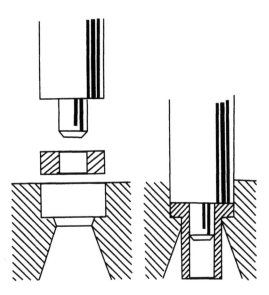

**FIGURE 3.68**  The forward impact process. (From
*Aluminum Impacts Design Manual,* the Aluminum
Association, New York, 1982. With permission.)

A *combination impact* is, as the name implies, a combination of forward and reverse metal flow. This method is used to produce complex-shaped parts (see Figure 3.69). The metal is confined inside the cavity between the upper and lower punches, forcing it to flow both up and down. If a solid slug is used, a web is left; if a hollow slug is used, a tubular part is formed.

By incorporating a cavity in the punch, the metal is allowed to flow upward into the punch, until the cavity is filled. Further punch movement causes forward extrusion of the remaining metal.

At the Royal Schelde Nuclear Equipment Division in the Netherlands, the author observed one of the useful examples of combining processes. There, an existing small extrusion press had been modified, to make a part for a nuclear power plant boiler water system. The tooling consisted of a formed male punch on the press ram and a formed female die in the die holder. The die had an orifice of approximately 10 mm, with provisions to extrude a tube with approximately 1-mm wall thickness in the center of the small bathtub fitting formed by the punch and die. The billet was heated, and the press speed increased to its maximum, which was less than the normal velocity for forward or combination impacting. The die was opened, the part removed, and the 5.0-m length of tube was carefully coiled to form the cooling coil. This tubing replaced a part that was previously made in two pieces and then electron-beam welded, and it eliminated any question of leaks, additional tests, welding, and so forth. The temperatures and pressures were controlled to obtain improved physical properties in the small housing formed on the end of the tubing. The process was not normal "extruding," nor was it really "impacting," but it was an excellent example of understanding processes and metal flow characteristics, in order to make a part that performed its required function.

## Impact Design Considerations

Good impact design practice includes full consideration of alloy selection, tool design and construction, impact production, lubrication, and possibly heat treatment of the impact following manufacture, in addition to good engineering design principles. It is essential that all contemplated impact designs falling outside recommended practices be discussed with an impact engineer. Basic design guidelines affecting cost, tool life, dimensional accuracy, and repeatability are:

1. Keep the design simple.
2. Avoid designs that are not symmetrical around the punch.
3. Make circular sections in planes perpendicular to the punch axis.
4. Avoid dimensional tolerance closer than necessary.
5. Use thin sections.
6. Use lowest-strength alloys applicable.

## Reverse Impacting

Press capacity and type of impact determine the maximum length that can be produced on a given press. In reverse impacting, length is determined by the inside diameter of the shell, the mechanical properties of the alloy, the reduction in area, and the stroke of the press. To avoid column failure of the punch, maximum shell length should not exceed

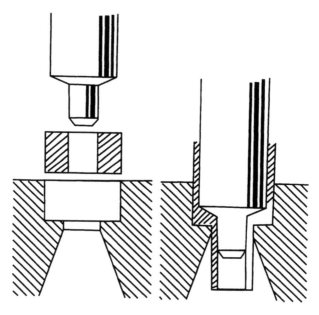

**FIGURE 3.69** Combination impact process. (From *Aluminum Impacts Design Manual,* The Aluminum Association, New York, 1982. With permission.)

eight times the inside diameter of the part (see Figure 3.70). While parts exceeding this 8:1 ratio have been reverse-impacted, it is usually more economical to add an ironing operation to obtain greater length.

A small outside corner radius or chamfer should be provided on the part bottom to promote an even flow of metal and to prevent dead metal in the corner of the die, which could cause slivers, poor surface finish, and possible separation in shear at the outside corners. See Figure 3.71 for examples. In Figure 3.71a, the inside radii, where the sidewall and base join, should be kept as small as possible. A sharp inside corner reduces friction during metal flow and improves surface finish and concentricity. If stress concentration at the junction of the base and sidewall is a major consideration, additional metal can be provided there without resorting to a large inside radius. In Figure 3.71b, the bottom's thickness at the base of the wall must be at least 15% greater than the thickness of the wall to prevent shear failure. Even distribution of metal flow is of great importance. The reverse impact on the left side of Figure 3.71b is correctly proportioned, while the drawing on the right side depicts a base that is thinner than the wall, causing defects. Practical minimum wall thicknesses for reverse impacts are shown in Figure 3.72.

### Forward Impacting

OD-to-ID ratio is not as great a limiting factor for forward impacting as it is for reverse impacting. The tube length is the main limitation. Irrigation tubing with 6-in. OD and an 0.058-in. wall has been produced in lengths up to 40 ft by hydraulic direct impacting. Tubing in 6061 alloy with 3/8-in. OD and an 0.035-in. wall has been made in 14-ft lengths.

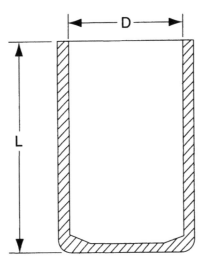

**FIGURE 3.70** Example of reverse impacting maximum length-to-diameter ratio. (From *Aluminum Impacts Design Manual*, The Aluminum Association, New York, 1982. With permission.)

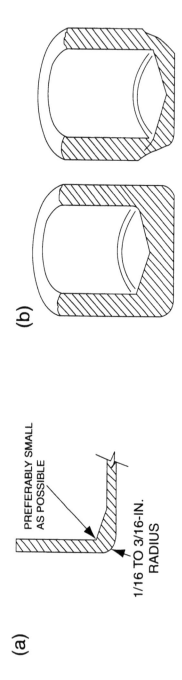

(a)

PREFERABLY SMALL
AS POSSIBLE

1/16 TO 3/16-IN.
RADIUS

(b)

*The inside radius between sidewall and base should be kept to a minimum. If additional strength is needed at this point it may be secured by tapering the bottom as shown. An outside radius should be provided to promote the even flow of metal.*

*Even distribution of metal flow is of great importance. The reverse impact at left is correctly proportioned while the drawing at the right depicts a base that is thinner than the wall, causing defects.*

**FIGURE 3.71** Examples of inside radii and bottom thickness of reverse impacts. (From *Aluminum Impacts Design Manual*, The Aluminum Association, New York, 1982. With permission.)

| Outside Diameter (inches) | Alloy and Wall Thicknesses (inch) | | | |
|:---:|:---:|:---:|:---:|:---:|
| | 1100 | 6061 | 2014 | 7075 |
| 1 | 0.010 | 0.015 | 0.035 | 0.040 |
| 2 | 0.020 | 0.030 | 0.070 | 0.080 |
| 3 | 0.030 | 0.045 | 0.105 | 0.120 |
| 4 | 0.040 | 0.060 | 0.140 | 0.160 |
| 5 | 0.050 | 0.075 | 0.175 | 0.200 |
| 6 | 0.060 | 0.090 | 0.210 | 0.240 |
| 7 | 0.075 | 0.110 | 0.245 | 0.280 |
| 8 | 0.100 | 0.130 | 0.280 | 0.320 |
| 9 | 0.110 | 0.145 | 0.315 | 0.360 |
| 10 | 0.125 | 0.165 | 0.350 | 0.400 |

Based on length-to-diameter ratio of 8:1 or less. Wall thicknesses less than those shown may be obtained by secondary operation.

**FIGURE 3.72** Practical wall thicknesses for reverse impacts. (From *Aluminum Impacts Design Manual,* The Aluminum Association, New York, 1982. With permission.)

## Secondary Operations

In some instances it is desirable to form or machine a part that has already been impacted. This process can be done without penalty. Metal so shaped has all the desirable characteristics: good grain flow, dimensional accuracy, smooth surface, and strength. Figure 3.73 illustrates two fairly common punch-and-die designs used for secondary operations. In Figure 3.73a an upset, head, or bulge is shown. In Figure 3.73b, a necking or nosing operation of a heavy wall impact is shown. Figure 3.74 is a drawing of the ironing operation used to reduce thickness of the impact's walls, increase its length, and ensure a smooth, uniform surface throughout. This result is accomplished by making the clearance space between the punch and the die slightly less than the thickness of the shell wall. The bottom of the shell retains its original thickness. This technique is often used to control tolerances after heat-treating the impact.

## Tolerances

Practically all factors involved in any fabricating process affect tolerances; hence, each is design-related either directly or indirectly. For example, eccentricity (a design

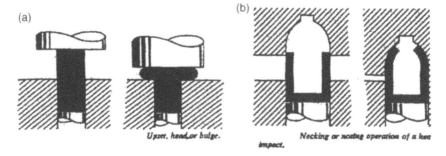

**FIGURE 3.73** Examples of secondary operations of upsetting and bulging of impacts. (From *Aluminum Impacts Design Manual*, The Aluminum Association, New York, 1982. With permission.)

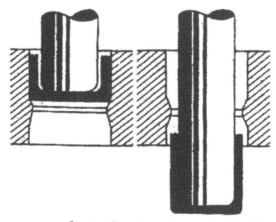

*Ironing: Drawing or draw wipe. Used for sizing and obtaining parts with a larger ratio of length to diameter.*

**FIGURE 3.74** Secondary process of ironing an impact. (From *Aluminum Impacts Design Manual*, The Aluminum Association, New York, 1982. With permission.)

consideration) helps reduce tool wear (a production factor). Some of the principal factors affecting tolerances are:

1. Tool wear
2. Irregular metal flow
3. Size of part
4. Shape of part
5. Alloy and temper
6. Wall thickness

Figure 3.75a shows commercial tolerances in general use for base, web, or flange thickness of aluminum impacts. Typical surface finishes are shown in Figure 3.75b. The finish depends on the alloy, lubricant, configuration, and tool design. Wall tolerances, as a general rule for 1100 alloy, can be held to $\pm10\%$ of wall thickness for walls 0.200 in. and $\pm7\%$ for walls between 0.250 and 1.00 in.

### Mechanical Properties

Figure 3.76 gives typical mechanical properties of some aluminum impacts, while Figure 3.77 shows the minimum mechanical properties of aluminum impacts that can be expected in production at 75°F.

While all of the discussion has been with reference to aluminum impacts, other nonferrous metals can also be impact extruded. These metals include lead, tin, zinc (heated to 300°F), magnesium (heated to 400°F), and so forth. For special applications, impacts in fine silver have been successfully made.

## 3.4   CASTING AND MOLDING PROCESSES

### 3.4.1   Introduction to Casting and Molding Processes

Up to this point we have discussed the fabrication of parts or products by forming sheet metal, machining, forging, extrusion, and the like. In some cases these processes seem to have become blended or modified or combined, making it somewhat difficult to classify the production process clearly. An example is the lathe that can perform milling and vice versa. Another is the forging, pressing, impacting, and extrusion area, where innovative combinations appear in all kinds of industries, using most types of materials. We have shown examples and limitations of several processes, all using a form of solid material. In some cases, the material was heated just below the liquid temperature; however, the processes started and finished with a solid metal.

Previously; we talked about the primary means of casting liquid metals into molds, in order to be able to influence the product design in the direction of a producible configuration. As we now introduce other important processes using liquid or powder materials or alloys as the input material, it becomes even more difficult to classify some of the processes. Any metal that can be melted and poured can be cast, and the size and

| (a) | BASE, WEB OR FLANGE | |
|---|---|
| Diameter (inches) | Tolerance (± inch) |
| Up to 3 | 0.012 |
| 3 to 5 | 0.015 |
| 5 to 7½ | 0.020 |
| Over 7½ | 0.030 |

| (b) | TYPICAL, AS-IMPACTED PRODUCTION FINISHES | |
|---|---|---|
| Alloy | Interior (micro inch) | Exterior (micro inch) |
| 1100 | 16–32 | 32–63 |
| 6061 | 63–125 | 125–250 |
| 7075 | 125–250 | 125–250 |

**FIGURE 3.75**  Commercial tolerances and finishes of impacts. (From *Aluminum Impacts Design Manual*, The Aluminum Association, New York, 1982. With permission.)

| Alloy and Temper | Tensile Strength [1]-KSI | | Elongation in 2 in. Percent | Brinell[2] Hardness |
|---|---|---|---|---|
| | Ultimate | Yield | | |
| 1100-F | 24.0 | 22.0 | 5 | 44 |
| -0 | 13.0 | 5.0 | 35 | 23 |
| -H112 | 24.0 | 22.0 | 5 | 44 |
| 2014-T4 | 62.0 | 42.0 | 20 | 105 |
| -T6 | 70.0 | 60.0 | .11 | 135 |
| 2618-T61 | 64.0 | 54.0 | 10 | 125 |
| 3003-F | 29.0 | 27.0 | 4 | 55 |
| -0 | 16.0 | 6.0 | 30 | 28 |
| -H112 | 29.0 | 27.0 | 4 | 55 |
| 6061-F | 28.0 | 25.0 | 5 | 52 |
| -0 | 18.0 | 8.0 | 25 | 30 |
| -H112 | 28.0 | 25.0 | 5 | 52 |
| -T4 | 35.0 | 21.0 | 22 | 65 |
| -T6 | 45.0 | 40.0 | 12 | 95 |
| -T84 | 43.0 | 39.0 | 6 | 85 |
| 6066-T6 | 57.0 | 52.0 | 10 | 115 |
| 6070-T6 | 55.0 | 51.0 | 10 | 115 |
| 7001-76 | 94.0 | 85.0 | 7 | 147 |
| 7075-T6 | 83.0 | 73.0 | 10 | 145 |
| -T73 | 73.0 | 62.0 | 12 | 135 |

1 Specimen axis parallel to direction of grain flow.
2 Brinell hardness is usually measured on the surface of an impact using a 500 kg load and a 10-mm penetrator ball.

NOTE: *These are not guaranteed values and therefore should not be used for design.

**FIGURE 3.76** Typical mechanical properties of some aluminum impacts at 75°F. (From *Aluminum Impacts Design Manual*, The Aluminum Association, New York, 1982. With permission.)

| Alloy and Temper | Thickness | Tensile Strengh [1]-KSI Ult Yield Min. Min. | | Elongation % min. in 2 in. or 4D[2] | Brinell Hardness 500 kg Load 10-mm ball Minimum |
|---|---|---|---|---|---|
| 1100-F* | All | ..... | | | |
| -0 | All | 11.0 | 3.0 | 15 | ----- |
| -H112 | All | 15.0 | 7.0 | 2 | ----- |
| 2014-T4 | All | 55.0 | 32.0 | 10 | 100 |
| -T6 | All | 65.0 | 55.0 | 6 | 125 |
| 2618-T61 | All | 58.0 | 45.0 | 4 | 115 |
| 3003-F* | All | ..... | | | |
| -0 | All | 14.0 | 5.0 | 14 | ----- |
| -H112 | All | 19.0 | 9.0 | 2 | ----- |
| 6061-F* | All | ---- | | | |
| -0 | All | | | 10 | ----- |
| -H112 | All | 22.0 | 16.0 | 2 | ----- |
| -T4 | All | 26.0 | 16.0 | 12 | 50 |
| -T6 | All | 38.0 | 35.0 | 6 | 80 |
| -T84 | .040-.093 | 35.0 | 30.0 | 4 | 75 |
| | .094 & over | 38.0 | 35.0 | 4 | 75 |
| 6066-T6 | All | 50.0 | 45.0 | 4 | 100 |
| 6070-T6 | All | 48.0 | 43.0 | 4 | 95 |
| 7001-T6 | All | 88.0 | 78.0 | 5 | 135 |
| 7075-T6 | All | 75.0 | 65.0 | 5 | 135 |
| -T73 | All | 66.0 | 56.0 | 5 | 125 |

1 Specimen axis parallel to direction of grain flow.
2 "D" equals specimen diameter. Elongation requirements do not apply to material thinner than 0.062 inch (nominal).
* For wrought products in F temper, there are no mechanical property limits.

**FIGURE 3.77** Minimum mechanical properties of some aluminum impacts at 75°F. (From *Aluminum Impacts Design Manual*, The Aluminum Association, New York, 1982. With permission.)

range of parts regularly produced by presently available methods are greater than that of any other process affording similar results. Generally, casting methods may be divided into those using nonmetallic molds and those using metallic molds. However, to permit a logical breakdown regarding design, these processes can best be subdivided according to the specific type of molding method used in casting, such as sand, centrifugal, permanent mold, and so forth.

We can discuss patterns, mold making, shrink rates, pouring, cooling or chilling, and good and bad product design features. Powder metal and sintering is a separate field (or is it?), which involves heating after solidifying rather than solidifying after heating, as in conventional casting processes. It has a relationship to forging, but is really not part of the normal forging industry. Making parts of rubber and synthetic rubber falls in somewhere, but again, it is really not a close family member of the processes discussed up to this point. Are rubber parts "castings," "forgings," "sintered," or none of the above? Ceramics are playing an increasingly important part in the world of manufacturing, and are related to the processing of metal parts. A combination of pure oxide ceramic and a metal constituent, both of which have good high-temperature properties (as opposed to the older clay-based ceramics), has brought about the development of highly specialized technical ceramics suited to unusual industrial applications.

All of the above processes have a place in today's complex industrial manufacturing arena and deserve at least a proper introduction in a handbook of manufacturing engineering.

### 3.4.2  Sand Casting

Sand casting is the oldest and most familiar method of casting. In this process, the castings are made by pouring molten metal into either green sand molds or molds made of baked sand. Refer to Chapter 2 for a further description of the sand casting process.

### 3.4.3  Centrifugal Casting

Centrifugal casting entails pouring a measured quantity of molten metal into a mold that is then rotated rapidly. The rotation of the mold forces the molten metal outward to give intimate contact between the metal and the mold. Spinning is continued until all the metal poured into the mold has solidified. Pouring must be done quickly to prevent chilling and laps. There are three standard types of centrifugal casting methods in general use.

*True centrifugal casting* is where the mold is rotated about its own axis without using a central core. If the mold is partially filled, a hole appears along the center of rotation of the casting, the diameter of the hole determined by the amount of metal used. The weight of casting produced to that of metal poured approaches 1:1.

*Semicentrifugal casting* is where central cores are used to give irregular shapes to the central hole. A measured amount of metal is poured so the mold space between the core and the outer wall is completely filled. Proper design of the mold is essential so that directional solidification of the metal is retained. Sand or plaster cores are usually employed. Differentially heated or cooled outer molds may be needed to control the direction of solidification.

*Centrifuged casting* is where irregular shapes can be obtained that would not be possible if the parts were rotated on their own axes. In this process, a number of molds are arranged about a central sprue, similar to the spokes of a wheel. Molten metal is fed into the castings through radial gates. The process is similar to semicentrifugal casting, except that several castings are produced at once, and the molds are not spun on their own axes. Directional solidification is a problem in this process, but it can be solved by proper location of the castings, proper gating, and mold temperature control.

### Process Data

Centrifugal casting molds may be made of a variety of materials, including steel, cast iron, sand, and graphite. Various wall coatings can be used, such as a mixture of graphite and sodium silicate in water.

Most aluminum casting alloys suitable for other processes can be used for centrifugal casting. The alloys should be poured at about 100°F less than with static casting. Alloys with short solidification ranges are preferable to those with wide freezing ranges.

## 3.4.4   Permanent Mold Casting

In permanent mold casting, the molten metal is poured by gravity into heated metal or graphite (permanent) molds. Sand casting is essentially a batch process, while permanent mold casting is suitable for quantity production of a continuous nature. This production-line approach requires a different arrangement of foundry equipment, metal-handling methods, and production procedures compared with those used in sand casting. A simple permanent mold usually consists of two halves that, when closed, form the mold cavity. Either metal or sand cores can be used; the process is designated as semipermanent mold casting when sand cores are used. The mold is heated before pouring, and held at a constant temperature during pouring. Some castings require either heating or cooling of the mold between pouring operations; in others, the molten metal keeps the mold at the desired temperature. Permanent molds must be prepared for use by coating with refractory material. The refractory coating serves two purposes:

1. The solidification rate in different sections of the mold can be controlled by varying the coating thickness.
2. The mold metal is protected from contact with the molten metal.

Solidification will start at those sections where the coat is thinnest, due to the faster heat dissipation at those points. The refractory coating is heavily applied along thin sections and through the gates and risers. Thin coats are used on areas of wide cross section to promote faster solidification. By properly applying the coating (by brushing or spraying), the rate of solidification throughout the casting will be uniform. The coating can be formulated from almost any refractory material finely ground in a water suspension with a suitable binding agent.

Permanent mold castings present different problems of gating and risering. The longest dimension of a permanent mold casting is usually vertical, as contrasted to horizontal positioning in sand casting. Sprues, risers, gates, and runners must all be designed as part of the mold structure. The difficulty and expense of making radical changes in metal molds make it essential that the entire feeding system of the mold be finalized before the mold is made. It is wise to build a mold with initially undersized gates and risers. Then, by making experimental castings, gradually increase the sizes of the feed channels until the best possible casting is obtained. This method obviates the possibility of having oversized channeling to begin with, and keeps to a minimum the amount of metal that will be poured for each casting.

With permanent mold casting, a carefully established and rigidly maintained sequence of operations is essential. Every step in the foundry, from charging the furnace to removal of the cast piece from the mold, must by systematized. If any of the factors (pouring temperature, mold temperature, pouring rate, solidification rate) are thrown out of balance, the resultant castings may end up as scrap.

### 3.4.5 Die Casting

In die casting, the molten metal is forced into a metal mold under considerable pressure, which is applied to the metal either pneumatically or hydraulically. Die casting gives low-cost production of large numbers of thin-sectioned parts. Close tolerances and extremely smooth surfaces can be produced without subsequent machining and finishing; also, small complex coring is possible, saving many drilling operations.

Such coring is not possible with sand or permanent mold casting. Intricate parts, not practical with other casting methods, can be easily produced by die casting. Satisfactory die casting depends on:

1. A suitable die-casting machine
2. A properly designed die

The die-casting machine consists of a substantial, rugged frame designed to support and open and close the die halves in perfect alignment. Usually the fixed, or ejector, half of the die is mounted on a stationary platen. The other half, or cover, is mounted on a movable platen. The two halves of the die must move together accurately and must be locked together with sufficient force to overcome the separating force developed as the metal is injected. Either a toggle linkage arrangement or hydraulic rams are used to lock the dies.

Besides the basic die halves, most dies have moving cores and other features that allow the production of complex castings. In operation, the cores, slides, and other moving die parts are operated by hydraulic action synchronized with the opening and closing of the main die halves. These added features complicate die design but enable the production of variegated surfaces.

Metal injection into the die cavity is by means of either the goose neck or the cold-chamber method. In goose-neck injection, the molten alloy is forced into the mold by means of pneumatic pressure. In cold-chamber injection, a hydraulically actuated plunger forces the molten metal from a cylindrical "shot" sleeve into the mold. The process is designated as "cold chamber" because the molten alloy is ladled into the shot sleeve just before it is forced into the mold. Pressures in the two methods vary; for the goose-neck method it is usually about 750 psi, while the cold-chamber method runs at 3,000 to 20,000 psi.

Metal enters the die in the cold-chamber process in a semimolten condition and forces the air out ahead of the metal. In the goose-neck system, the metal enters the mold in a completely molten state and tends to mix with air in the die cavity. This tendency to produce porous castings has resulted in goose-neck equipment becoming obsolete; it is now largely replaced by cold-chamber machines.

## Die Considerations

Heat-treated alloy steel dies are needed for die casting aluminum. These dies start to check after long usage due to the thermal shock from the molten metal. Initial tooling cost is high due to the skilled labor needed to make a die. However, the productivity of a die-casting machine is high and will bring the unit cost down when the production run is long, and where the part has sufficient complexity and need for precision to warrant die casting. To best utilize this process, consult die-casting experts such as those available through the American Die Casting Institute, 366 Madison Avenue, New York, NY 10017.

Frequently, die-cast parts improve products through production simplification and more sales appeal. Inserts of other materials can easily be placed in the die cavity and permanently molded in the finished piece. Die-cast parts can be bulk finished by barrel finishing, automatic polishing and buffing, continuous chemical treatment, or painting. Die castings offer freedom of shape and an unlimited range of surface ornamentation with but a slight increase in tooling or finishing cost.

## 3.4.6 Plaster Molding

The most common specialty casting process employs plaster molds. This process is a refinement of sand casting in that the sand is replaced by plaster, giving the finished casting a smoother surface and allowing greater accuracy in the dimensions of the molded part. A plaster mold is used for just one casting, since it is necessary to destroy the mold to remove the casting. The process is usually confined to castings under 2 lb. Gypsum plasters are the type most often used to make plaster molds. It is essential that all water be removed from the molds before casting.

The equipment required for plaster molds is more expensive than that for sand molds. Metal match plates and metal core boxes of extreme accuracy are used. Most plaster castings are poured by gravity, but moderate pressure gives improved casting detail.

The aluminum alloys used with plaster molds must be carefully selected. The refractory nature of the plaster results in a slow solidification time, with resultant lowering of mechanical properties. This refractory quality, however, enables thin and intricate sections to be cast. Due to their excellent fluidity, aluminum-silicon alloys are best for plaster casting.

### 3.4.7 Investment Casting

Investment casting, based on the "lost wax" process, allows for the intricacy of design of sand casting and the precision of die casting. To make an investment casting mold, a refractory-type plaster is poured around an expendable wax (or low-melting-temperature plastic) pattern. As the plaster sets, it is dried in a oven and the wax pattern is melted out. As in plaster casting, the molds are used only once, because they must be broken to remove the casting. More often, in today's shops, the refractory material is sprayed over the wax patterns and dried in an oven, melting out the wax pattern as above. This method allows the use of a robot to spray the refractory material the same way each time, improving repeatability and lowering costs.

A master mold is required to make the wax patterns. The number of castings required determines the permanency of the master mold. When large numbers of castings are to be made, the master molds should be made of metal, and an injection molding process should be used for the production of the wax patterns. This process is expensive because of the number of steps required, the need for skilled operators at each step, and the slow production rate compared to other casting processes. The use of plaster or other refractory material limits the choice of materials to those suited to plaster casting.

The molten metal may be poured under pressure, or poured with the mold in a vacuum chamber. Extremely sharp details can be obtained in the cast piece. The accuracy of this process is very high—even greater than that achieved with die casting because there are no moving parts in the mold. Naturally, the ultimate accuracy of the method depends on the accuracy of the master mold used to make the wax patterns. Because of the slow cooling of this type casting, it is imperative that the metal be thoroughly fluxed with chlorine gas in order to eliminate pinhole porosity.

### 3.4.8 Power Metal

For lack of a better, more concise term for the pressing and sintering of metal powders into machine parts of all varieties, the generally recognized powder metallurgy (PM) is normally employed. The first porous metal bearings were marketed following World War I, and with the outbreak of World War II, the vast potential of powder metallurgy began to be realized for its value not only as a method of fabricating parts whose physical characteristics are impossible to produce otherwise, but also as a large-volume mass-production process having excellent speed and material economy. Beyond the well-known oilless or self-lubricating bearings and similar parts, there is a tremendous field of machine parts in production. Some of these include clutch friction facings, internal and external splines, rollers, external and internal gears, ratchets, piston rings, bushings, magnets, and the like.

## Metallurgy

Several general classes of metal-powder structure can be set up. In the first, consolidation during sintering is primarily a particle-to-particle cohesion or contact fusion of particles containing a melting constituent. In the second, one of the powders acts as a melting medium, bonding or cementing together a high-melting-point constituent. In the third class, consolidation of a fairly high-melting-point metal is achieved as in the first category, but lacking high density, the compact is impregnated with low-melting-point metal.

## Production Steps

The procedures employed in producing PM parts are generally as follows:

1. Selection of the powder or powders best suited for the part being designed as well as for the most rapid production
2. Wet or dry mixing of powders where more than one powder is to be used
3. Pressing in suitable dies
4. Low-temperature, short-time sintering, usually referred to as presintering, for increasing strength of fragile parts, removing lubricants or binders, and soon
5. Machining or otherwise forming of presintered parts
6. Sintering green compacts or presintered parts to obtain the desired mechanical properties such as proper density, hardness, strength, conductivity, and soon
7. Impregnating low-density sintered compacts, usually by dipping in molten metal so as to fill all pores or by allowing the impregnant to melt and fill the pores during sintering
8. Coining or sizing operations, cold or hot, when necessary, to attain more exacting dimensional tolerances and also improve properties
9. A hot-pressing operation to replace the usual pressing and sintering

## Compacting

Pressing or compacting of PM parts generally requires anywhere from 5 to around 100 tons of pressure per square inch. Common garden-variety parts are produced with 20 to 50 tons/in.$^2$ pressure. It is interesting to observe, however, that the final density of a PM product is not determined only by the pressure under which it is cold-pressed or briquetted. Rate of pressure application, particle size, type of material, sintering time and temperature, occluded gas, and so on, also have effects on final density and size.

Press stroke also presents certain broad limitations as to part size. The compression ratio between the volume of powder in a die before and after pressing is dependent on loading weight, particle size, form and composition, metal hardness, and pressure used. With most common metals and alloys, this ratio is usually 3 to 1, but it may vary from 5 to 1 up to 10 to 1 with fine powders, and 2 to 1 up to 4 to 1 with medium-size powders. Combined with the compression ratio is the general limitation that diameter-to-length ratio of parts be restricted to a maximum of 3 to 1.

## Equipment

Presses may be mechanical, hydraulic, or a combination of both. Small parts that can be made at high speed with relatively low pressures are best produced in mechanical automatic presses with single- or multiple-cavity dies. Such presses for average parts usually are built in pressure capacities ranging from 100 to 150 tons, although sizes to 1500 tons are available. Presses are generally of the single-punch type or of the rotary multiple-punch type. Single-punch presses are either of the single-action type that compresses with the top punch only, or of the double-action type that employs movement of the lower punch simultaneously with the upper to obtain more uniform compacting and automatic ejection. As many as two upper telescoping punch movements and three or four lower punch telescoping punch movements are used along with side core movements for complex designs. However, it is generally difficult to press powder into reentrant angles, sharp corners, or undercuts. An average die produces 50,000 pieces before wear necessitates refitting or replacement.

Where the ordinary process for handling PM parts does not result in a satisfactory density or the pressures required are extremely high, *hot pressing* can often be used. In this method, the powders are heated in the dies and pressure is applied to form the part. The hot-pressing method yields high and nearly ideal density and greater strength at relatively low pressures. Carbide parts up to 100 in.$^2$ in cross section, the greatest dimension of which can be 18 in. with a length of 8 in., have been produced by hot pressing, especially parts too large for regular cold pressing and sintering, and thin-wall parts that tend to go out of round.

Ordinarily, to obtain desirable density and precision, parts of materials other than carbides are *coined* or *sized* after sintering, but naturally this step adds to the cost. Compacts up to 10 in.$^2$ in cross section are readily coined in hydraulic presses at a rate of about 4 to 6 pieces/min with hand feeding, and up to 10 pieces/min with automatic feeding. Dies may produce 100,000 to 200,000 pieces before replacement.

## High-Density Powder Metallurgy

High-density powdered metals can be produced by a number of techniques, including powder forging, powder rolling, liquid metal infiltration, and liquid-phase sintering. Although powder forging has a relatively long history, it is only recently that the process has become important commercially. A typical flow diagram for the hot forging process is shown in Figure 3.78. This is a diagram of the Federal Mogul process called Sinta-Forge. Material improvements have included development of clean special grades of low-alloy powder. Iron-carbon and iron-copper-carbon alloys are another addition to the supply. Figure 3.79 is a compilation of representative mechanical properties and combinations of alloys available.

The properties of powder-forged steel are intermediate between the properties found for forgings between the horizontal and transverse direction of the forging.

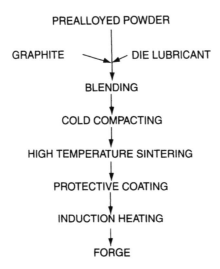

PREALLOYED POWDER

GRAPHITE ⟍ ⟋ DIE LUBRICANT

BLENDING

COLD COMPACTING

HIGH TEMPERATURE SINTERING

PROTECTIVE COATING

INDUCTION HEATING

FORGE

**FIGURE 3.78** Typical flow diagram of the hot-forging process for PM (Sinta-Forge process of Federal Mogul Corp. With permission.)

| Material group | Density (g/cm³) | Tensile strength (N/mm²) | Elon- gation (%) | Examples of applications |
|---|---|---|---|---|
| Iron and low-alloy com- | 5.2-6.8 | 5-20 | 2-8 | Bearings and light-duty struc-tural components |
| pacts | 6.1-7.4 | 14-50 | 8-30 | Medium-duty structural parts, magnetic components |
| Alloyed steel compacts | 6.8-7.4 | 20-80 | 2-15 | Heavy-duty structural parts components |
| Stainless steel compacts | 6.3-7.6 | 30-75 | 5-30 | Components with good corrosion resistance |
| Bronzes | 5.5-7.5 | 10-30 | 2-11 | Filters, bearings, and machine components |
| Brass | 7.0-7.9 | 11-24 | 5-35 | Machine components |

**FIGURE 3.79** Representative mechanical properties and combination of PM alloys available. (From R. Bolz, *Production Process*, Industrial Press, New York, 1963. With permission.)

This is illustrated in Figure 3.80. The advantage of the powder-forged alloy is the consistent properties in longitudinal and transverse directions. This has been found particularly useful in powder-forged connecting rods and gear wheels. It appears that from a material cost saving and manufacturing process simplification, powder forg-ing will continue to grow and replace machined parts and drop forgings.

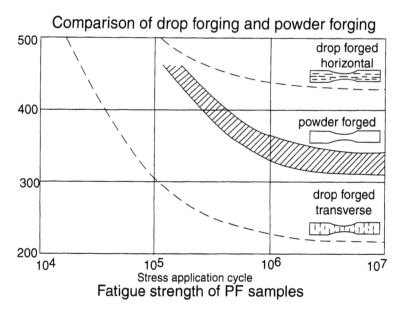

**FIGURE 3.80** Properties of powder-forged steel in comparison with horizontal and longitudinal properties of conventional forgings. (From R. Bolz, *Production Processes*, Industrial Press, New York, 1963. With permission.)

## Hot Isostatic Pressing of Metal Powders

Use of hot isostatic pressing (HIP) in conjunction with PM processing has produced parts with 100% density and properties equivalent to those of wrought alloys. Although the principal use of the HIP process has been for producing large billets for subsequent mechanical working and improving the properties of castings, it is expected that it will be used for much smaller parts on a high-volume production basis. The principal barrier to the widespread use of HIP is the necessity for obtaining a closed surface rather than the porous interconnected condition that exists in conventional as-pressed parts.

## Injection Molding of Metal Powders

An alternative to the use of pressure techniques to form a green compact is the use of injection molding. This technique uses the technology of plastic injection molding by combining metal powders with a polymer binder to produce a slurry-type mixture that can be injected into a complex die. Unlike plastics, however, the size of the parts produced by this method is somewhat restricted. Complex geometries including undercuts, holes, and reentrant angles may be produced using appropriate slide tooling on the die. Following injection, a binder removal operation at relatively low temperatures (400–600°F) is done. Next, a sintering operation is conducted that is much the same as that for conventional PM processing. In injection molding of PM, shrinkage between the original injection molded dimensions and final sintered dimensions may be as much as 20%; therefore, careful tool design is needed. The polymer-PM blend must produce a predictable shrinkage perform.

The range of materials applicable to this process is limited only by the use of any fine powder material that can be sintered. These include steels, both stainless and carbon steel, and alloy steels as well as superalloys, tungsten carbides, and so on. The technique has also been used with ceramic materials such as aluminum oxide.

## 3.4.9   Ceramics/Cermets

Among the earliest objects fashioned by man, ceramic products have withstood the test of centuries and still find applications in numerous fields. Exceptional properties of ceramics have increased their use and their availability and have brought about the development of highly specialized technical ceramics suited to unusual industrial applications. Properly chosen and suitably designed, technical ceramics can fill a real need in the design of industrial equipment.

Failure to consider the use of ceramics in mechanical equipment generally stems from rather meager knowledge of these materials and lack of data with which to design such parts to ensure not only better performance but lower production costs.

### Types of Ceramics

Traditionally, ceramic materials have been composed largely of naturally occurring clays, alone or in admixture with various amounts of quartz, feldspar, and other nonmetallic minerals.

*High-clay ceramics* account for the largest tonnage of manufactured ceramics. For convenience, these may be said to have a clay content in excess of 50%. The ceramics are characterized by high shrinkage—approximately 25% by volume—during drying and firing and therefore exhibit the widest size variation. To a large extent, the production tolerances are a direct function of the clay content. Parts produced include almost all structural clay products, clay-based refractories, most chemical porcelains and stoneware, and electrical and mechanical porcelains.

*Low-clay ceramics* include steatite and other low-loss dielectrics, special and super-refractories, and special porcelains. Steatite normally carries more than 80% talc (hydrated magnesium silicate), which is bonded with ceramic fluxes. Shrinkage is low, and close tolerances can be held.

Various methods are used for forming clay-type ceramic bodies prior to firing. Practical production design must take cognizance of not only the tolerance effect of the method required but also the general range and limitations imposed. The various forming methods are either the wet or dry process types, shown in Figure 3.81. The *plastic wet process* is probably the oldest and most diversified general process, consisting of extrusion, casting, pressing, throwing, or jiggering. Figure 3.82a shows four of the processes, and Figure 3.82b shows an extruded blank that was machined after extruding and drying, but before firing. The *dry process* is different only in that a semimoist granular powder is used, and results in parts with much less dimensional variation. Pressing in metal dies is the only method of forming used; this method is

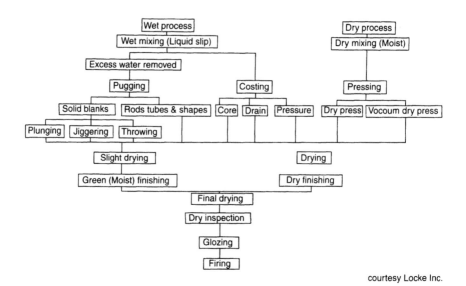

courtesy Locke Inc.

**FIGURE 3.81**  Forming methods for clay-type ceramic bodies prior to firing.

adaptable to high production rates with automatic presses. The process is generally limited to parts of 2 to 4 in., but parts up to 14 in. have been made.

*Clay-free ceramics* are entirely devoid of clay. Included in this group are the pure oxide types and the so-called cermets, which have properties between metals and ceramics. A combination of pure oxide ceramic and a metal constituent, both of which have good high-temperature properties, the metal-ceramics have been developed to meet the need for a material capable of withstanding operating temperatures to 2400°F or more. The metal constituent is employed to provide thermal conductivity and shock resistance, while the ceramic provides resistance to deformation under high stress at high temperatures.

Some of the most recent innovations in cutting tool technology have been taking place in ceramics and cermets. Ceramic cutters, which first became available in the late 1940s, are characterized by high wear resistance, good hot hardness, and a low coefficient of friction. Additionally, ceramics are inherently inexpensive materials providing a cost advantage compared to carbide materials. As with all cutting tool innovations, increased productivity resulting from higher material removal rates and from less frequent tool changes is the driving force behind implementation. The use of ceramic cutting tools has been fastest in Japan, where 5–7% of the cutting tools are ceramic, compared to about half this usage in the United States in 1985.

Ceramic cutters are not without limitations. The poor mechanical and thermal shock resistance of ceramics limits their usefulness to applications with low feed rates and noninterrupted cuts, and to machine tools capable of great rigidity and high speeds.

Three types of ceramic cutting tools predominate today: aluminum oxide ($Al_2O_3$), sialon (a combination of silicon, aluminum, oxygen, and nitrogen), and silicon nitride ($Si_3N_4$). All three types are manufactured by cold pressing the ceramic powder and then sintering at high temperature in an oven.

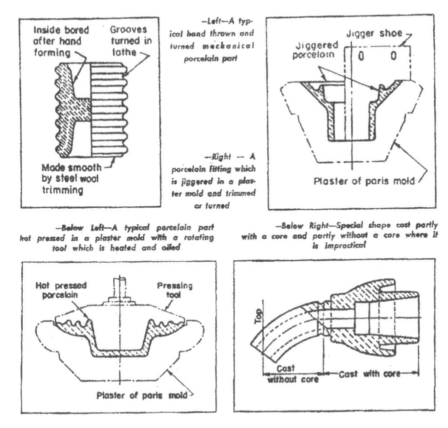

**FIGURE 3.82** Four of the common forming methods using the plastic wet process for clay-type ceramics. (Courtesy of Westinghouse Electric Corp. With permission.)

*Aluminum oxide,* the most common ceramic cutting tool material, has been commercially available since the 1950s. It is most often applied when machining cast iron softer than 235 BHN and steel softer than Rc34. It also performs best for finishing, semifinishing, and noninterrupted cuts. When applied to cast iron, single-point cutting tools are able to achieve production removal rates of 1442 mm³/min at a cutting speed of 610 mm/min, a depth of cut of 5.85mm, and feed rate of 0.41 mm per revolution.

*Sialon,* which has been available since 1981, offers a high degree of resistance to mechanical shock and exhibits very good hot hardness. These properties, in turn, provide uniform wear and lower chances of chipping failure compared to aluminum oxide. Unlike aluminum oxide, sialon can withstand severe thermal shock resulting from the use of coolants. It cannot be used on carbon steels because of chemical incompatibility.

*Silicon nitride* (Si$_3$N$_4$) cutting tools exhibit some of the same attributes of sialon, but with the addition of increased toughness, the ability to withstand higher chip loads, and higher cutting speeds. Used most often for the high-speed milling and roughing of cast iron and nickel-based alloys, silicon nitride is able to machine gray cast iron at speeds in excess of 1500 m/min.

*Cermet* tools are made of approximately 70% $Al_2O_3$ and 30% titanium carbide (TiC). Unlike ceramic tools, cermets are manufactured by hot pressing, that is, simultaneously pressing and sintering powder materials. The resulting cermets are much tougher and move fracture resistant than ceramics, allowing them to be used for machining cast iron materials harder than BHN 235 or steels harder than Rc34. The high toughness and hardness inherent with cermets also provide additional advantages of cutting speeds 600% faster than carbides, together with the ability to perform interrupted cuts.

A new class of cutting tool materials, known as ceramic/ceramic composites, are composed of an $Al_2O_3$ matrix reinforced with thousands of tiny silicon carbide whiskers measuring only 0.02μm in diameter by 1μm in length. The addition of these reinforcing fibers acts to distribute cutting forces throughout the matrix and to assist in carrying heat away from the cutting edge. This action results in a 200% increase in fracture toughness compared to cermets, and this process is suitable for applications requiring either roughing or finishing cuts.

## 3.4.10 Rubber Products

Like the plastics, rubber and rubberlike synthetics provide almost possibilities as components of machine assemblies. Again, it is a case where specific properties, largely unattainable with any other materials, are a design necessity and some knowledge of rubbers, their advantages and limitations, and their manufacture into parts is invaluable. As a rule, it is possible to design or redesign a rubber part to closely approach the ideal conditions for economical production without affecting or destroying the intended functional characteristics.

### Materials

The rubber materials that can be used for molded parts now range from natural stocks from various parts of the world to a full array of rubberlike synthetics. The crude or synthetic-base material is compounded with curing agents, antioxidants, accelerators, lubricants, and so on, and is thoroughly kneaded on mixing mills. It emerges as large slabs for storage pending subsequent manufacturing operations. This "green" or uncured compounded rubber can vary from a soft, gummy state to a hard, leathery condition. In no case is it liquid. (We are omitting discussions of some of the silicone compounds, which can be cast in a liquid or semiliquid state—making them more in the adhesive or plastic material field, rather that grouped with rubbers.) The nature of this uncured stock largely determines just how it needs to be shaped or prepared to fit a mold cavity for proper flow. A soft compound can be roughly shaped and laid in or adjacent to the mold cavity, whereas a hard stock requires careful tailoring and direct placement in the cavity. Characteristics of rubber and synthetic materials are shown in Figure 3.83.

### Preliminary Processing

To obtain uncured stock of controlled size and shape for molding, the compounded material is processed by one of several preliminary operations: (1) slabbing,

| Material | Desig-nation (Gov't) | Tensile Strength | | Hardness (Durom-eter) | Mixing Eff. (%) | Extrud-ing (%) | Calen-dering (%) | Cotieston | Molding |
|---|---|---|---|---|---|---|---|---|---|
| | | Pure Gum | Black Rein-forced | | | | | | |
| Natural | ----- | 3000 | 4500 | any | 100 | 100 | 100 | Excellent | Excellent |
| Buna S | GRS | 400 | 3000 | any | 85 | 90 | 90 | Fair | Fair |
| Buna N | GRA | 600 | 3500 | any | 50 | 50 | 50 | Poor | Excellent |
| Butyl | GRI | 3000 | 3000 | 28 to 85 | 90 | 90 | 100 | Good | Good |
| thiokol | GRP | 300 | 1500 | 35 to 90 | 75 | 75 | 50 | Poor | Good |
| Neoprene | GRM | 3500 | 3500 | 25 to 90 | 100 | 75 | 90 | Good | Good |

**FIGURE 3.83** Characteristics of rubber and synthetic rubber materials. (From R. Bolz, *Production Processes,* Industrial Press, New York, 1963. With permission.)

(2) calendaring, or (3) extruding (termed "tubing" in the rubber field.) In slabbing, the compounded rubber is loaded into a mill, consisting of two large steel rolls that operate at slightly different speeds, which kneads and heats the compound and reduces its plasticity. After sufficient milling, the rubber is cut off in slabs, the gauge of the stock being determined by the roll spacing. Cutting templates for "slabbing off" pieces to proper contour for molding are often employed.

The calendar prepares the rubber compound in thin sheets that can be held to close tolerances. In calendaring, a ribbon of rubber is fed from a warmup mill to the calendar rolls, from which it emerges as a sheet or strip. Rubberizing or "frictionizing" of fabric is also done in a calendar. Extruding or tubing is carried out in a screw-type extruder, the shape and size of the stock being determined by the contour of the die. After cooling in a water tank, the extruded raw stock is cut to necessary lengths for molding.

## Molding Methods

There are four practical molding processes: (1) compression molding (most widely used), (2) transfer injection molding, (3) full injection molding, and (4) extrusion molding.

*Compression molding* consists of placing a piece or pieces of prepared stock in the heated mold cavity, bringing the halves of the mold together under pressure of 500 to 1000 psi, and curing (see Figure 3.84). Heat for curing is usually supplied by the heated platens of the press used. Depending on size, it is possible to mold from one to as many as 360 pieces per mold.

*Transfer-injection molding* permits the use of a single piece of prepared compound. Intricately shaped parts can be molded with improved efficiency over the compression method. The prepared piece is placed in a charging cavity in the mold and forced at high pressure through runners or channels into the mold cavity (see Figure 3.85). Usually the mold is opened and closed by hydraulic pressure, a separate plunger being used for injection.

*Mull injection molding* uses an extrusion head as an integral part of the molding machine. Lengths of extruded compounded rubber stock are fed directly into the extruder, which in turn injects or forces the material into the mold cavity or cavities. These units (see Figure 3.86), are entirely automatic in operation with the exception of stripping the finished parts from the mold. Cavities are laid out so that the stock

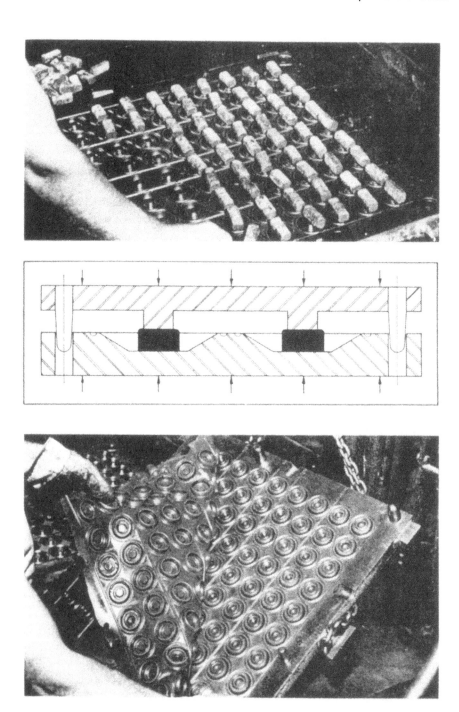

**FIGURE 3.84** Cross section of compression mold and views showing mold loading and removal of cured parts. (Courtesy Goodyear Tire & Rubber Co. With permission.)

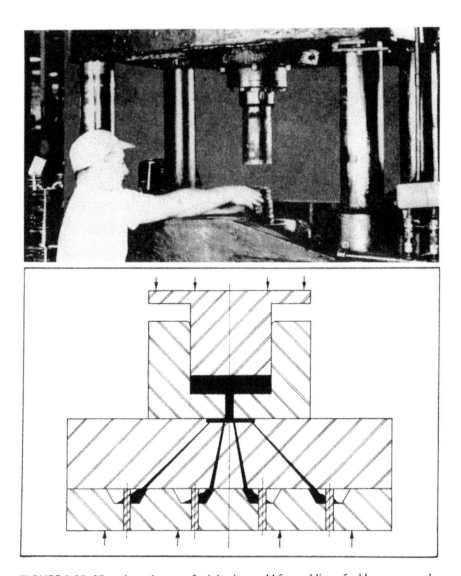

**FIGURE 3.85**  View through a transfer injection mold for molding of rubber compounds. (Courtesy Goodyear Tire & Rubber Co. With permission.)

is injected into a central canal that branches into two or more feeders and thence to the cavities. High pressure and turbulence developed during injection result in high temperature, reducing the curing period materially. Resultant savings in molding justify the use of this equipment in many instances.

*Extrusion molding* can be used to extrude a wide variety of uniform-cross-section parts to the desired shape. Very intricate sections are practicable. Screw-type feed is employed for forcing the stock through the die, as shown in Figure 3.87. Unlike the

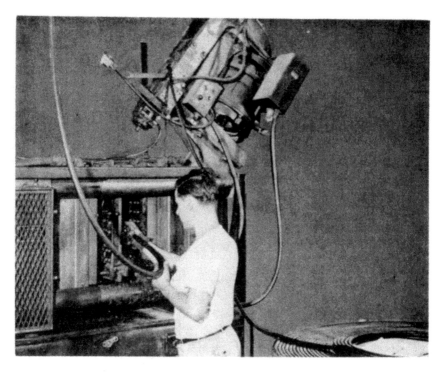

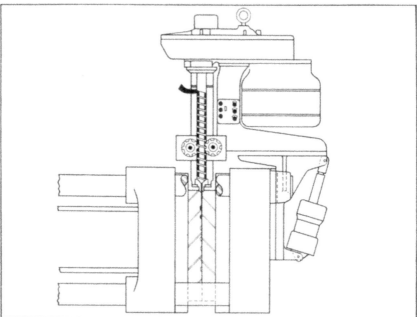

**FIGURE 3.86** Screw-type injection molding of lengths of extruded compounded rubber stock, and cross section through extrusion head and die. (Photo and drawing courtesy Goodyear Tire & Rubber Co. With permission.)

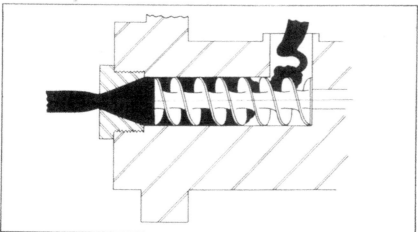

**FIGURE 3.87** Extrusion molding of uniform cross-section parts. (Photo and drawing courtesy Goodyear Tire & Rubber Co. With permission.)

other molding methods, extrusion does not permit curing during the cycle. After extruding and removal from the water cooling tank, extruded stock is cured under temperature and pressure by means of steam, and after cooling, it is cut to the necessary lengths. Extruded shapes require fair volume production for economy, because at least 100 lb of stock are required to put an extruder into operation. As example, a tube of 6 in. diameter and 1/8 in. wall can be produced in lengths to about 50 ft; sections under 3/16 in. in diameter can be extruded to lengths of about 500 ft.

### Parting Lines

As in all molding methods, placement of the mold parting or partings is extremely important, not only for ensuring simplest possible mold design and operation, but also for simplifying flash removal and finishing. Any particular location on a part that for design purposes should be free from flash, should be so indicated on the drawing. Because nonfills result in rejects, molded rubber parts invariably have overflow flash, and molds are designed to accommodate this condition. Circular flash, as shown in Figure 3.88a, is readily removable, automatically and cheaply. The part in Figure 3.88b is difficult to trim, whereas the one molded vertically in Figure 3.88c is much more economical.

### Metal Inserts

Where inserted pieces are to be completely imbedded in rubber, it is difficult to ensure positive location. As a rule, rubber flow in the mold varies and exact position of full floating inserts is impossible to predict. Thus, where inserts are employed, the parts should be designed so that the mold ensures positive positioning (see Figure 3.89).

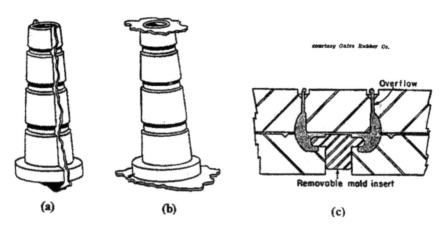

courtesy Gates Rubber Co.

Overflow

Removable mold insert

(a)                              (b)                              (c)

**FIGURE 3.88** Parts molded in (a) are difficult to trim; circular flash is readily removed in (b); while (c) is much more economical.

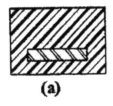

**(a)** **(b)** **(c)**

*courtesy Sirvene Div., Chicago Rawhide Mfg. Co.*

*courtesy Gates Rubber Co.*

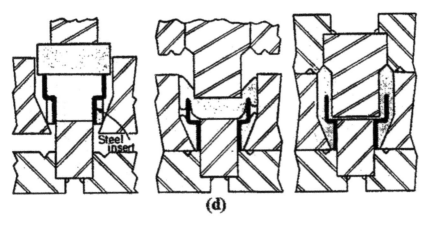

**(d)**

**FIGURE 3.89** Totally imbedded metal inserts (a) are difficult to position. Design (b) is difficult to trim, and (c) is preferred for economy. Imbedded inserts should be designed to be positioned by the dies positively as shown in (d), where the insert is located by the center pin.

## 3.5 BIBLIOGRAPHY

*ABC's of Aluminum,* Reynolds Metals Company, Richmond, VA, 1962.

Alting, Leo, *Manufacturing Engineering Processes,* Marcel Dekker, New York, 1982.

*Aluminum Forging Design Manual,* The Aluminum Association, New York, 1975.

*Aluminum Impacts Design Manual,* The Aluminum Association, New York, 1980.

Bolz, Roger, *Production Processes,* Industrial Press, New York, 1963.

*Casting Aluminum,* Reynolds Metals Company, Richmond, VA, 1965.

*Design Manual for Aluminum Precision Forgings,* Aluminum Company of America, Pittsburgh, PA, 1950.

*Designing for Alcoa Die Castings,* Aluminum Company of America, Pittsburgh, PA, 1955.

*Designing with Aluminum Extrusions,* Reynolds Metals Company, Louisville, KY, 1952.

*Facts and Guideline Tolerances for Precision Aluminum Forgings,* Forging Industr Association, Cleveland, OH, 1975.

*Forging Industry Handbook,* Forging Industry Association, Cleveland, OH, 1970.

*Forging Product Information,* Kaiser Aluminum, Oakland, CA, 1959.

*Forming Alcoa Aluminum,* Aluminum Company of America, Pittsburgh, PA, 1973.

Gillett, H. W., *The Behavior of Engineering Metals,* John Wiley, New York, 1951.

Ginzburg, Vladimir B., *High-Quality Steel Rolling,* Marcel Dekker, New York, 1993.

*How to Design Metal Stampings,* Dayton Rogers Manufacturing Company, Minneapolis, MN, 1993.

*Investment Casting Handbook,* Investment Casting Institute, Chicago, IL, 1968.

*Modern Steels and Their Properties,* Bethlehem Steel Company, Bethlehem, PA, 1964.

Tanner, John P., *Manufacturing Engineering,* Marcel Dekker, New York, 1991.

Tijunelis and McKee, *Manufacturing High Technology Handbook,* Marcel Dekker, New York, 1987.

# 4 Nontraditional Machining Methods

*Lawrence J. Rhoades*

with

*V. M. Torbillo*

## 4.0 INTRODUCTION TO NONTRADITIONAL MACHINING METHODS

The term *nontraditional machining* refers to a variety of thermal, chemical, electrical, and mechanical material-removal processes. The impetus for the development of nontraditional machining methods has come from the revolution in materials, the demand for new standards of product performance and durability, the complex shapes of products engineered for specific purposes, and considerations of tool wear and economic return. Nontraditional machining methods have also been developed to satisfy the trend toward increased precision and to create improved surface conditions. Because nontraditional machining processes can provide new ways of satisfying the demands of nascent technological advances in many areas, design engineers need not limit ideas to traditional machining methods. A new horizon of choices has opened up for the design of products.

There are several reasons for choosing nontraditional machining methods. One of the most important of these is the trend toward using engineered materials that are difficult to machine with conventional processes. High-temperature and high-performance metal alloys are often difficult to machine using traditional methods, but nontraditional machining processes generally work well with these materials. Because of the varying degrees of hardness they present, composites can be extremely difficult to machine using conventional methods. However, due to the selective machining and low applied forces of some nontraditional machining methods, composites can be machined with accuracy. Nontraditional machining methods are also ideal for machining ceramics and a host of other difficult materials.

Another reason for choosing nontraditional machining methods is that the features to be machined are often difficult or impossible to do with traditional methods. Complex geometric shapes, or shapes derived by the needs of flow, can present special machining problems. Nontraditional machining methods deal with these problems, offering full-form machining as opposed to single-point machining and providing the ability to reach normally inaccessible places. Nontraditional machining methods can create and machine special holes. These methods can machine shape cross sections, tapered holes, and holes where there is a high length-to-diameter ratio. Nontraditional machining methods can also provide special profile cutting for thin slots and small internal corner radii.

Burr-free machining, which may be essential in some applications, can be achieved by nontraditional machining methods. Low applied forces can prevent damage to the workpiece that might occur during traditional machining. Surface conditions left by some of the processes that fall into the category of nontraditional machining can ensure that the workpiece meets specific demands. The selective machining provided by nontraditional machining methods can be particularly interesting in the milling of composite materials in which there are combinations of components and some of the materials may be required to be left emerging from the field of a different material. Tool wear advantages that are offered by some of the nontraditional machining processes allow for continuous machining with, at least in theory, zero tool wear. All of these unique advantages of nontraditional machining methods open almost limitless possibilities of design and application. Figures 4.1 through 4.5 show the range and applications of nontraditional machining processes. A closer look at some of these processes will show their strengths in general and for specific tasks.

| CUTTING MECHANISM | PROCESS | 3D FORM SINKING | HOLE DRILLING | CUTTING | NON-TRADITIONAL EQUIVALENTS OF TRADITIONAL MACHINING (PROFILE MILLING, TURNING, GRINDING) |
|---|---|---|---|---|---|
| THERMAL REMOVAL | EDM | RAM EDM | SMALL HOLE EDM | WIRE CUT EDM | EDM PROFILING<br>ELECTRICAL DISCHARGE GRINDING |
| | LASER | | LASER DRILLING | LASER CUTTING | LASER TURNING<br>LASER MILLING |
| | PLASMA ARC | | | PLASMA ARC CUTTING | PLASMA ARC TURNING |
| DISSOLUTION | CHEMICAL | CHEM MILLING | CHEM MILLING<br>PHOTO ETCHING | CHEM MILLING<br>PHOTO ETCHING | |
| | ELECTROLYTIC | SALT WATER ELECTROLYTES | STEM ELECTROSTREAM CAPILLARY DRILLING | | |
| HYBRID BOTH MECHANICAL ABRASION AND ELECTROLYTIC REMOVAL | ELECTROLYTIC GRINDING | | | ECG "CUT OFF" | ECG FORM GRINDING |
| MECHANICAL ABRASION | ULTRASONIC | ULTRASONIC IMPACT GRINDING | ULTRASONIC DRILLING | ULTRASONIC "KNIFE" CUTTING | USM PROFILE MILLING<br>ROTARY ULTRASONIC MACHINING |
| | ABRASIVE WATER JET | | AWJ DRILLING | AWJ CUTTING<br>ABRASIVE SUSPENSION JET CUTTING | AWJ TURNING<br>AWJ MILLING |

**FIGURE 4.1**    Nontraditional equivalents of traditional machining.

| | EDM | RAM EDM |
|---|---|---|
| THERMAL REMOVAL | LASER | |
| | PLASMA ARC | |
| DISSOLUTION | CHEMICAL | CHEM MILLING |
| | ELECTROLYTIC | SALT WATER ELECTROLYTES |
| HYBRID | ELECTROLYTIC GRINDING | |
| MECHANICAL ABRASION | ULTRASONIC | ULTRASONIC IMPACT GRINDING |
| | ABRASIVE WATER JET | |

**FIGURE 4.2**    Summary of nontraditional 3-D form sinking processes.

| | EDM | SMALL HOLE EDM |
|---|---|---|
| THERMAL REMOVAL | LASER | LASER DRILLING |
| | PLASMA ARC | |
| DISSOLUTION | CHEMICAL | CHEM MILLING PHOTO ETCHING |
| | ELECTROLYTIC | STEM / ELECTROSTREAM / CAPILLARY DRILLING |
| HYBRID | ELECTROLYTIC GRINDING | |
| MECHANICAL ABRASION | ULTRASONIC | ULTRASONIC DRILLING |
| | ABRASIVE WATER JET | AWJ DRILLING |

**FIGURE 4.3**    Summary of nontraditional hole drilling.

| | EDM | WIRE CUT EDM |
|---|---|---|
| THERMAL REMOVAL | LASER | LASER CUTTING |
| | PLASMA ARC | PLASMA ARC CUTTING |
| DISSOLUTION | CHEMICAL | CHEM MILLING<br>PHOTO ETCHING |
| | ELECTROLYTIC | |
| HYBRID | ELECTROLYTIC GRINDING | ECG "CUT OFF" |
| MECHANICAL ABRASION | ULTRASONIC | ULTRASONIC "KNIFE" CUTTING |
| | ABRASIVE WATER JET | AWJ CUTTING<br>ABRASIVE SUSPENSION JET CUTTING |

**FIGURE 4.4**    Summary of nontraditional cutting processes.

| | EDM | EDM PROFILING<br>ELECTRICAL DISCHARGE GRINDING |
|---|---|---|
| THERMAL REMOVAL | LASER | LASER TURNING<br>LASER MILLING |
| | PLASMA ARC | PLASMA ARC TURNING |
| DISSOLUTION | CHEMICAL | |
| | ELECTROLYTIC | |
| HYBRID | ELECTROLYTIC GRINDING | ECG FORM GRINDING |
| MECHANICAL ABRASION | ULTRASONIC | USM MILLING<br>ROTARY ULTRASONIC MACHINE |
| | ABRASIVE WATER JET | AWJ TURNING<br>AWJ MILLING |

**FIGURE 4.5**    Summary of nontraditional profile milling, turning, and grinding processes.

## 4.1   THERMAL REMOVAL PROCESSES

In thermal removal processes, high-intensity heat is focused on a small area of the workpiece, causing it to melt and vaporize. One kind of thermal removal process is electrical discharge machining (EDM). In EDM, sparks between the electrode and the workpiece perform the material removal (see Figure 4.6). Two other types of thermal removal processes are laser machining and plasma arc machining. In both of these processes, a directed energy beam performs the cutting (see Figure 4.7). In all of the thermal removal processes, not all of the material removed is vaporized—much of it is melted. Most of the melted material is expelled from the

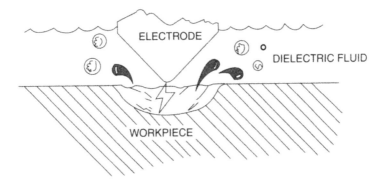

**FIGURE 4.6**   Schematic of the thermal removal process of EDM. Sparks between the electrode and the workpiece perform the material removal.

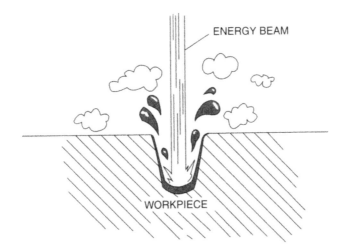

**FIGURE 4.7**   Schematic of laser and plasma arc thermal removal processes. The directed energy beam performs the cutting.

cut by the turbulence of the adjacent vaporization or by the flow of an assist gas used in the process. Some material remains and resolidifies on the surface, cooling rapidly as heat is transferred to the subsurface material. The remaining recast layer is likely to have microcracks and residual tensile surface stresses, encouraging those cracks to widen when the material is fatigued. Beneath the recast layer there is typically a heat-affected zone, where the material's grain structure may have been altered (see Figure 4.8 and Figure 4.9). A closer look at thermal removal processes will show in detail how they work.

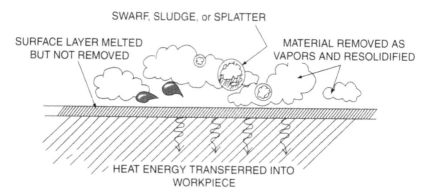

**FIGURE 4.8**    In thermal removal processes, some material remains and resolidifies on the surface.

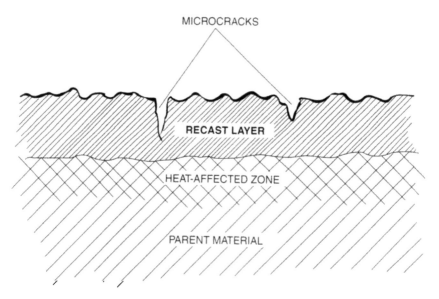

**FIGURE 4.9**    Microcracks appear in the recast layer formed in thermal removal processes.

## 4.1.1 Electrical Discharge Machining

In EDM, carefully controlled sparks (electrical discharges) are generated between an electrode and the workpiece. The electrode material chosen and the characteristics of the generated sparks are designed so that much more material is removed from the workpiece than from the electrode. The EDM machine typically has a built-in servo control system, and it is this control system that manipulates the electrode and the workpiece relative to one another to maintain a controlled spark gap. A flow of dielectric fluid is maintained in the spark gap. This fluid is used to provide a nonconductive barrier. It is also used to cool and resolidify the removed material into swarf particles and to flush this swarf from the machining gap to prevent uncontrolled arcing.

In RAM EDM, the form of a preshaped electrode is eroded into the workpiece material. Sparks occur between the electrode and the workpiece at the smallest gap, removing microscopic amounts of material with each spark. In the 3-D form sinking of RAM EDM, often the shaped electrode is manipulated under servo control. This manipulation, along with the high frequencies at which the voltage pulses are discharged, enables sparking to occur along the entire face of the electrode. The electrode then progressively advances into the workpiece, generating a uniform spark gap around itself, and the shape of the electrode is gradually reproduced on the workpiece (see Figure 4.10).

Nevertheless, in EDM the electrode wears as it removes material from the workpiece, and the electrode may need to be replaced or reshaped to machine multiple workpieces. Often more than one electrode is required to rough and finish machine a single workpiece with deep features. The servo-controlled manipulation of the electrode within the cavity being machined improves flushing and provides more uniform electrode wear. Flushing is important in EDM, especially when deep, complex shapes have to be reproduced, for the spark gap must be kept free from machining debris, which can cause uncontrolled arcing.

Another kind of EDM is small-hole EDM. When EDM is used to produce holes, typically a small-diameter wire, rod, or tube is held in an electrode holder

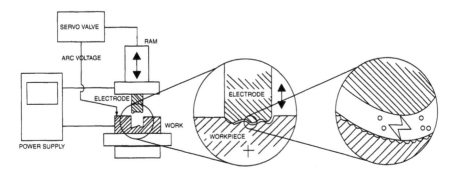

**FIGURE 4.10** In RAM EDM, the form of a preshaped electrode is eroded into the workpiece material.

and advanced into the workpiece (see Figure 4.11). Electrode wear is compensated by advancing or refeeding the electrode material through the holder to reestablish proper electrode length. If necessary, the electrode ends are trimmed to restore the desired tip shape. To enhance productivity, small-hole EDM is employed using a number of electrodes held simultaneously to drill multiple holes precisely. If a single hole is to be drilled, the electrode can be rotated during machining to improve flushing and distribute wear uniformly. Using tube electrodes permits flushing through the tube. Small-hole EDM allows precise drilling operations in a variety of materials.

Wire EDM has proven to be a major development in EDM. In wire EDM, the electrode used for cutting is a small-diameter wire, usually brass or copper with a diameter of 0.05 to 0.25 mm. Servo control systems are employed for guiding the movement of the tool and the workpiece relative to one another and for controlling the direction of the machining. The small-diameter wire is held taut between two spools. Fresh wire is continuously fed into the cut to make up for electrode wear and encourage flushing. The dielectric fluid used in traveling wire EDM is typically deionized water because of low viscosity and high dielectric constant. The dielectric fluid is usually injected into the machining zone coaxially with the wire. Cutting speeds have increased by a factor of more than ten since wire EDM was introduced, and those speeds are now in the range of 140 $cm^2$/hr. Accuracies can routinely be held to ±0.01 mm. Wire EDM is the most widely used unattended machining operation, with overnight "lights out" operation a routine practice even in small manufacturing companies in this country and throughout the world (see Figure 4.12).

EDM can also be applied to milling and grinding operations (see Figure 4.13 and Figure 4.14). EDM can be used as a profile milling operation similar to conventional

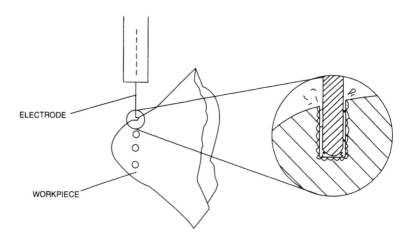

**FIGURE 4.11** Example of using the EDM process to produce small-diameter holes.

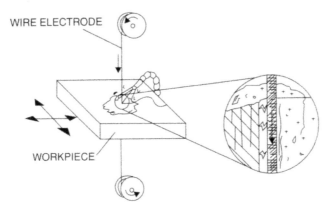

**FIGURE 4.12**  Process of cutting shapes using wire EDM.

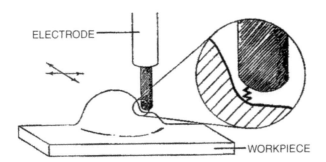

**FIGURE 4.13** EDM profile milling where the electrode is used like a milling cutter, with the machine manipulating the tool and workpiece relative to one another under servo control.

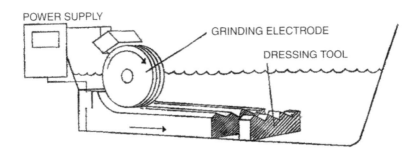

**FIGURE 4.14** EDM grinding where the electrode is used like a grinding wheel, with the machine manipulating the tool and workpiece relative to one another. The dressing tool shown can be used similar to redressing a conventional grinding wheel.

milling. The electrode is used like a milling cutter. The servo control of the milling machine manipulates the tool and the workpiece relative to one another for accurate milling. The process is not very fast, but it is quite flexible. Another EDM process is EDM grinding. In EDM grinding, the electrode is a wheel, typically graphite, with power connected to it. The electrode is rotated while the workpiece and the wheel are moved relative to one another under servo control. The electrode wear is distributed across the entire wheel diameter, and redressing can easily be performed, as in conventional grinding. Generally, the work is done submerged within the dielectric fluid.

EDM is a versatile and precise thermal removal process. It has a wide range of applications in machining a variety of materials. It can be controlled with an accuracy that many traditional machining methods lack, and as the trend toward increased precision and hard materials grows, EDM will become even more integral to manufacturing.

## 4.1.2 Laser Beam Machining

Another thermal removal process is laser beam machining. In laser beam machining, an intense beam of collimated, single wavelength, in-phase light is focused by an optical lens onto the workpiece point to be machined. The light absorbed by the workpiece is converted to heat, which melts and vaporizes the workpiece material. Molten material is evacuated from the cut by the adjacent vaporization turbulence that typically occurs in drilling operations, or by the use of an assist gas in cutting operations (see Figure 4.15). Laser beam machining uses a directional, coherent, monochromatic beam of light to achieve precision in cutting and drilling. The intensity of this light produces a tremendous amount of heat at the point of application to the workpiece, and laser beam machining can take place at relatively high speeds.

A number of different types of lasers are used in laser beam machining, each with certain advantages for different operations or applications. The most commonly used lasers for machining are Nd:YAG lasers, which have certain advantages for hole drilling due to their higher pulse energy, and $CO_2$ gas lasers, which have certain advantages in cutting since they are capable of delivering much higher average power. Lasers may be operated in either pulsed or continuous-wave (CW) modes. The most powerful $CO_2$ lasers, however, are operable only in CW mode. $CO_2$ lasers can have output power generally ranging from 100 to 2000 W when pulsed, and from 250 to 5000 W in CW mode. Some lasers are capable of an output power of 25,000 W.

A 1250-W $CO_2$ laser can cut mild steel at speeds ranging between 40 and 140 $cm^2$/min, depending on material thickness. At a thickness of 12 mm, the cut can be 40 $cm^2$/min. At a thickness of 2 mm, a 1250-W $CO_2$ laser can cut at a rate of 140 $cm^2$/min. Aluminum is generally cut at about one half the speed of carbon steel with a $CO_2$ laser because of the high thermal conductivity of the aluminum. In cutting applications, the laser beam may be transmitted and switched by using

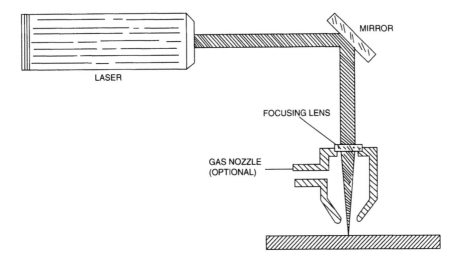

**FIGURE 4.15** Schematic of laser beam machining using a directional, coherent, monochromatic beam of light to achieve precision in cutting and drilling.

mirrors to manipulate the beam, or with Nd:YAG lasers by the use of a combination of fiber optic cables and switches. Below is the typical average power range and maximum pulse energy for the two types of lasers most commonly used in laser beam machining.

|                          | Nd:YAG    | $CO_2$                             |
|--------------------------|-----------|------------------------------------|
| Typical average power range | 100-400 W | 250-5000 W CW<br>100-2000 W pulse |
| Maximum pulse energy     | 80 J      | 2 J                                |

In the application of laser beam machining, it is possible to perform the operations of drilling and cutting with speed and precision. For instance, percussion (i.e., repeated pulse) drilling of Inconel 718 with a 250-W Nd:YAG laser can produce accurate holes 12 mm deep in under 10 sec and holes 25 mm deep in 40 sec. Length-to-diameter ratios are limited to about 30 or 40 to 1 with conventional Nd:YAG lasers. However, other laser technologies allow the length-to-diameter ratio to be higher. Frequently, gas is used to assist laser beam machining. A coaxial columnar flow of gas (oxygen, air, or inert gas) at pressures ranging from 1 to 6 bar expels molten metal from the cut. Oxygen assists in cutting steel and other materials at an increased rate because of the oxidation reaction with the metals.

Laser beam machining uses thermal energy to drill and cut with speed and precision that traditional machining methods often cannot duplicate. Its applications are becoming more varied as new techniques are developed.

## 4.1.3 Plasma Arc Machining

The final thermal removal process to be discussed is plasma arc machining. Like EDM and laser beam machining, it employs thermal energy to accomplish its work. In plasma arc machining, gas heated to very high temperatures by a high-voltage electric arc partially ionizes and consequently becomes electrically conductive, sustaining the arc. When gas is heated to the degree that electrons become ionized (electrically charged), the gas is called a plasma. Primary gases used for plasma arc machining may be nitrogen, argon–hydrogen, or air. The gas is forced at a high rate of speed through a nozzle and through the arc. As the gas travels, it becomes superheated and ionized. The superheated gas reaches temperatures of 3,000 to 10,000 K. A hot tungsten cathode and a water-cooled copper anode provide the electric arc, and the gas is introduced around the cathode. It then flows out through the anode. The size of the orifice at the cathode determines the temperature, with small orifices providing higher temperatures. The ionized particle stream is consequently a high-velocity, well-columnated, extremely hot plasma jet, supporting a highly focused, high-voltage, "lightning like" electric arc between the electrode and the workpiece. With such high temperatures, when the plasma touches the workpiece, the metal is rapidly melted and vaporized. The high-velocity gas stream then expels molten material from the cut.

In plasma arc machining, frequently a swirling, annular stream of either water or a secondary gas is injected to flow coaxially with the plasma arc. The use of water can serve several purposes. It increases the stability of the arc and increases cutting speeds. Water injected coaxially can also cool the workpiece and reduce smoke and fumes. It can also increase nozzle life. Sometimes a secondary gas is introduced, surrounding the plasma stream. The choice of the secondary gas depends on the metal being cut. Hydrogen is often used as a secondary gas for the machining of stainless steel or aluminum and other nonferrous metals. Carbon dioxide gas can be used successfully with both ferrous and nonferrous metals. Oxygen is often introduced as a secondary gas surrounding the plasma stream, adding the heat from the exothermic oxidation reaction with steel and other materials to assist in the cut (see Figure 4.16).

Plasma arc machining is widely used for sheet and plate cutting. It is incorporated in many CNC sheet metal punching machines. With plasma arc machining, cutting speeds of 700 $cm^2$/min and higher can be achieved. Accuracies are limited, however, to ±0.1 mm at best, and a taper of 2° or more is normally generated. A recast layer and heat-affected zone of roughly 0.5 mm depth are typical. Nevertheless, plasma arc machining is a fast and effective method of machining in many applications.

EDM, laser beam machining, and plasma arc machining with differentials of an order of magnitude offer precision and machining speeds that many conventional processes do not (see Figure 4.17). In cutting, for instance, wire EDM provides cutting speeds of about 40 $cm^2$/hr. Laser beam machining provides cutting speeds on average of 70 $cm^2$/min. Plasma arc machining provides cutting speeds of about

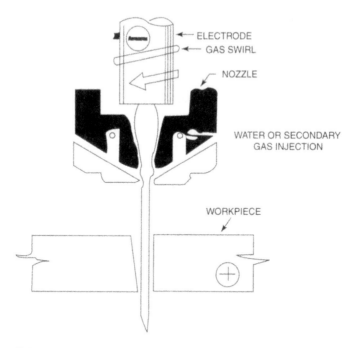

**FIGURE 4.16** In plasma arc machining, gas heated to very high temperatures by a high-voltage electric arc partially ionizes and consequently becomes electrically conductive, sustaining the arc.

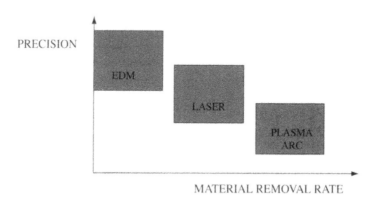

**FIGURE 4.17** Relation of precision and material removal rate of the major thermal removal processes.

700 cm$^2$/min. The precision, however, ranges from about $\pm 0.1$ mm at best for plasma arc machining to $\pm 0.01$ mm as typical for wire EDM. The trade-off of speed for precision might be a determining factor in choosing the right kind of thermal removal process for the specific application.

## 4.2 CHEMICAL MACHINING

Chemical milling is accomplished simply by dipping the workpiece into a tank with strong etchants. Nitric acid, hydrochloric acid, and hydrofluoric acid are the most commonly used etchants in chemical machining. The areas where no machining is desired are masked, typically with an elastomer like silicone rubber or with an epoxy. Periodic interim steps are used to monitor remaining material or, if necessary, to remask or to cover undercut areas. Progressive masking and etching can produce complex shapes in difficult-to-form materials, even very large ones, with potentially no applied forces (see Figure 4.18).

Photo etching is another variant of chemical machining. Photo etching uses a photo-resistant maskant to generate intricate 2-D patterns in thin, flat metal sheets. The metal sheets used in photo etching have a thickness range from 0.01 to 1.5 mm. The process of photo etching uses a relatively mild etchant such as ferric chloride, which is typically applied by spraying as conveyorized parts pass through a spray chamber. Photo etching is widely used to produce circuit boards and other sheet materials. It can produce thousands of holes at once with a high degree of accuracy.

### 4.2.1   Electrochemical Machining

Electrochemical machining (ECM) removes metal by anodic dissolution. ECM electrolytes are normally safe-to-handle common salt-water solutions. However, the sludges produced in the ECM of certain materials, notably chromium, can be

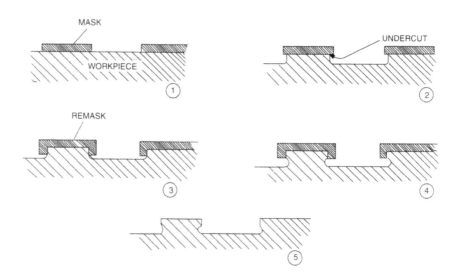

**FIGURE 4.18** Chemical milling is accomplished simply by dipping the workpiece into a tank with strong etchants. The areas where no machining is desired are masked, typically with an elastomer like silicone rubber or with an epoxy. Progressive steps of masking and etching are shown as 1 through 5 for a complex part.

poisonous and must be processed or disposed of carefully and responsibly. ECM uses high amperage (500–20,000 A, typically) and low voltage (10–30 V) DC power with relatively high electrolyte flow rates through the machining gap (e.g., 3000–6000 cm/sec). Metals removed by ECM quickly become a metal hydroxide, insoluble sludge, which is washed from the machining gap by the electrolyte and is removed by some filtration process—settling, centrifuge, or some other type of filtration. The form machined is a near mirror image of the cathode (tool), with variations resulting from electrolyte flow, hydroxide concentration, and temperature changes.

There are compelling advantages to ECM processes. One is that there is virtually no tool wear. Another advantage to ECM is that work can be done at relatively high speeds. For example, a 15,000-A machine can cut 25 cm$^3$/min, regardless of material hardness. ECM normally provides excellent surface conditions with no thermal damage. Nevertheless, there are some drawbacks to using ECM. The cost of equipment is relatively high because corrosive-resistant materials are required for the corrosive salt environment in which ECM operations take place. Machining tight accuracies can often require extensive tool-shape development. This results from the changing conductivity of the electrolyte as it passes across the machining gap due to the heat and metal hydroxide added from the machining process, causing the electrolytic machining gap to vary as much as 0.02 to 0.2 mm. Pulsed power ECM reduces this variance. Finally, sludge handling and removal, particularly when chromium alloys are used, can be a concern.

ECM can provide both speed and precision in many applications. ECM is well suited to machining the complex shapes and difficult materials used in turbine engines, and it is commonly used to machine compressor blades. It is also widely used, with stationary electrodes, as a deburring method. There is great potential in the ECM process to influence design and production.

## 4.2.2   Stem, Capillary, and Electrostream Drilling

Some commonly applied variations of ECM include stem, capillary, and electrostream drilling. These processes use strong acids as electrolytes. Stem drilling uses titanium tubes with conventional low voltage, generally 8–14 V. Electrolyte is pumped down the central bore of the titanium tube and out through the gap formed between the wall of the tube and the hole being dissolved in the workpiece. Capillary drilling uses straight glass tubes with a platinum wire inside each to conduct electricity to the electrolyte stream. In capillary drilling, the voltage is higher, generally in the range of 100–300 V. Again it is the electrolyte that dissolves the hole in the workpiece as it is pumped through the glass tubes and out through the gap. Electrostream drilling uses glass nozzles, shaped glass tubes without wires. A much higher voltage, generally in the range of 600–900 V, allows the electrolyte stream to cut without a conductive tool.

With stem, capillary, and electrostream drilling, speed and precision are key benefits. Cutting speeds range from 0.75 to 3 mm/min with precision, and multiple holes can be simultaneously machined. All of these processes can produce long, small holes.

|            | Diameter (mm) | Depth (mm) |
|------------|---------------|------------|
| Stem       | 0.5           | 75         |
|            | 6.0           | 1000       |
| Electrostream | 0.15       | 6          |
| Capillary  | 0.5           | 25         |

## 4.2.3   Electrochemical Grinding

Electrochemical grinding (ECG) is a hybrid process combining electrolytic dissolution and mechanical grinding. In ECG, a grinding wheel with conductive bonding material and protruding nonconductive abrasive particles does the cutting. Electrolyte is carried across the machining gap with the surface of the wheel (see Figure 4.19). The grinding wheel is the cathode (tool), and the workpiece is the anode. The nonconducting particles protruding from the wheel act as a spacer between the wheel and the workpiece to allow a constant gap for the flow of the electrolyte. The electrolyte is fed through a tube or nozzle to flow into the machining gap.

In ECG, the feed rate and voltage settings determine the relative roles of the electrolytic and mechanical action. With high electrolytic action—in other words, higher voltage and lower feed rates—wheel wear is minimized and more of the cutting is accomplished by electrochemical action. With lower voltage and higher feed rates, there will be more mechanical cutting. Although higher rates of mechanical

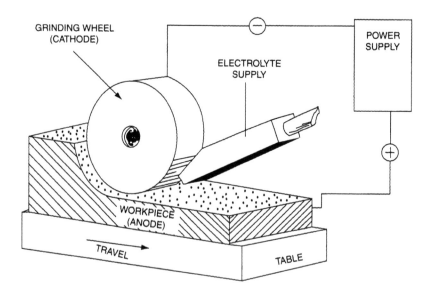

**FIGURE 4.19** ECG is a hybrid process combining electrolytic dissolution and mechanical grinding.

cutting produce more wheel wear, they also improve machining accuracy. In ECG operations, generally 90% of the stock is removed electrolytically. Then low-voltage finish passes can be made to produce high accuracy and sharp edges that electrolysis alone cannot achieve.

There are real advantages to ECG. ECG can machine dissimilar materials and composites well, since the process itself is a composite process. ECG uses electrolytes and machining conditions that are prone to surface passivation, thus minimizing unwanted stray etching, since grinding action will mechanically remove the nonconductive passivation layer, thereby exposing only the desired areas to continued electrolytic action. ECG can machine without burrs, can have very low grinding wheel wear since most of the removal is by electrolytic action, is capable of extremely high accuracy, and leaves stress-free surfaces.

## 4.3 MECHANICAL PROCESSES

Nontraditional machining methods employ mechanical processes just as traditional methods do. These processes include ultrasonic machining, abrasive water jet machining, and abrasive flow machining. Each of these processes has certain advantages in the machining of different materials and shapes.

### 4.3.1 Ultrasonic Machining

In ultrasonic machining, a tool is vibrated along its axis at its resonant frequency. The material, the size, and the shape of the tool are typically chosen to have a resonant frequency in the range of 20 kHz, which is above the human hearing range and may therefore be deemed ultrasonic. The vibration amplitude of the tool is typically 0.01 to 0.05 mm. The ultrasonic machining process erodes holes or cavities in hard or brittle material by means of the cutting action of an abrasive medium. Ultrasonic impact grinding uses an abrasive/water slurry. This slurry is drawn into the machining gap between the vibrating tool and the workpiece. The abrasive particles are propelled or hammered against the workpiece by the transmitted vibrations of the tool. The particles then microscopically erode or chip away at the workpiece.

Rotary ultrasonic machining uses an abrasive surfaced tool that is simultaneously rotated and vibrated. The combination of rotating and vibrating action of the tool makes rotary ultrasonic machining ideal for drilling holes and performing ultrasonic profile milling in ceramics and brittle engineered materials that are difficult to machine with traditional processes. Ultrasonic assisted machining adds ultrasonic vibrations to conventional drilling, turning, and milling operations.

Ultrasonic machining is ideal for certain kinds of materials and applications. Brittle materials, particularly ceramics and glass, are typical candidates for ultrasonic machining. Ultrasonic machining is capable of machining complex, highly detailed shapes and can be machined to very close tolerances ($\pm 0.01$ mm routinely) with properly designed machines and generators. Complex geometric shapes and 3-D contours can be machined with relative ease in brittle materials. Multiple holes, sometimes hundreds, can be simultaneously drilled into very hard materials with great accuracy.

Ultrasonic machining can be used to form and redress graphite electrodes for EDM. It is especially suited to the forming and redressing of intricately shaped and detailed configurations requiring sharp internal corners and excellent surface finishes. Low machining forces permit the manufacture of fragile electrodes too specialized to be conventionally machined. Redressing can be accomplished quickly, typically in 2 to 10 min, often eliminating the need for multiple electrodes. One electrode can be used for roughing, redressed for semifinishing, and redressed again for final finishing. Because of this advantage to ultrasonic machining of electrodes, EDM parameters can be selected for speed and finish, without regard to electrode wear. Minutes spent in ultrasonic electrode redressing can save hours of EDM time while also improving final finish and accuracy.

A variation of ultrasonic machining is ultrasonic polishing. Ultrasonic polishing can uniformly polish and remove a precise surface layer from machined or electrical discharge machined workpieces by using an abradable tool tip. This process uses a sonotrode (see Figure 4.20 for machining and Figure 4.21 for polishing) that has a special tip that is highly abradable, such as graphite. By vibrating the abradably tipped tool into the workpiece, the tool tip takes the exact mirror image of the workpiece surface and uniformly removes the surface layer from the workpiece material, improving the surface finish and removing undesirable surface layers. The polishing action occurs as fine abrasive particles in the slurry abrade the workpiece surface,

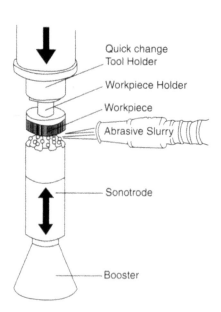

**FIGURE 4.20** Example of ultra sonic machining with a sonotrode and abrasive slurry.

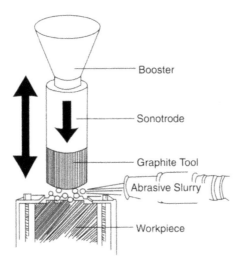

**FIGURE 4.21** Example of ultrasonic polishing with a graphite tool on the sonotrode with
an abrasive slurry.

typically removing only slightly more material than the surface roughness depth
(e.g., 0.01 mm from a 1-μm Ra electrical discharge machined surface). The extent
of polishing required is determined by the initial surface roughness and the finish
required after polishing. Typical surface improvements range from 3:1 to 5:1. A variety of materials including tool steels, carbides, and ceramics can be successfully
processed with ultrasonic polishing.

## 4.3.2   Abrasive Water Jet Machining

Another of the mechanical processes of nontraditional machining methods is abrasive water jet machining (AWJ). AWJ cuts by propelling high-velocity abrasive particles at the workpiece. The propulsion is done by entraining abrasive particles into a
small-diameter, high-pressure water stream. The process is widely used for cutting,
with limited use for drilling, turning, and milling.

In AWJ, high-pressure water is fed into a tube and through a small-diameter
water orifice, producing a high-velocity water jet. Abrasive particles are fed into
the mixing chamber of the nozzle through another tube. The abrasive is entrained
into the high-velocity water stream as they both pass through a mixing tube to the
workpiece (see Figure 4.22). Water-pumping pressures of 2000–3000 bar are typically used, with orifice exit velocities of about Mach 3 in a 0.2- to 0.3-mm diameter
stream. Introducing about 8% (by weight) of garnet abrasive into this stream accel-
erates the abrasive particles to perform the cutting. However, it also disperses the
stream and wears both the mixing tube and the abrasive particles. The following table
shows the speeds of AWJ.

| MAXIMUM CUTTING SPEED (mm/min.) | | | | |
|---|---|---|---|---|
| **Thickness (mm)** | **Aluminum** | **Carbon Steel** | **Stainless Steel** | **Alloy 718** |
| 3 | 1250 | 750 | 600 | 550 |
| 6 | 750 | 500 | 400 | 300 |
| 12 | 450 | 300 | 250 | 150 |
| 25 | 200 | 150 | 100 | 40 |
| 50 | 150 | 75 | 55 | 5 |
| 100 | 100 | 25 | 25 | — |

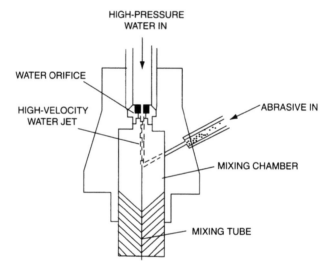

**FIGURE 4.22** Principle of AWJ.

AWJ is a relatively fast and precise mechanical process of nontraditional machining. Its primary use is for sheet-cutting operations. It is a method of choice for creating prototypes. AWJ has many applications in the aerospace industry. It also has several applications in the automotive and appliance industries.

Another variation of AWJ currently under development is abrasive suspension jet machining (ASJ). In ASJ, the abrasive and the suspension carrier are propelled together through a diamond nozzle. ASJ's more uniform stream velocity and stream coherency promise improved precision over AWJ, and cutting efficiency is high enough to use pressures 75% lower than with AWJ. Abrasive materials can be continually reused with ASJ, dramatically reducing waste generation. When it is perfected, ASJ will broaden the applications for abrasive jet machining.

### 4.3.3 Abrasive Flow Machining

Abrasive flow machining (AFM) hones surfaces and edges by extruding a pliable abrasive-filled medium through or across the workpiece. In this process, two vertically opposed cylinders extrude abrasive media back and forth through passages formed by the workpiece and the tooling (see Figure 4.23). Abrasive action occurs wherever the media enter, and passes through the most restrictive passages. The major elements of the process include the tooling, which confines and directs the media flow to the appropriate areas; the machine, which controls the media extrusion pressure, flow volume, and, if desired, the flow rate; and the media, which determine the pattern and aggressiveness of the abrasive action that occurs. By selectively permitting and blocking flow into or out of workpiece passages, tooling can be designed to provide media flow paths through the workpiece. These flow paths restrict flow at the areas where deburring, radiusing, and surface improvements are desired. Frequently, multiple passages or parts are simultaneously processed.

The machine controls the extrusion pressure. The range of useful pressures extends from low pressures, down to 7 bar, to high pressures, in some cases over 200 bar. Increasing extrusion pressure generally increases process productivity. However, there may be reasons to choose a lower extrusion pressure. If the workpiece is fragile, lower extrusion pressures may be necessary. Lower extrusion pressures might also be beneficial for lower tooling costs, or when there is the desire to machine multiple parts or large-area parts.

The machine also controls the volume of the media flow. The volume of flow per stroke (in cubic inches or centiliters) can be preset, as can the number of cycles. Each cycle is typically two strokes, one up and one down. Following the dimensional change caused by smoothing the rough surface peaks, stock removal is directly proportional to media extrusion volume. This permits precise control of the minute

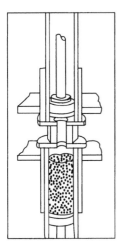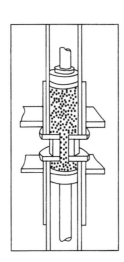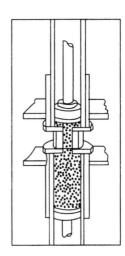

**FIGURE 4.23** Principle of AFM.

enlargement of flow passages while maintaining shape within a fraction of the surface layer removed.

AFM can controllably remove undesirable surface layers remaining from thermal machining processes such as EDM and laser beam machining. AFM can also improve finishes by an order of magnitude. For example, 1-μm Ra electrical discharge machined surface is improved to 0.1-μm, or a 2.5-μm Ra finish is improved to 0.25-μm (see Figure 4.24). Required dimensional change to achieve this improvement is slightly greater than the original total roughness per surface. For example, an EDM surface with a 1-μm Ra surface typically has a total roughness (Rt) of about 8-μm. Apparent stock removal (dimensional change) will be about 10-μm (0.01 mm).

AFM is used in a variety of applications. In aircraft turbine engines, it is used to machine compressor blades, blisks, impellers, turbine blades and vanes, disks, casings, and other components, with more applications being discovered frequently. AFM is ideal for dies and molds. It can be used for extrusion dies, cold-heading dies, tableting and compacting dies, forging dies, and die-casting dies. It can also be used for plastic injection molds and glass molds. Other applications include electronic components, medical components, and high-precision pumps, valves, and tubes. The use of AFM in finishing diesel and automotive components is rapidly growing.

## 4.4  BURNISHING

V. M. Torbilo, professor, Ben-Gurion University, Beer-Sheva, Israel

### 4.4.1  Introduction to Burnishing

Burnishing is one of the methods of finish machining, yielding significant improvements in the service properties of machined parts. It provides efficient machining of parts made of most of the engineering metallic materials, including the high-strength alloys of practically any hardness. This method is used in machine production, especially for finishing of precision and critical parts.

Burnishing is a method of finishing and hardening machined parts by plastic deformation of the surface. The plastic deformation of the processed surface is accomplished by the pressure of a sliding tool (burnisher) with a rounded working surface (Figure 4.25). During burnishing, the surface roughness caused by the previous machining is flattened and leveled, and the surface acquires a mirrorlike finish.

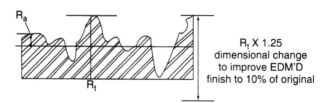

**FIGURE 4.24** Figure AFM can improve surface finishes by an order of magnitude (e.g.. micron Ra surface is improved to 0.1 micran Ra).

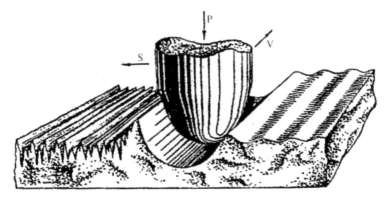

**FIGURE 4.25** Principle of burnishing.

The surface layer strength increases and compressive residual stresses are generated. After burnishing, the surface becomes smooth and clear of metallic splinters or abrasive grains that usually occur during abrasive machining. Combination of properties of the burnished surface determines its high working specifications, including wear resistance, fatigue strength, and so forth.

## 4.4.2 The Burnishing Process

The burnishing method is rather simple in its basic operation. The burnishing tool consists of a working element and a metallic holder. The working element is made of superhard materials and has a rounded form, usually spherical. The tool is held in a tool holder on a regular metal-working machine. Burnishing is most often performed on turning machines (Figure 4.26). Burnishing can also be performed on machines of other types, such as milling, boring, or drilling machines. CNC machine tools and machining centers can also be used in the burnishing process.

Burnishing has the following main features: use of superhard materials as a deforming element (most often diamonds), a small radius on the deforming element (0.5–4.0 mm), and sliding friction between the deforming element and the workpiece being processed. The high hardness of diamonds and other superhard materials provides an opportunity to burnish almost all metals susceptible to plastic deformation in the cold state, whether they are relatively soft or hardened as high as Rockwell C 60–65.

Due to the small rounding radius of burnishers, the contact area between them and a workpiece is small (less than 0.1–0.2 mm$^2$). This allows the creation of high pressures exceeding the yield limit of the processed material in the contact area at comparatively small burnishing tool forces (50–260 N). It also reduces the requirements for the rigidity of manufacturing equipment. However, the small contact area in burnishing combined with slow feed rates limits process efficiency. Sliding

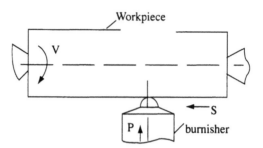

**FIGURE 4.26** Principle of burnishing on a lathe.

friction between a burnisher and a workpiece increases the quality of the burnished surface.

Since there is no metal cutting in burnishing, and the burnisher's hardness is much higher than that of the processed material, the burnishing process can be regarded as a process of motion of a rigid rounded deforming element (indentor), forced into the processed surface and deforming it. In the area of tool–workpiece contact, some physical processes occur. The main ones are surface-layer deformation, friction between the tool and the workpiece, heating of the tool and the workpiece, and wear of the former. These physical processes establish the quality of the surface layer and determine the efficiency and economy of this manufacturing process, considering the machinability of different metals and the associated tool life of the burnisher.

## 4.4.3  Materials Suitable for Burnishing

Almost all industrial metals and alloys that are subject to plastic strain in the cold condition can be burnished. The results of burnishing depend to a great extent on a material's type, properties, and machinability. The following criteria characterize machinability at burnishing: (1) surface smoothability; (2) hardenability; and (3) wearability, denoting the material's ability to wear the tool's working surface.

Almost all types of steels (of different chemical composition, structure, and hardness up to HRC 62–64), including nickel-based alloys, burnish well. Aluminum and copper alloys have good machinability from the standpoint of burnishing. Gray cast iron is generally known to be a low-plastic material; therefore, its burnishing is inefficient. But some types of cast iron, such as high-strength cast iron and alloy cast iron, can be burnished.

Electrodeposited coatings (chrome, nickel, etc.) can be treated by burnishing. There are some specific features in burnishing of coatings. Burnishing of titanium alloys is usually not practical because of the great adhesive interaction between the processed material and the burnisher's working surface.

### 4.4.4 Tools, Toolholders, and Machines

A burnisher is a metallic holder in which a working element made of superhard material (diamond, synthetic corundum, carbide) is fixed. Several shapes of the burnisher's working part are utilized: the spherical surface, the side surface of a cylinder, the surface of a circular torus, and the cone surface burnisher. The spherical shape is the most versatile. It allows one to burnish outer and inner round surfaces and flat surfaces.

Burnishing can be performed on standard universal and special machine tools—turning, boring, planing, milling, and so forth—with normal and high precision. Higher precision is required for burnishing with rigid fixing of a burnisher to the machine. Burnishing on lathes is most common. At this point, particular attention should be paid to the value of the spindle's radial concentricity (not more than 0.01–0.02 mm runout), the rigidity of the support, and vibration resistance. A number of feeds, beginning with about 0.02 mm/rev, should be provided on a machine tool. The machine should also be equipped with a lubricating–cooling device.

Burnishing tools are mounted on metal-cutting machines with the help of holders. Two main types of holders are distinguished by their method of a tool attachment, either rigid or elastic. Holders with a rigid or fixed tool have a very simple design and differ little from mounting a cutting tool on a lathe cross-slide, as shown in Figure 4.27. Several designs of holders using an elastic element to maintain pressure have been developed, including hydraulic and electromagnetic. This tool has more flexibility of use, since it tends to hold the required pressure between the surface of the part and the burnishing tool face, even though the part may have some eccentricity or other minor surface deviations. See Figure 4.28 for a common example using a spring as the force element to hold pressure.

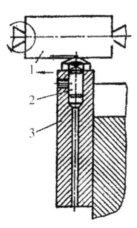

**FIGURE 4.27** Example of rigrid burnishing (1) workpiece, (2) burnisher, (3) rigid holder.

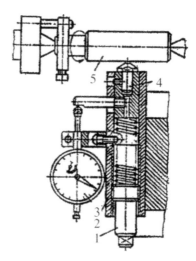

**FIGURE 4.28** Example of elastic burnishing: (1) screw, (2) spring, (3) indicator, (4) burnisher, (5) workpiece.

### 4.4.5 Basic Methods of Burnishing

Burnishing with rigidly fixed tools creates a solid kinematic link between the tool and the workpiece, as in turning, for example. The burnisher is fixed on a machine tool in the same way as a cutter, and its position in relation to the workpiece is determined only by the machine's kinematics and the elasticity of the manufacturing system. With rigid burnishing, the burnisher is indented into the surface for a predetermined depth, which varies from several microns to several hundredths of a millimeter. The depth depends on the plasticity of the material, its surface roughness, and the burnishing tool radius. The advantage of burnishing with a rigidly fixed tool is that it offers an opportunity to increase the precision of dimensions and shape of the workpiece by redistribution of volumes of plastically strained metal. By rigidly fixing the burnisher, however, the value of the burnisher's indentation, and hence the burnishing force, can vary considerably because of the workpiece's beat. Therefore, the machine tool (for example, the lathe) must have precision spindle runout and very good rigidity in the bearings, frame, ways, cross-slide, and work holder. The part must also be rigidly mounted, and the burnishing tool precisely and rigidly mounted. Burnishing with a rigidly fixed tool is recommended for processing of especially precise machined parts for high-precision machines.

Burnishing with an elastic tool mount is a simpler and more convenient method. In this method of burnishing, the tool is pressed elastically to a workpiece with the help of a spring, or in some other way. The force pressing the burnisher depends on the plasticity of the material, its surface roughness, and the burnisher's radius. It is easily controlled and should be kept constant during burnishing. In this process there is no rigid kinematic link between the workpiece and the tool, and the position of the latter in relation to the workpiece is determined only by the surface itself. With an elastically fixed tool, the errors of the shape are copied and are not corrected. Only smoothing of the surface roughness and surface hardening take place. Simplicity of setting and comparatively moderate requirements for precision and rigidity of machine tools and workpieces are the advantages of this method of burnishing.

### Vibroburnishing

In vibroburnishing, the burnisher is imparted an oscillating movement in the direction of feed in addition to the normal direction and movement of feed. Oscillations can be imparted mechanically, by imparting oscillating movement to the burnisher, or by generating ultrasonic oscillations. With vibroburnishing, a net of sinusoidal grooves forms on the piece surface. The microrelief of the burnished surface can be regulated within wide limits by changing the conditions of burnishing (the speed of the workpiece rotation, feed, amplitude and frequency of the tool's oscillations, burnishing force, and the burnisher's radius). Vibroburnishing allows one to create a regulated surface microrelief and to raise the wear resistance of machine parts by improving the lubrication of contacting surfaces and preventing their seizure.

## 4.4.6 Burnishing Tool Life and Wear

The wear of burnishers is influenced by a combination of factors. Brittle damage (chipping) is the main type of damage to the working surface of the tools in burnishing ground, hardened steels. Hard particles of carbides in steel, and also the abrasive particles left in the surface after grinding, push the burnisher's surface and chip the particles. Besides this main process of damaging the burnisher surface, friction-fatigue damage also takes place. The fatigue damage is worsened by the presence of defects in the structure. Microcracks, pores, cavities, and so forth play the role of stress concentrators and generate fatigue damage. The accumulation of fatigue damage leads to the cracking and chipping of the surface, which amplifies the abrasive wear of the latter. Friction-fatigue wear, not saturation by the abrasive particles, seems to be the main type of wear in the burnishing of soft materials.

Thermal damage due to graphitization of the diamond surface is important only at high temperatures ($> 500$–$600°C$). Under the usual burnishing conditions, when the contact temperature is $200$–$400°C$, diamond thermal damage seems to be scarcely noticeable. Table 4.1 shows the burnishing tool life in processing materials of different hardnesses.

## 4.4.7 Burnishing of Various Surface Forms

Outer cylindrical surfaces are most frequently burnished. Burnishing is used for finishing of cylindrical surfaces of shafts, bars, pistons, piston and crank pins, bearing rings, and many other pieces. Burnishing of outer cylindrical surfaces is usually performed on turning machines.

High burnishing speed and high efficiency in burnishing cylindrical continuous pieces such as shafts, piston pins, and so forth can be attained by centerless burnishing.

TABLE 4.1  Tool Life of a Diamond
Burnisher

| Processed materials | Burnishing length to a blunting L, in km |
|---|---|
| Aluminum | up to 1000 |
| Bronzes | 150–200 |
| Nonhardened steels | 150–250 |
| Hardened steels | 50–120 |
| Cast iron | 50–80 |
| Hard metal coatings | 10–20 |
| Cemented carbides | 0.2–2.0 |

The rigidity of the manufacturing system in centerless burnishing is much higher than with ordinary burnishing on centers, and allows one to increase the burnishing speed significantly. Centerless burnishing is performed on special automatic machines.

## Holes

A technology for burnishing holes 20–200 mm in diameter and up to 500 mm deep has been developed. The burnishing of the holes in pieces made of high-strength and hardened steels, which are difficult to process by other finishing methods, is especially effective. Burnishing of the holes can be performed on turning machines, boring and drilling machines, and in machining centers.

## Flat Surfaces

One can perform burnishing of flat face surfaces of round workpieces and flat linear surfaces. In the first case, the workpiece rotates (processing on turning machines). In the second case, a tool (or a workpiece) has forward motion (processing on a planer), or the tool has a rotational motion (on milling machines).

## Contoured Surfaces

Burnishing of contoured and conic round surfaces is most frequently used. The diamond burnisher can also roll on the workpiece surface grooves, the radius of which is determined by the diamond's working part. Roller paths of bearing rings can also be processed by burnishing.

## Gears

The high quality of teeth of critical gears can be provided by gear burnishing. It can be used to improve the quality and service life of gears.

## Threads

Burnishing is sometimes applied for finishing of trapezoidal threads. The lateral sides of the profile of a trapezoidal thread are burnished by tools with a cylindrical working surface. Burnishing is applied as well for finishing the threads of semicircular profiles that are applied in lead-screw nuts, which are widely used in machine tools and actuators for the aircraft industry.

## 4.4.8 Surface Finishes (Roughness)

A burnished surface differs from surfaces processed by other finishing methods by its structure. After burnishing, an even and solid surface is formed that is distinguished by a mirror luster. The roughness of a burnished surface comprises a combination of irregularities that were formed by the process of burnishing, whose spacing equals the feed, with crushed initial irregularities. Surfaces of a similar roughness height (Figure 4.29) that were received by different finishing methods differ in the shape of the irregularities and in their service properties.

A ground surface has irregularities in the shape of sharpened projections and peaks, the rounding radius of them is equal to $R = 0.07–0.10$ mm. Polished and superfinished surfaces have a more blunted shape of irregularities with $R = 0.2–0.4$ mm. For a burnished surface, a smoothened, rounded shape of irregularities is typical with $R = 1.0–3.5$ mm.

The bearing area curve gives an indication of the bearing capability of the surface. It characterizes the filling of the profile irregularities along the height and the bearing capability of the surface at different levels of profile height. In Figure 4.30 the bearing area curves for surfaces processed by different finishing methods are

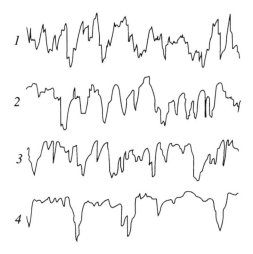

**FIGURE 4.29** Profile of Ra=0.1μm, processed by (1) grinding, (2) polishing, (3) superfinishing, and (4) burnishing.

presented. When the surface finish height is similar, surfaces processed by different methods have different bearing capabilities. The burnished surface has the highest, and the ground one the lowest.

A quantitative description of the bearing capability is the bearing length ratio $t_p$, where $p$ is the level of the profile section. The values of the parameters $t_{10}$, $t_{20}$, and $t_{30}$, which describe the properties of the upper part of the surface roughness layer, are the most important ones from the point of view of the service properties of the surface. A comparison of the values of the parameter $t_p$ for some finished surfaces (see Table 4.2) shows the considerable advantage of the burnished one.

The main factors that affect the surface roughness of a burnished surface are the properties of the burnisher (the material, shape, and condition of the work surface, and its radius), the properties of the workpiece (hardness, surface roughness, and stability), the processing regimes (force, feed, velocity), and the kind of applied lubricant. These factors predetermine the nature and intensity of the contact processes (deformation, friction, heating) that form the surface.

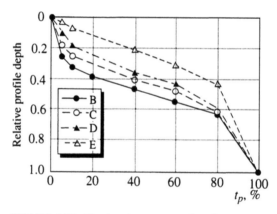

**FIGURE 4.30** The bearing curves of surfaces Ra = 0.1 μm: (B) grinding, (C) polishing, (D) superfinishing, (E) burnishing.

TABLE 4.2  Values of Bearing Length
Ratio $t_p$ for Some Finishing Methods

| Finishing method | $t\,10$, % | $t\,30$, % |
|---|---|---|
| Burnishing | 5–10 | 30–50 |
| Superfinish | 5–8 | 25–45 |
| Polishing | 4–8 | 20–42 |
| Lapping | 5–6 | 22–27 |
| Grinding | 2–3 | 12–23 |

The hardened steels are burnished effectively when the initial surface roughness does not exceed Ra = 1.5–2.0 µm. When the initial surface roughness is Ra = 1.5–2.0 µm, a stable reduction of the surface roughness occurs and a surface roughness of Ra = 0.03–0.30 µm is achieved, respectively. When soft materials such as copper and aluminum alloys are burnished, surfaces with an initial surface roughness of Ra = 0.5–5.0 µm can be burnished effectively. In this process a surface roughness of Ra = 0.04–0.30 µm can be achieved.

## Accuracy

The accuracy of the burnishing depends on the processing method. Elastic fixation allows the burnisher to copy the errors of the blank shape. Because the variation in the burnishing force caused by the beating of the processed blank is small, the reduction and the change of its size occurs uniformly under elastic fixation of the tool. The size of the blank changes due to crushing of initial irregularities and redistribution of the metal in the surface layer.

After the burnishing, the outer diameter of the processed workpieces decreases and the diameter of the holes increases. The change in the dimensions depends mostly on the initial roughness of the processed surface and is usually found by the equation

$$<d = (1.2–1.4)\, Rz\ in$$

where $<d$ is the change in the workpiece diameter, and $Rz\ in$ is the height of the irregularities before the burnishing.

The necessity of assigning an allowance for the burnishing operation depends on the tolerance value for the processed surface. As a rule, the tolerance for the final dimension is significantly higher than the value of dimension change during burnishing. Therefore, there is usually no need to assign a special allowance. When processing with micrometer tolerances, the allowance for dimension change that can be calculated by the above-mentioned equation has to be predetermined.

When burnishing with a rigid tool, forced correction of shape errors occurs in the transverse as well as the longitudinal sections of the workpiece. All the error indexes are reduced two to four times. The value of the possible shape correction is within the limits of the initial irregularities crushing (micrometers and parts of micrometers), but rigid burnishing it imposes increased demands on the accuracy and rigidity of the machine tool and the accuracy of the tool and workpiece setting.

## 4.4.9  Changes in the Surface Layer

During burnishing, the surface layer of the metal undergoes momentary deformation and heating; in other words, it is subjected to a kind of thermomechanical processing. As a result, structural and phase changes may occur that affect the strength and the service properties of the surface layer. The temperature that develops during burnishing is usually lower than the critical ones that cause structural and phase transformations; therefore, deformation plays the leading role in chang-ing the properties of the surface layer.

After burnishing, the shattering of the grains and formation of a disperse struc-ture takes place in the thin surface layer, as well as stretching of grains located near the surface in the direction of the deformation and creation of an oriented structure or texture. Thus, in the process of burnishing hardened low-tempered steels of a martensite structure, significant property changes of the metal's thin surface layer take place. Under the effect of plastic deformation, the dislocation density increases significantly. Creation of disperse carbides, crushing of grains and blocks (coher-ent scattering areas), their deorientation under the effect of grid distortions, and microstresses that occur during dissociation of the residual austenite create addi-tional barriers for dislocation movement, increase the resistance of surface layer to the plastic deformation, and strain-harden it. The thickness of the deformed layer is usually 0.02–0.04 mm.

When steels are martempered, normalized, or annealed, phase transformations do not occur in the surface layer after burnishing. This is explained by the high stability of the ferrite–pearlite and sorbite structure. Brushing and stretching of surface layer grains is evident. The deformation rate of the grains is maximal at the surface and decreases with the depth. The thickness of the deformed layer is usually 0.2–0.4 mm.

## Strain Hardening of the Surface Layer

The structure-phase transformations that occur in the surface layer during burnish-ing cause it to strengthen (strain harden); the hardness, strength, and yield strength increase, but the plasticity decreases. Usually the strain hardening of the surface is estimated by the increment of its hardness. The main indexes of strain hardening are (1) the strain hardening rate,

$$\delta = (H_b - H_{in}) / H_{in}$$

where $H_b$, $H_{in}$ are the hardness of the burnished and initial surfaces, respectively, and (2) the thickness of the strengthened (strain-hardened) layer, $h_s$.

The main factors that affect the strengthening of a burnished surface are the ini-tial properties of the processed workpiece (the material, its structural condition, the hardness), the tool properties (the radius of the burnisher, its wear rate), burnishing conditions (force, feed, speed), and the type of lubricant.

The structural condition of the material strongly affects its strain hardening. For example, Figure 4.31 shows the variation in microhardness with the thickness of the surface layer after burnishing for the carbon steel AISI 1045 in different structural conditions. Maximal strain hardening rate occurs during the burnishing of steels with ferrite, austenite, and martensite structures, and minimal strain hardening rate occurs when the steels have a sorbite and troostite structure. Table 4.3 shows the strain hard-ening rate of steels and alloys in different structural conditions as achieved during burnishing.

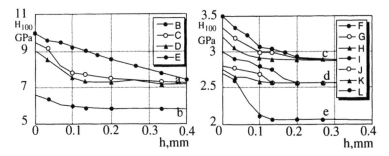

**FIGURE 4.31** Surface layer microhardness $H_{100}$ versus steel structural condition and burnishing force. Structural condition: (a) martensite, (b) troostite, (c) sorbite, (d) ferrite + pearlite, (e) ferrite. The burnishing force $P=100$ N (D, H, *K*); $P=200$ N(C, G, J); $P=300$ N(B, E, F, I, L). Workpiece material is carbon steel AISI 1045.

TABLE 4.3  Strain Hardening Rates of Different Materials

| Material | Structure | Relative pressure $P_o$ | Strain hardening rate $\delta$ |
|---|---|---|---|
| Steel | Austenite | 0.95–1.00 | 0.50–0.60 |
| Steel | Ferrite + pearlite | 0.95–1.00 | 0.35–0.45 |
| Steel | Pearlite | 0.85–0.95 | 0.25–0.35 |
| Steel | Sorbite | 0.80–0.90 | 0.20–0.30 |
| Steel | Troostite | 0.80–0.90 | 0.20–0.30 |
| Steel | Martensite | 0.80–0.90 | 0.35–0.45 |
| Aluminum alloys | — | 0.90–1.00 | 0.15–0.20 |
| Copper alloys | — | 0.90–1.00 | 0.15–0.20 |

Pressure is the second main factor that affects strengthening. The value of the pressure during burnishing determines the intensity of plastic deformation and affects the surface layer strain hardening properties to the greatest extent. As can be seen from the graphs in Figure 4.31, the surface hardness increases significantly in relation to the initial value, and it increases when the force of burnishing increases.

When the pressure rises, the strain-hardened layer thickness increases almost linearly. As this takes place, strain hardening of deeper and deeper metal layers occurs. In the general case, the thickness of the strain-hardened layer is 0.1–0.3 mm for hardened steels and 0.2–0.5 mm for nonhardened steels and nonferrous alloys. The lower limit of the listed figures is for low pressures and the upper one is for higher ones.

**Residual Stresses**

During burnishing, the surface layer of the processed workpiece is subjected to intensive plastic deformation, but its heating is not high. When hardened steels of a martensite structure are burnished, the phase transformations that take place are reduced to almost complete residual austenite dissociation with its conversion to martensite. These effects cause the development of compressive residual stresses in the surface layer after burnishing. Figure 4.32 shows typical diagrams of residual stresses that are formed after burnishing.

Significant compressive stresses, which are close to the elasticity strength of the material, are formed in the surface layer. The depth of their occurrence is 0.15–0.35 mm, depending on the material and the burnishing conditions. Maximal tangential stresses occur not on the surface but at a certain depth. The axial stresses are maximal at the surface and gradually decrease as one moves away from it. At a depth greater than 0.1 mm, the tangential and axial residual stresses become practically identical.

## 4.4.10 Surface Improvements Due to Burnishing

### Wear Resistance

The burnished surface is characterized by a combination of the following properties: low roughness (RA = 0.04–0.32 μm), large bearing capacity ($t_p$ up to 60%), hardening with great value (20–50%) and depth (0.2–0.4 μm), residual compressive stresses, and absence of abrasive particles charged into the surface. Such a surface is likely to have good working properties, particularly in friction conditions.

Figure 4.33 gives some results of comparative wear tests on ground, polished, and burnished sample rollers (Ra = 0.10–0.14 μm). The burnished rollers were the least worn. Full-scale field tests of certain wares with burnished machine parts

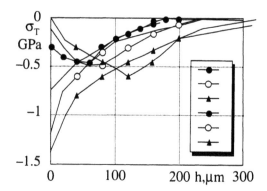

**FIGURE 4.32** Tangential (B, C, D) and axial (E, F, G) residual stresses in surface layer $P=50$ N(B, E); 100 N(C, F); 200 N(D, G). Alloy steel AISI 5140, HRC 52.

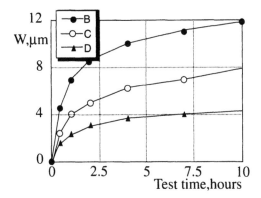

**FIGURE 4.33** Wear of ground (B), polished (C), and burnished (D) surfaces. Workpiece material carbon steel AISI 1045, HRC 60.

(in particular, truck compressors with burnished piston pins and crankshafts, and ball bearings of a turbodrill) demonstrate that burnishing reduces wear significantly (an average of 35–45%). Especially effective is the burnishing of sealing surfaces.

## Seizure Resistance

Seizure of friction metallic surfaces is an unfavorable phenomenon, deteriorating the work of a friction pair and resulting in machine parts damage. The seizure resistance of friction surfaces depends greatly on the method of finishing of a harder piece. The finishing methods create various microreliefs of the friction surface, differing in the asperities' shape, bearing capacity, and so forth. Investigation results show that burnishing provides a smooth-enough surface and significantly improves the seizure resistance of a friction surface.

## Fatigue Strength

Fatigue strength at cyclic loadings depends mostly on the condition of the surface layer, as the surface is usually loaded to a greater extent and fatigue failure most often begins from the surface. Fatigue strength is known to depend on the surface roughness, since the cavities between the projections, especially those with a sharp bottom, are the stress concentrators. Burnishing reduces the surface roughness, makes it smoother, and flattens some surface defects such as marks, scratches, and so forth. Fatigue strength also depends on the strength and structural metal strength, and eliminates or reduces the action of structural stress concentrators—burns, microcracks, and so forth. Finally, fatigue strength depends very strongly on residual stresses. It is well known that the compressive stresses are favorable from the point of view of the material fatigue strength. As was said above, significant residual compressive stresses are created by burnishing in the surface layer. Comparative tests of a number of carbon and alloyed steels show that the fatigue limit increased 20–40% as a result of burnishing.

### Corrosion Resistance

Many machined parts in their working environment are subjected to the corrosive action of air or liquid media. Numerous tests have shown that burnishing, by providing good surface smoothness and compressive residual stresses, improves machined parts' corrosion resistance.

### Contact Rigidity

Contact rigidity is the ability of the surface layers of machined parts, in contact with others, to resist strain under the action of loading forces. Contact displacements under the action of work loading greatly influence the precision working of machines and instruments, the precision of the machine's processing and assembly, and so forth. Contact rigidity is determined by the properties of the material and the surface quality of the conjugate parts. The surface hardness and roughness are the most important parameters for contact rigidity. The higher the surface hardness and bearing capacity of roughness, the smaller the contact strain and the higher the contact rigidity. Comparative tests of contact rigidity show that burnishing reduces the contact strain by half. Thus burnishing, by improving the quality of the surface layer, greatly improves contact rigidity.

## 4.5   CONCLUSIONS

Nontraditional machining methods meet the needs of emerging technologies by providing many new choices for design engineers and manufacturers. They can be applied in a variety of ways, just as traditional machining processes can. These methods offer advantages in dealing with certain kinds of materials and in the performance of intricate milling, machining, and drilling operations. Nontraditional machining methods offer an arsenal of new tools that can, in turn, offer new manufacturing strategies to harness the benefits promised by advanced engineered materials. These nontraditional machining methods can also be important in generating and machining the complex shapes and features of tomorrow's products. Design, material selection, and manufacturing options that are constrained by traditional machining limitations can be overcome by the special capabilities of nontraditional machining methods.

## 4.6   BIBLIOGRAPHY

Farrar, F., How diamond burnishing can help the engineer looking for improved surface finishes, *Ind. Diamond Rev.* 28:552 (1968).
Hull, E. H., Diamond burnishing, *Machinery* 5:92 (1962).

Khvorostukhin, L. A., *Increase of Machine Parts Bearing Capacity by Surface Hardening*, Mascinostroenie, Moscow, 1988.

Schneider, Y. G., *Service Properties of Machine Parts with Regular Microrelief*, Mashinostroenie, Leningrad, 1982.

Torbilo, V. M., *Diamond Burnishing*, Machinostroenie, Moscow, 1972.

# 5 Nonmetals: Plastics

*Bruce Wendle*

## 5.0 INTRODUCTION TO PLASTICS

This chapter on nonmetallics provides some guidelines on the right way to develop plastic applications. It will not make you a plastics engineer, but it will give you a place to start. Additional references are listed at the end of the chapter.

One of the biggest mistakes is to try to replace metal applications with plastics on a one-for-one basis. Plastic cannot be substituted on the drawings for a metal part in the hope of reducing cost, weight, or obtaining the same structural properties. The metal part may be overdesigned and loaded with labor-intensive functions that are not needed in a plastic part. A good metal design is normally made up of several parts fastened together to accomplish a specific function. A good plastic design, on the other hand, can usually be combined into fewer parts and is often able to accomplish other functions at the same time.

A well-designed plastics application is usually lighter in weight, more functional, sometimes transparent, and often lower in cost than a metallic application designed for the same function. In addition, plastics can be modified in any number of ways to provide a higher degree of functionality. Glass fibers can be added to provide additional stiffness and reduce shrinkage. Pigments can be added to the part to achieve almost any color desired. Molybdenum sulfide can be added to provide natural lubricity, and any number of fillers can be added to reduce costs.

The variety of ways that plastics can be processed or formed also increases their usefulness. Small- to medium-sized parts can be made by injection molding, large sheets can be formed by vacuum forming, or medium to large parts can be produced by hand layup or spray techniques to produce parts such as boat hulls. Other processes such as blow molding and rotational molding provide ways to make hollow parts. Profiles can be produced by extrusion. Nearly any shape or size of part can be manufactured using one of these techniques (see Figure 5.1 and Figure 5.2 for some examples).

In this chapter we will discuss the various properties of plastics that make them unique, touch on the technology of tooling, and cover some of the design rules needed for a successful plastic project. In addition, an effort will be made to provide additional sources of information to assist in the design process.

Some of the problems with plastics will be discussed as well. Such areas as volume sensitivity, high tooling costs, and property deficiencies will be covered. The important thing to realize is that plastics, when used correctly, can provide you a functional, inexpensive part that will get the job done.

**FIGURE 5.1**    The first carbon matrix composite vertical stabilizer for a Boeing 777 jet liner is hoisted into position. (Courtesy of Boeing Commercial Airplane Company. With permission.)

## 5.1   DEFINITIONS

The world of plastics has its own language. As with any technical field, it has a unique jargon. We will define the important terms and describe why each is important. For a more complete list we recommend *Whittington's Dictionary of Plastics* (Technomic Publishing Co.). To understand any list of plastic words and definitions, you must first know that plastic materials come in two distinct forms: thermoplastic and thermosetting materials (see Figure 5.3). A *thermoplastic* is any resin or plastic compound that in the solid state is capable of repeatedly being softened and re-formed by an increase in temperature. A *thermosetting* material, on the other hand, is a resin or compound that in its final state is substantially infusible and insoluble. Thermosetting resins are often liquids at some state in their manufacture or processing and are cured by heat, catalysis, or other chemical means. After being fully cured, thermosets cannot be resoftened by heat. Some plastics that are normally thermoplastic can be made thermosetting by means of cross-linking. Some common terminology defining the processing of plastic follows:

> *Injection molding:* A method of molding objects from granular or powdered plastics, usually thermoplastic, in which the material is fed from a hopper to a heated chamber, where it is softened. A ram or screw then forces the

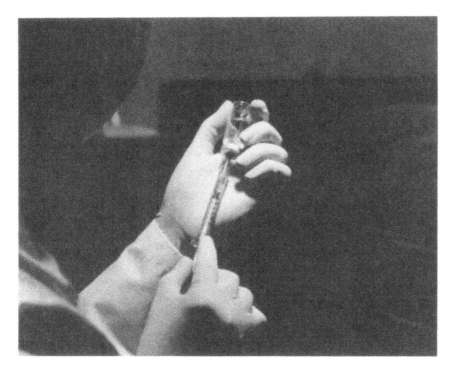

**FIGURE 5.2**   High-pressure syringe made from GE Lexan GR resin. (Courtesy of General Electric Company. With permission.)

material into a mold. Pressure is maintained until the mass has hardened sufficiently for removal from the cooled mold (see Figure 5.4.).

*Extrusion:* The process of forming continuous shapes by forcing a molten plastic material through a die.

*Thermoforming:* A method of forming plastic sheet or film into a three-dimensional shape, in which the plastic sheet is clamped in a frame suspended above a mold, heated until it becomes softened, and then drawn down into contact with the mold by means of a differential pressure to make the sheet conform to the shape of a mold or die positioned below the frame.

*Vacuum forming:* A form of thermoforming in which the differential pressure used is a vacuum applied through holes in the mold or die.

*Blow molding:* The process of forming hollow articles by expanding a hot plastic element against the internal surface of a mold. In its most common form, the plastic element used is in the form of an extruded tube (called a parison). Many variations of the process exist, including using two sheets of cellulose nitrate bonded together instead of the parison. Sometimes the parison is a preform made by injection molding.

*Rotational molding:* The process whereby a finely divided, sinterable powdered plastic is sintered, then fused against the walls of a mold. The process

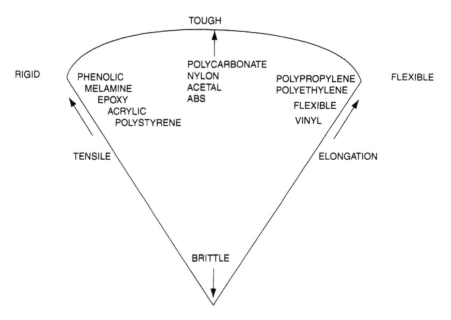

**FIGURE 5.3**  Physical characteristics of plastic materials. (From R. D. Beck, *Plastic Product Design*, Van Nostrand Reinhold, New York, 1980. With permission.)

**FIGURE 5.4**  Large, 1500-ton clamp injection molding machine capable of shooting 50 lb of plastic. (Courtesy of Cincinnatti Milacron. With permission.)

forms hollow articles by rotating the mold containing the powdered material about one or more axes at a relatively slow speed until the charge is distributed on the inner mold walls by gravitational forces and hardened by heating and then cooling the mold.

*Layup molding:* A method of forming reinforced plastic articles by placing a web of the reinforcement fibers, which may or may not be preimpregnated with a resin, in a mold or over a form. When a dry fiber is used, fluid resin is applied to impregnate or coat the reinforcement. This is followed by heating and curing the resin. When little or no pressure is used in the curing process, the process is sometimes called contact pressure molding. When pressure is applied during curing, the process is often named after the means of applying pressure, such as bag molding or autoclave molding. A related process is called spray-up, in which a chop gun is used to apply the reinforcement and resin at the same time. The techniques above are used with thermoset resins.

*Tooling:* In the case of plastics, the hollow form into which a plastic material is placed and which imparts the final shape to the finished article.

*Mold:* A form of tooling, usually made from metal or composite. It normally consists of a concave, or female, section shaped to the outside of the part (known as the cavity), and a convex, or male, section shaped to the inside of the part (known as the core). For some molding processes, only the cavity or the core is utilized. (See Figure 5.5.)

*Sprue:* In injection or transfer molding, the main channel that connects the mold's filling orifice with the runners leading to each cavity gate. The term is also used for the piece of plastic material formed in this channel.

*Runner:* In an injection or transfer mold, the feed channel that connects the sprue with the cavity gate. This term is also used for the plastic material formed in this channel.

*Cavity:* The female portion of a mold. This is often the side into which the plastic material is injected. In injection molding, this is usually the movable side of the mold, which opens after the part has solidified. (See Figure 5.6.)

*Core:* Usually the male side of a mold. This is often the side of the tool that is fixed to the injection molding machine and from which parts are ejected. However, cores can also be used to create undercuts and are often moved mechanically or hydraulically in planes different from the normal open and close directions of the mold.

*Platen:* A steel plate used to transmit pressure to a mold assembly in a press. In some cases, heat is often transferred through this plate as well.

*Nozzle:* In injection or transfer molding, the orifice-containing plug at the end of the injection cylinder or transfer chamber that contacts the mold sprue bushing and conducts molten resin into the mold. The nozzle is shaped to form a seal under pressure against the sprue bushing. Its orifice is tapered to maintain the desired flow of resin, and sometimes contains a check valve to prevent backflow, or an on–off valve to interrupt the flow at any desired point in the molding cycle.

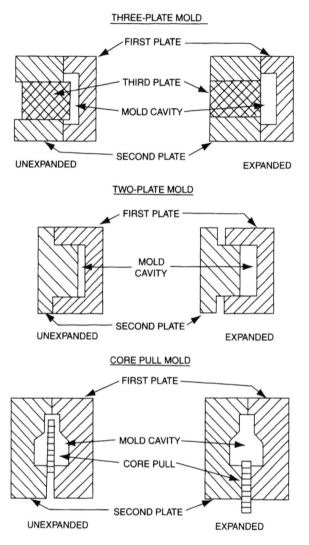

**FIGURE 5.5** Typical injection mold design concepts.

*Die:* A steel block containing an orifice through which plastic is extruded, shaping the extrudate to the desired profile; or the recessed block into which plastic material is injected or pressed, shaping the material to the desired form. The term *cavity* is more often used.

*Polymer:* The product of a chemical reaction (polymerization) in which the molecules of a simple substance (monomer) are linked together to form large molecules. The terms *polymer, resin,* and *plastic* are often used synonymously, although the latter also refers to compounds containing additives.

**FIGURE 5.6**     A tool maker works on the cavity of a large mold. (Photo courtesy of SPL.)

These terms and many others make up the vocabulary of the plastics industry. As you gain experience in the field, you will find others and will no doubt make up some of your own.

## 5.2   DESIGNING WITH PLASTICS

Regardless of whether you are working with thermosets or thermoplastics, many of the design rules are the same. Thermoplastics are more forgiving materials and do not cross-link when exposed to heat or pressure. Cross-linking generally is built into thermosetting polymers by the addition of branching or added cross-structure to the long molecular chains that make up the polymer's backbone, much like adding rungs to a ladder.

Thermoplastic materials may not be as stiff or have as high a modulus as thermosets. Thermoplastics are generally more pliable and have higher impact strengths. Thermosets often vary depending on which fiber (glass, carbon, etc.) is used with them. (See Figure 5.7.)

### 5.2.1   Design Ground Rules

Rule 1: Wherever possible, keep the wall section constant—the cardinal rule for all plastics design. Unless you are designing a foamed product, this is a must for good plastics design. Use only the thickness necessary to get the job done. Too much material generates higher material costs and increases cycle time. It also increases the chance

| PLASTIC | SPECIFIC GRAVITY | MOLD SHRINKAGE IN/IN | TENSILE STRENGTH 10 PSI | FLEXURAL MODULUS 10 PSI | DEFLECTION TEMPERATURE 164 PSI(DEG F.) | THERMAL EXPANSION 10 IN/IN (F) |
|---|---|---|---|---|---|---|
| ASTM | D792 | D995 | D638 | D790 | D648 | D696 |
| ABS | 1.05 | 0.006 | 6 | 0.32 | 195 | 5.3 |
| W/30% GLASS | 1.28 | 0.001 | 14.5 | 1.1 | 220 | 1.6 |
| ACETAL | 1.42 | 0.02 | 8.8 | 0.4 | 230 | 4.5 |
| W/30% GLASS | 1.63 | 0.003 | 19.5 | 1.4 | 325 | 2.2 |
| NYLON 6/6 | 1.14 | 0.018 | 11.6 | 420 | 170 | 4.5 |
| W/30% GLASS | 1.37 | 0.004 | 26 | 170 | 490 | 1.8 |
| PPO | 1.06 | 0.005 | 9.5 | 490 | 265 | 3.3 |
| W/30% GLASS | 1.27 | 0.002 | 21 | 1.3 | 310 | 1.4 |
| POLYCARBONATE | 1.2 | 0.006 | 9 | 0.33 | 265 | 3.7 |
| W/30% GLASS | 1.43 | 0.001 | 18.5 | 1.2 | 300 | 1.3 |
| POLYESTER TP | 1.31 | 0.2 | 8.5 | 0.34 | 130 | 5.3 |
| W/30% GLASS | 1.52 | 0.003 | 19.5 | 1.4 | 430 | 1.2 |
| POLYETHYLENE | 0.95 | 0.02 | 2.6 | 0.2 | 120 | 6 |
| W/30% GLASS | 1.17 | 0.003 | 10 | 0.9 | 260 | 2.7 |
| POLYPROPYLENE | 0.91 | 0.018 | 4.9 | 0.18 | 135 | 4 |
| W/30% GLASS | 1.13 | 0.004 | 9.8 | 0.8 | 295 | 2 |
| POLYSTYRENE | 1.07 | 0.004 | 7 | 0.45 | 180 | 3.6 |
| W/30% GLASS | 1.28 | 0.001 | 13.5 | 1.3 | 215 | 1.9 |
| POLYSULFONE | 1.24 | 0.007 | 10 | 0.4 | 340 | 3.1 |
| W/30% GLASS | 1.45 | 0.003 | 18 | 1.2 | 365 | 1.4 |

**FIGURE 5.7** Table illustrating the change in properties of a thermoplastic material when glass fiber is added. (From Beck, R. D., *Plastic Product Design,* Van Nostrand Reinhold, New York, 1980. With permission.)

of warping problems. In plastics, if a cross section does not cool uniformly, chances are that the thicker, slower cooling area will warp and throw the part out of tolerance. Sometimes coring ribs or pins are called for just to core out these thick sections. (See Figure 5.8 for minimum thicknesses.)

Rule 2: Avoid undercuts where possible. The plastics tooling industry has come up with many clever ways to create undercuts, but keep in mind that they all cost money. The best way to get around the problem is just to design the undercuts out, unless you must have them for function.

Rule 3: Add generous fillets (radii) to all inside corners. All plastics materials are notch sensitive in one way or another, so to avoid areas of high stress, keep the sharp corners to a minimum. Generous fillets also help the material to flow more evenly around sharp corners in the tool. Figure 5.9 and Figure 5.10 show design standards for high-density polyethylene and polycarbonate.

Rule 4: Allow for sufficient draft on all vertical walls. Sometimes, because of fit or function, this is difficult to do. However, a part that locks in the mold will not help either you or your molder. Most molders would like to have as much draft as possible. The designer, on the other hand, would like to design all parts with zero draft. Often a compromise is in order. Anywhere from 1/2° to 2° is common. Be sure that if the part is going to be textured, there is enough draft on the vertical walls to remove the part. A degree of draft is required for every 0.001 in. of texture depth.

Rule 5: Mold metal inserts into thermoplastic parts, as shown in Figure 5.11.

| SUGGESTED WALL THICKNESS FOR PLASTIC MOLDING MATERIAL | | | |
|---|---|---|---|
| THERMOPLASTIC MATERIALS | | MINIMUM/IN. | MAXIMUM/IN |
| ACETAL | | 0.016 | 0.125 |
| ABS | | 0.03 | 0.125 |
| ACRYLIC | | 0.025 | 0.25 |
| CELLULOSICS | | 0.025 | 0.187 |
| FEP Fluoroplastic | | 0.01 | 0.5 |
| NYLON | | 0.015 | 0.126 |
| POLYCARBONATE | | 0.04 | 0.375 |
| POLYESTER TP | | 0.025 | 0.5 |
| POLYETHYLENE(LD) | | 0.02 | 0.25 |
| POLYETHYLENE(HD) | | 0.035 | 0.25 |
| EVA | | 0.02 | 0.125 |
| POLYPROPYLENE | | 0.025 | 0.3 |
| POLYSULFONE | | 0.04 | 0.375 |
| MODPPO | | 0.03 | 0.375 |
| POLYSTYRENE | | 0.03 | 0.25 |
| SAN | | 0.03 | 0.25 |
| PFV(RIGID) | | 0.04 | 0.375 |
| POLYURETHANE | | 0.025 | 1.5 |
| THERMOSETTING MATERIALS | | | |
| ALKYD-GLASS FILLED | | 0.04 | 0.5 |
| ALKYD-MINERAL FILLED | | 0.04 | 0.375 |
| DIALLYL PHTHALATE | | 0.04 | 0.375 |
| EPOXY GLASS | | 0.03 | 1 |
| MELLAMINE-CELLULOSE | | 0.035 | 0.187 |
| UREA-CELLULOSE | | 0.035 | 0.187 |
| PHENOLIC-GENERAL PUR. | | 0.05 | 1 |
| PHENOLIC-FLOCK FILLED | | 0.05 | 1 |
| PHENOLIC-GLASS | | 0.03 | 0.75 |
| PHENOLIC-FABRIC | | 0.062 | 0.375 |
| PHENOLIC-MINERAL | | 0.125 | 1 |
| SILICONE GLASS | | 0.05 | 0.25 |
| POLYESTER PREMIX | | 0.04 | 1 |
| | | | |

**FIGURE 5.8**    Suggested wall thicknesses for plastic moldings. (From Beck, R. D., *Plastic Product Design,* Van Nostrand Reinhold, New York, 1980. With permission.)

Rule 6: Let the tool maker and molder know where the critical surface areas on the parts are located. If some areas are going to be seen by the consumer, indicate these areas on the drawings. If a gate vestage is not acceptable, indicate this to the molder. Textures and special treatment areas should be so designated. If an area is going to require close tolerances, be specific. There are many ways that a molder can give specific treatment to an area, but these areas must be known before the tool is built.

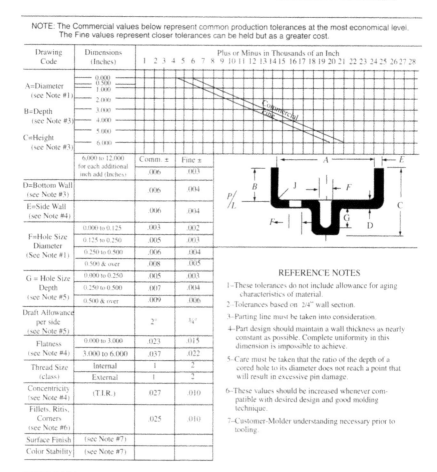

| STANDARDS AND PRACTICES OF PLASTICS MOLDERS | | | | Engineering and Technical Standards HIGH DENSITY POLYETHYLENE | | |

NOTE: The Commercial values below represent common production tolerances at the most economical level. The Fine values represent closer tolerances can be held but as a greater cost.

| Drawing Code | Dimensions (Inches) | | | Plus or Minus in Thousands of an Inch  1  2  3  4  5  6  7  8  9  10 11 12 13 14 15  16 17 18 19 20 21  22 23 24 25  26 27 28 | | |
|---|---|---|---|---|---|---|
| A=Diameter (see Note #1) | 0.000 0.500 1.000 2.000 | | | | | |
| B=Depth (see Note #3) | 3.000 4.000 | | | | | |
| C=Height (see Note #3) | 5.000 6.000 | | | Commercial Line | | |
| | 6,000 to 12,000 for each additional inch add (Inches) | Comm. ± | Fine ± | | | |
| | | .006 | .003 | | | |
| D=Bottom Wall (see Note #3) | | .006 | .004 | | | |
| E=Side Wall (see Note #4) | | .006 | .004 | | | |
| F=Hole Size Diameter (See Note #1) | 0.000 to 0.125 | .003 | .002 | | | |
| | 0.125 to 0.250 | .005 | .003 | | | |
| | 0.250 to 0.500 | .006 | .004 | | | |
| | 0.500 & over | .008 | .005 | | | |
| G = Hole Size Depth (see Note #5) | 0.000 to 0.250 | .005 | .003 | | | |
| | 0.250 to 0.500 | .007 | .004 | | | |
| | 0.500 & over | .009 | .006 | | | |
| Draft Allowance per side (see Note #5) | | 2° | ¾° | | | |
| Flatness (see Note #4) | 0.000 to 3.000 | .023 | .015 | | | |
| | 3.000 to 6.000 | .037 | .022 | | | |
| Thread Size (class) | Internal | 1 | 2 | | | |
| | External | 1 | 2 | | | |
| Concentricity (see Note #4) | (T.I.R.) | .027 | .010 | | | |
| Fillets, Ritis, Corners (see Note #6) | | .025 | .010 | | | |
| Surface Finish | (see Note #7) | | | | | |
| Color Stability | (see Note #7) | | | | | |

REFERENCE NOTES

1–These tolerances do not include allowance for aging characteristics of material.

2–Tolerances based on 2/4" wall section.

3–Parting line must be taken into consideration.

4–Part design should maintain a wall thickness as nearly constant as possible. Complete uniformity in this dimension is impossible to achieve.

5–Care must be taken that the ratio of the depth of a cored hole to its diameter does not reach a point that will result in excessive pin damage.

6–These values should be increased whenever compatible with desired design and good molding technique.

7–Customer-Molder understanding necessary prior to tooling.

**FIGURE 5.9**  Engineering and technical standards for high-density polyethylene. (From *Standards and Practices of Plastics Molders*. With permission.)

## 5.2.2  Design Checklist

A list of design rules follows:

1. Strive for wall sections that are constant and of a minimum thickness.
2. Avoid undercuts wherever possible.

| STANDARDS AND PRACTICES OF PLASTICS MOLDERS | | | Engineering and Technical Standards POLYCARBONATE |
|---|---|---|---|

NOTE: The Commercial values shown below represent common production tolerances at the most economical level. The Fine values represent closer tolerances can be held but as a greater cost.

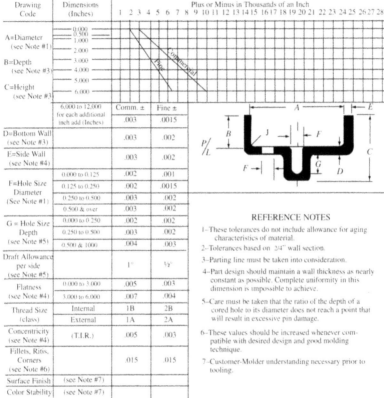

| Drawing Code | Dimensions (Inches) | | Plus or Minus in Thousands of an Inch 1 2 3 4 5 6 7 8 9 10 11 12 13 14 15 16 17 18 19 20 21 22 23 24 25 26 27 28 |
|---|---|---|---|
| A=Diameter (see Note #1) | 0.000 0.500 1.000 2.000 | | |
| B=Depth (see Note #3) | 3.000 4.000 5.000 | | |
| C=Height (see Note #3) | 6.000 | | |

| | | Comm. ± | Fine ± |
|---|---|---|---|
| 6.000 to 12,000 for each additional inch add (Inches) | | .003 | .0015 |
| D=Bottom Wall (see Note #3) | | .003 | .002 |
| E=Side Wall (see Note #4) | | .003 | .002 |
| F=Hole Size Diameter (See Note #1) | 0.000 to 0.125 | .002 | .001 |
| | 0.125 to 0.250 | .002 | .0015 |
| | 0.250 to 0.500 | .003 | .002 |
| | 0.500 & over | .003 | .002 |
| G = Hole Size Depth (see Note #5) | 0.000 to 0.250 | .002 | .002 |
| | 0.250 to 0.500 | .003 | .002 |
| | 0.500 & 1000 | .004 | .003 |
| Draft Allowance per side (see Note #5) | | 1° | ½° |
| Flatness (see Note #4) | 0.000 to 3.000 | .005 | .003 |
| | 3.000 to 6.000 | .007 | .004 |
| Thread Size (class) | Internal | 1B | 2B |
| | External | 1A | 2A |
| Concentricity (see Note #4) | (T.I.R.) | .005 | .003 |
| Fillets, Ribs, Corners (see Note #6) | | .015 | .015 |
| Surface Finish | (see Note #7) | | |
| Color Stability | (see Note #7) | | |

REFERENCE NOTES

1–These tolerances do not include allowance for aging characteristics of material.

2–Tolerances based on 2/4" wall section.

3–Parting line must be taken into consideration.

4–Part design should maintain a wall thickness as nearly constant as possible. Complete uniformity in this dimension is impossible to achieve.

5–Care must be taken that the ratio of the depth of a cored hole to its diameter does not reach a point that will result in excessive pin damage.

6–These values should be increased whenever compatible with desired design and good molding technique.

7–Customer-Molder understanding necessary prior to tooling.

**FIGURE 5.10** Engineering and technical standards for polycarbonate. (From *Standards and Practices of Plastics Molders*. With permission.)

3. Add generous fillets (radii) to all inside corners.

4. Avoid sharp transitions in wall design.

5. Allow for draft wherever possible.

6. Indicate desired surface—that is, textured or polished—and ejection pin location.

7. Consider the interface of all joining walls. Thickness at joining rib wall should be 75% of the wall it joins.

| METAL INSERTS IN THERMOPLASTIC MATERIALS | | | | TEST DATA | |
|---|---|---|---|---|---|
| | | | ABS | POLYCARBONATE | |
| THREAD SIZE | LENGTH | ROTATION | TENSILE | ROTATION | TENSILE |
| | INCH | INCH-LBS. | LBS. | INCH-LBS. | LBS. |
| NO. 0 | 0.115 | 50 IN-OZ * | 79 | 70 IN-OZ * | 135 |
| | 0.188 | 65 IN-OZ * | 162 | 52 IN-OZ * | 258 |
| NO. 2 | 0.115 | 67.5 IN-OZ | 79 | 110 IN-OZ * | 135 |
| | 0.118 | 61 IN-OZ | 162 | 168 IN-OZ * | 258 |
| NO. 4 | 0.135 | 13 | 147 | 20 * | 230 |
| | 0.219 | 17 | 258 | 23 * | 417 |
| NO. 6 | 0.15 | 23 | 220 | 33 * | 341 |
| | 0.25 | 25 | 370 | 45 | 661 |
| NO. 8 | 0.185 | 37 | 304 | 52 | 538 |
| | 0.312 | 45 | 469 | 88 * | 910 |
| NO. 10 | 0.225 | 51 | 448 | 86 | 773 |
| | 0.375 | 68 | 726 | 125 | 1388 |
| NO. 12 | 0.265 | 60 | 508 | 96 | 937 |
| | 0.438 | 78 | 809 | 140 | 1520 |
| 1/4" | 0.3 | 102 | 700 | 157 | 1283 |
| | 0.5 | 116 | 1157 | 231 | 2073 |
| 5/16" | 0.335 | 155 | 739 | 259 | 1555 |
| | 0.562 | 214 | 1435 | 345 | 3128 |
| 3/8" | 0.375 | 220 | 940 | 383 | 2065 |
| | 0.625 | 229 | 1743 | 520 | 3638 |

* Indicates Screw Failure

**FIGURE 5.11**    Metal inserts in thermoplastic materials. (Courtesy of Helicoil, Inc. With permission.)

8. Core out all thick sections to minimize cycle time and material.
9. Design for the particular process selected.

Above all, listen to the advice of the molder and the tool builder. They may not know much about your widget, but they do know how to save money when building a plastic part. There are very few configurations that cannot be built in plastics, but some may cost you more money and time than an alternative design.

## 5.3   MATERIAL AND PROCESS SELECTION

It is difficult to determine the material and subsequently the process needed for a given application. Plastic products are often a compromise between design goals and technical feasibility from the manufacturing point of view. Before approaching a plastics molder, a consultant, or a plastics engineer, the following questions should be answered:

### Product Requirements Checklist

1. End-use temperature—What will the application see in general use, and what is the highest temperature to which it will be exposed? Any low-temperature exposure should be indicated as well.

2. Environment—Where will the application be used? Will it be exposed to any chemicals and for how long? Will it be exposed to sunlight? How long?
3. Flammability—What are the flammability requirements of the application? Fire-retardant packages are available, but their addition can cause loss of critical properties.
4. Part strength—What loads will the part need to withstand? Are they static or dynamic? Does the part get impacted?
5. Life cycle—How long is the application expected to last?
6. Light transmission—Is the application transparent, translucent, or opaque?
7. Part combination—Is the part under consideration next to another part that could be combined with it?
8. Part usage—What volume of parts do you expect to use? This is important because the cost of the tooling may be prohibitive if the volume is too low.
9. Government regulations—Does the part have to meet any government or agency regulations?

With the answers to these questions, a determination of the optimum material and process to use can be made.

## 5.3.1 Polymer Selection

The choice of which material to use in a product application is always difficult. The first question is whether to use plastics at all, or is some other material better suited? Economics plays a major part in this decision.

Tooling for any plastic material is a large part of the cost. With metal or wood, the part can be manufactured in small numbers with little or no tooling, but this can be very labor intensive. Parts can be machined from plastics as well, and this option should be considered, especially for prototypes or models and for low-volume production. (See Subchapter 5.4.)

Part size will often dictate which process and which material to choose. Very large parts are usually made from a fiber/matrix composite using thermoset material and produced using a layup technique. This choice will provide a large, rigid part with excellent physical properties.

If the part has a complex shape or contains ribs and holes, then the choice is probably going to be one of the molding techniques. With a hollow part such as a container, blow molding or rotational molding will probably be the choice.

When you get into thermoplastics, a variety of forming and molding techniques are available. The large variety of polymers, each with its own set of properties, makes the decision difficult.

If the part is going to be transparent, a material such as polycarbonate, styrene, or acrylic in highly polished tooling is often the only choice. Forming the part from transparent sheet is also a possibility, depending on the configuration.

The environment in which the part is going to perform will also affect your decision. High-heat applications, where the continuous-use temperature may get above 300°F, will dictate the use of materials such as polyetherimide or polyarylsufone. High-impact applications will force you to consider a material such as polycarbonate and will probably force you to reconsider any glass fibers or other fillers and additives.

## 5.3.2   Modification of Plastic Polymers

There are 50 or so basic polymers available to use in plastics applications. In recent years, the introduction of numerous combinations of these polymers, and the addition of fillers, additives, and modifications to basic polymers, has been nothing short of miraculous.

*Copolymerization,* the combining of basic polymers into families of copolymers, has been proceeding at a rapid rate. The synergistic effect often seen with this approach has given us materials with outstanding and unusual properties.

The filling of materials with additives such as glass and carbon fibers, calcium carbonate, mica, lubricants, pigments, and numerous others has changed the physical properties and appearance of many of our basic resins. The addition of glass or carbon fibers to a thermoplastic usually increases the stiffness of the polymer but often reduces the impact strength. The length of fiber is also a factor, with the longer length (up to 0.250 in.) about the longest that will go through an injection molding machine without breaking up.

The art of coloring plastics has progressed, and now it is very much a science in its own right. Color analysis computers are now available that can read colors accurately, and color labs can reproduce colors to near-perfect standards. Environmental problems with some colors have caused the industry to examine and reevaluate some of the pigments commonly used. As an example, color pigments containing lead and cadmium are being reformulated because of their toxic nature.

It should also be noted that putting pigments into polymers often reduces physical properties. Sometimes, the natural darker color of the basic polymer needs to be covered up with pigment to match a lighter color, causing heavy loading and a loss in physical properties.

Flammability and smoke generation are also properties demanding careful evaluation. These properties are getting close attention from government agencies responsible for health and safety in many different fields. As an example, commercial aircraft builders are under a strict mandate to eliminate polymers that burn rapidly or give off toxic fumes when combusted. This problem is often eliminated with flame-retardant additives that are blended into the plastic raw materials.

## 5.3.3   Secondary Operations

For all types of plastic materials, the secondary operations performed on them can be expensive and time consuming. Most of the processes available produce a part requiring residual secondary work to be done. Some, such as injection molding, leave relatively little flash to clean or surface treatment to apply to the part. Others, such as vacuum forming, normally produce products that need a great deal of secondary work before they are ready for use in production.

When bonding or joining is required, a choice of whether to use mechanical fastening or adhesives must be made. These actions require specialized treatment, and such things as compatibility and stress formation must be taken into consideration. Metal inserts can be added to provide threaded holes or threaded studs. (See Figure 5.11.)

Various adhesive systems work with specific plastics; others create problems due to solvent contamination and adhesion failures. It is always best to check with your material suppliers before using a specific adhesive system with a given material.

Machining of various plastic materials can usually be accomplished with the use of the recommended feed, speed, and cutters. Removing plastic materials by such techniques as sanding can be done, but may create problems due to the abrasive material surface rapidly loading up with melted plastic, or the plastic surface becoming embedded with abrasive particles.

Painting of plastic materials has developed into a science. Adhesion problems are quite common, especially with materials such as polyethylene and polypropylene. Contamination is usually caused by solvents in the paints, and may cause stress cracking. Some materials require a barrier coat before applying the color or final coat. In many cases the pigments are added to the molding resin and the parts are integrally colored. In some cases the parts are molded in the same color as the paint applied to them, to prevent show-through in the final finish.

There are many coatings available for coating transparent products such as sunglass lenses that give the clear plastic better abrasion properties and some protection against chemical attack. These systems are a direct result of space technology and have increased the markets for plastic products considerably.

## 5.4  TOOLING

The production of any plastic application requires many difficult decisions. The choice of tooling is one of those areas. It is normally the most expensive part of any application and is filled with pitfalls. It is often said that tooling is the most important part of any plastic project. If the tooling is not right, the part that comes from it cannot be right.

The tooling for almost any part design can be designed and built, but the cost may be prohibitive. Undercuts and hard-to-mold areas are easier to eliminate in the part design stage than they are after the metal has been cut. Redesigning a part after the tool has been made is expensive and time consuming. The best approach is to contact a good tool maker and listen to this individual. He or she can save you untold dollars, time, and gray hairs.

There are different classes of tooling. Because of this, you need to know the volume of parts you plan to build with the tool in question. Aluminum tooling is usually good enough to produce a limited quantity of parts, but damaged tooling will not produce any parts. Hardened tool steel is more expensive but will be good for larger volumes.

The important rule here is to get good advice when it comes to tooling and then listen to it. Also, it is important to give your mold maker enough time to give you a first-class tool. Often the demand for a short lead time on a product dooms it to failure because the mold maker was not given enough time to do a good job.

Another good suggestion is to get your tool maker to give you weekly updates on progress. Nothing is so discouraging as to arrive at the date of mold delivery and find out that the tool maker is not finished. Also, leave time in the schedule for tweaking a tool after the first mold trial. A tool is like an expensive piece of machinery and will often need some modification before it produces perfect parts. Recent introduction of computer-aided design/computer-aided manufacturing (CAD/CAM) systems has improved the ability of tool makers to produce a good part the first time, but you can still expect problems to crop up before you have a part ready for production.

You may want to take advantage of CAD systems to simulate the filling of a plastic part. The information derived from one of these systems, such as Moldflow, can be very useful in assisting the mold designer in laying out the tooling before metal is cut. Other programs to help designers improve the cooling characteristics of a tool and help eliminate warping are also available. Moldflow is available from Moldflow Pty Ltd, Kalamazoo, Michigan.

Above all, treat the tooling arrangement as you would any business deal. Put everything in writing and have good documentation. Be sure the molder and the tool maker both know what you expect. Often, added features or changes are given via verbal orders, and this can only lead to problems. Update your drawings or data sets regularly, and don't be afraid to write everything down. It could save you both money and time.

## 5.5  BIBLIOGRAPHY

Many books are available to further your knowledge of the plastics industry. A short list follows.

### Tooling

Donaldson, Le Cain, and Goold, *Tool Design,* McGraw Hill, New York, 1957.
Menges, G., and Mohren, P., *How to Make Injection Molds,* 1986.
Michaeli, W., *Extrusion Dies for Plastics and Rubbers,* 2nd ed., 1992.

### Design

Beck, R., *Plastic Product Design,* Van Nostrand Reinhold, New York, 1980.
*Designing with Plastic, The Fundamentals,* Hoeschst Celanese Design Manual TDM-1, Hoeschst Celanese, 1989.
Dym, J. B., *Product Design with Plastics,* 1983.
*Plastic Snap-Fit Joints,* Miles, Inc., 1992.
Rosato, D. V., *Designing with Plastics and Composites, A Handbook,* 1991.

### Structural Foam

Wendle, B., *Structural Foam,* Marcel Dekker, New York, 1985.

### Extrusion

Richardson, P., *Introduction to Extrusion,* 1974.

### Blow Molding

Rosato, D. V., *Blow Molding Handbook,* D. V. Rosato, 1989.

## Thermosets

Whelan, T. and Goff, J., *Molding of Thermosetting Plastics*, 1990.

## General

Wendle, B., *What Every Engineer Should Know about Developing Plastic Products*, Marcel Dekker, New York, 1994.

Other information is available through organizations such as the Society of Plastic Industries and the Society of Plastic Engineers, as well as various trade publications.

## Important Phone Numbers

| | |
|---|---|
| Society of Plastic Industries | (202) 371-5200 |
| Society of Plastic Engineers | (203) 775-0471 |
| American Society for Testing and Materials | (215) 299-5400 |
| American Mold Builders Association | (708) 980-7667 |
| Directory of Moldmaker Services | (202) 371-0742 |
| Members Directory Canadian Association of Moldmakers | (519) 255-9520 |

# 6 Composite Manufacturing

*John F. Maguire*

with

*Don Weed*

and

*Thomas J. Rose*

## 6.0 INTRODUCTION AND BACKGROUND

It is difficult to find a truly satisfactory definition of composite material. The *American Heritage Dictionary* (Houghton Mifflin, Boston, 1981) comes close, with "a complex material, such as wood or fiber glass, in which two or more complementary substances, especially metals, ceramics, glasses, and polymers, combine to produce some structural or functional properties not present in any individual component." The problem with even a good definition, of course, is that it is all-encompassing, so that every material in the universe could in some sense be defined as a composite. This loss of exclusivity diminishes the usefulness of the definition. For our purposes, we shall restrict attention to that subset of materials known as fiber-reinforced advanced polymeric composites. In these materials, a reinforcing fiber is embedded in an organic polymeric resin. The fiber acts as a structural reinforcement and the resin binds the fibers together. This transfers loads and provides structural and dimensional integrity.

Advanced composite materials, developed in the latter half of the twentieth century, may well provide a key to enabling technology for the twenty-first century. These materials are strong, light, and corrosion resistant, offering considerable technical advantages in aerospace, automotive, offshore petrochemical, infrastructure, and other general engineering applications. Composite components may be made by laminating or laying up layers of composite material, each ply consisting of one or more patterns, which may be as large as 4 1/2 ft wide by 9 ft long. These patterns are cut from a continuous roll of cloth, or from sheets, with standard widths measuring up to 4 1/2 ft. Composite material wider than 12 in. is referred to as *broadgoods*. If a material is narrower, it is usually a *unidirectional tape.* In such a tape the fibers run in the longitudinal direction along the length of the tape. There are no fill fibers, and the material is stabilized with backing paper to permit handling prior to use.

There are many additional types of processes that utilize the fiber as strands, flakes, and so forth that will be discussed in this chapter. Today there are projects in automobiles, bridge construction, and a host of sporting goods applications. Figure 6.1 shows a representative cross section of products that are currently manufactured using composite materials [1].

The strength and stiffness of these materials far exceeds those of metals [2], as shown in Figure 6.2, and it is this combination of strength and stiffness coupled with light weight that lies at the heart of the performance advantage. While the need for lightweight materials of exceedingly high strength and stiffness is apparent in aerospace applications (see Figure 6.3), it is not quite so obvious that a strong, lightweight plastic might offer advantages, say, in the construction of a bridge or a submarine. The average density of a submarine must be equal to that of seawater, regardless of the material from which the pressure hull is constructed. However, the drag on a submarine increases dramatically when the radius of the pressure hull is increased, so that one arrives quickly at a situation where the payload advantage of a larger radius is more than offset by the need for a much bigger power plant. One very effective way to

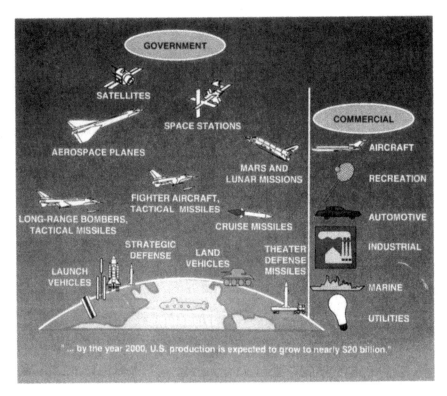

**FIGURE 6.1**    Representative cross section of products containing composites.

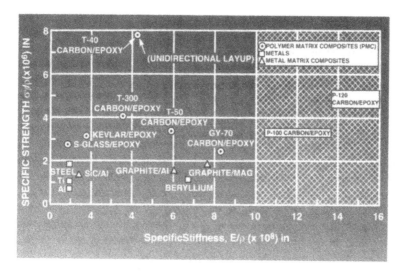

**FIGURE 6.2**   Strength and stiffness of composite materials and metals.

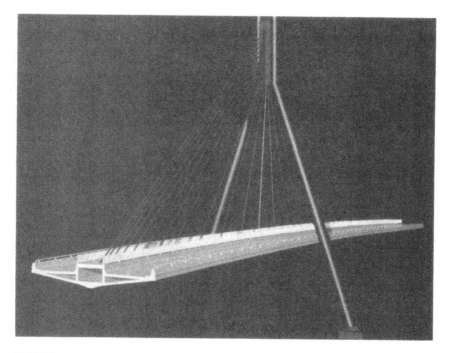

**FIGURE 6.3**   Design concept for an all-composite bridge.

increase the payload or endurance is to decrease the weight of the pressure hull while keeping the radius constant, and it is in this area that composite materials offer significant promise. Similarly, since most of the material in a bridge is needed to hold up the bridge itself rather than carry the loads that cross it, composite materials offer much more efficient structures with far greater spans. Figure 6.3 shows the design concept for an all-composite bridge. For automobile applications, reduced weight translates into increased fuel efficiency, reduced emissions, and greater payload—demonstrating the significant benefits to be accrued from greater utilization of composite materials [3].

Before contemplating the fabrication of detail components or major subassemblies using composites in a true production manufacturing environment, it is wise to quantify the advantages in terms of cost. If we define the savings in dollars per pound of fabricated structure across various industries, we arrive at a plot as shown in Figure 6.4. In this figure, the savings per pound are plotted as a function of the potential market. In satellite applications the savings are huge ($10,000 per pound), but the actual total amount of material used in the manufacturing is minuscule. On the other hand, in automobile or marine applications the savings are far more modest but the tonnage is enormous due to the production rates.

The level of sophistication required to make a given component or subassembly is not a constant, but depends to a large extent on the standards and operating practices of the particular industry. For example, in the aerospace industry the structural integrity is of the utmost importance and cosmetic considerations may be somewhat secondary. On the other hand, in the automobile industry, while the less stringent mechanical performance criteria must be met, there is the additional requirement

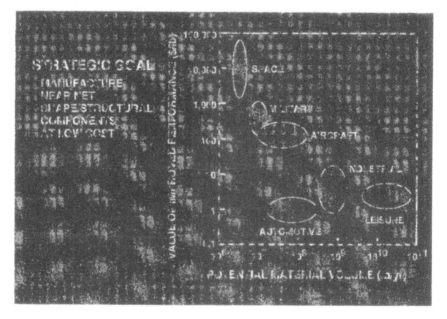

**FIGURE 6.4**   Savings in dollars per pound of fabricated structure.

of a very high degree of surface finish for cosmetic reasons. This requires consideration of aesthetic qualities that have little to do with mechanical or functional design requirements but may have a high impact on cost. The challenge, then, is not to be able to fabricate a given structure per se, but rather to design and develop a product and implement cost-effective manufacturing technologies that will allow the technical benefits afforded by composite structures to translate into competitive advantages in the commercial marketplace. This requires a two-pronged attack on current practices. First, it will be necessary to produce structural plastics that will have properties close to those of metals [4], and that can be processed using exceedingly low-cost manufacturing technologies such as injection molding [5]. This approach will require fundamental work in the structure of matter and will be of a long-term nature. Second, more effective manufacturing technologies and processes must be utilized in order to reduce the cost associated with current technologies.

## 6.0.1  Design/Manufacturing Interface

In today's competitive cost environment, the producibility of a new design is not merely important—it is the key to survival as a leader in the field of manufacturing. Many authorities would agree that perhaps as much as 90% of the ultimate cost of a product may be predicated by the design, and only 10% can be influenced by the manufacturing process. However, by considering the known production processes, the quality and reliability inherent in the process, and the material selections available within the constraints of product function, we can optimize a design early in the development of a new or revised product. This is certainly true in the field of composites.

In common with other manufacturing processes, the fabrication of composites may be conveniently discussed in terms of the labor, materials, tooling, and equipment requirements. Quite straightforward design changes can sometimes lead to great cost savings in manufacturing. Structures perfectly acceptable in metals may be difficult or impossible to fabricate in composites. For example, the flanges on the outer duct of an aircraft engine may be fabricated in metal using standard shop practices, but are a real challenge in carbon fiber-reinforced plastic. Here it is good to bear in mind that there may be little to be gained but a good deal to be lost by dogmatic adherence to a particular "all-composite" philosophy. If a particular structure or part of a structure would be easier to fabricate in metals, then the smart compromise is to combine the materials to meet design requirements at minimum cost. Failure to recognize this can lead to program delays and very substantial cost overruns. In the development of the world's first all-composite small transport airplane, the Lear Fan 2100, the rigid adherence to an "all-composite dogma" resulted in the cabin door hinges being designed and fabricated from carbon fiber-reinforced epoxy. These hinges failed twice during pressure testing of the fuselage, resulting in significant delays and cost to the program, all for the rather minimal advantages to be gained by the weight reduction due to the door hinges. Therefore, before locking in on a particular fabrication technology, one should ask whether a hybrid approach, possibly combining metals and composites, would lead to acceptable performance at lowest cost.

## 6.0.2  Materials

The composite materials most commonly used in the field can be made up of at least two constituents:

1. The *structural constituent,* usually the reinforcement, used to determine the internal structure of the composite, such as fibers, particles, lamina, flakes, and fillers
2. The *body constituent,* or matrix, used to enclose the composite structural constituent and give it its bulk form, such as epoxy, polyester, polyimide, vinyl ester, and bismalemides

### Fibers

Of all composite materials, the fiber type (specifically the inclusion of fibers in a matrix) has generated the most interest among engineers concerned with structural applications. The fibers, coated and uncoated, typically control the strength and stiffness characteristics, formability, and machining characteristics of the laminate. The more commonly used fibers today include glass, carbon, and Kevlar. See Figure 6.5 for stress–strain diagrams for some common fibers.

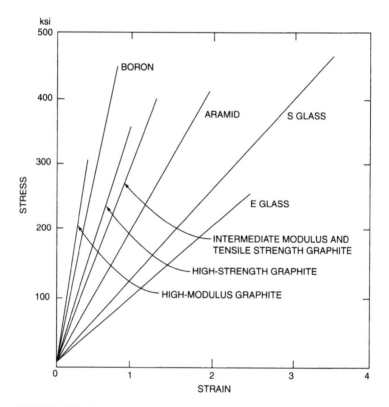

**FIGURE 6.5**   Stress–strain diagram for fibers used in hybrid construction.

## Glass

Glass is the most widely used reinforcing material, accounting for more than 70% of the reinforcement for thermosetting resins. Forms of glass fiber materials include roving (continuous strand), chopped strand, woven fabrics, continuous-strand mat, chopped-strand mat, and milled fibers (0.032–0.125 in. long). The longer fibers provide the greatest strength; continuous fibers are the strongest. Glass does not burn, and it retains good mechanical properties: up to approximately 50% of its strength up to 700°F, and 25% of its strength up to 1000T. Moisture resistance is excellent, and glass fibers do not swell, stretch, disintegrate, or undergo other chemical changes when wet. The sizing or chemical treatment applied to a glass fiber surface is designed to provide compatibility with the type of resin matrix used. It also improves handling characteristics of the glass fiber, such as the ability to control tension, choppability, and wet-out. See Figure 6.6 for glass fiber nomenclature.

E glass was the first glass developed specifically for production of continuous fibers. It is a lime–alumina–borosilicate glass designed primarily for electrical applications. It was found to be adaptable and highly effective in a great variety of processes and products, ranging from decorative to structural applications. It has become known as the standard textile glass. Most continuous-filament glass produced today is E glass.

S glass is a high-tensile-strength glass. Its tensile strength is 33% greater and its modulus almost 20% greater than that of E glass. Significant properties of S glass for aerospace applications are its high strength-to-weight ratio, its superior strength

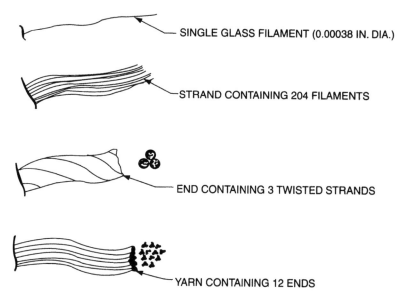

SINGLE GLASS FILAMENT (0.00038 IN. DIA.)

STRAND CONTAINING 204 FILAMENTS

END CONTAINING 3 TWISTED STRANDS

YARN CONTAINING 12 ENDS

**FIGURE 6.6**    Definition of forms of glass fibers in the process of making a yarn.

retention at elevated temperatures, and its high fatigue limit. S glass costs up to 20 times more than E glass, depending on the special form of the fibers.

## Carbon

The terms *carbon* and *graphite fiber* are frequently used interchangeably. The correct term in most cases is *carbon,* because commercially produced fibers do not exhibit the characteristic X-ray parameters of graphite. Due to a unique combination of properties, carbon fibers are the ideal reinforcement for lightweight, high-strength, and high-stiffness structures. High-performance carbon fibers are available in a range of properties, product forms, and prices. Continuous yams or tows contain from several hundred to several thousand filaments per strand, and generally fall into two categories:

1. High-strength $(350–500 \times 10^3 \, psi)$, intermediate-modulus $(30–50 \times 10^6 \, psi)$ fibers
2. High-modulus $(50–75 \times 10^6 \, psi)$, intermediate-strength $(250–350 \times 10^3 \, psi)$ fibers

The high-modulus fibers are generally more costly, and the higher-filament-coum tows are lower in price than yams containing fewer fibers.

## Aramid (Kevlar)

Introduced commercially in the 1970s, Kevlar aramid is an aromatic long-chain polyamide polymer, the fibers of which may be produced by spinning using standard textile techniques. The low-density, high-tensile-strength, low-cost fiber produces tough, impact-resistant structures with about half the stiffness of graphite structures. The fiber was originally developed to replace steel in radial tires and has found increasing use in the belts of radial car tires, and carcasses of radial truck tires, where it saves weight and increases strength and durability.

Kevlar 29 is the low-density, high-strength aramid fiber designed for ballistic protection, slash and cut resistance, ropes, cables, and coated fabrics for inflatables and architectural fabrics.

Kevlar 49 aramid fiber is characterized by low density, high strength, and high modulus. These properties are the key to its successful use as a reinforcement for plastic composites in aerospace, marine, automotive, sports equipment, and other industrial applications.

## Matrices

The matrix, usually resin as a binder, determines the transverse mechanical properties, interlaminar shear characteristics, and service temperature of the laminate. The matrix (or body constituent) serves two very important functions: (1) it holds the fibers in place, and (2) under an applied force, it deforms and distributes the stresses to the high-modulus fiber constituent. Both influence the selection of shop processes and tool design. The matrix may be a thermoplastic or thermosetting type of resin.

Thermosets include the epoxies, bismaliemides, and polyimides as well as some of the lower-cost resins such as the phenolics, polyesters, and vinylesters. Generally, they require the addition of a catalyst in order to cure. The type of catalyst influences the pot life of the mix and whether heat is required to achieve full cure strength. Once cured, they cannot be resoftened by heat for re-forming. Although thermosetting plastics do not have much greater tensile strength than the thermoplastics, they are much more effective as the temperature environment increases.

Thermoplastics are softened by heating, permitting the forming of sheet material that retains its formed shape after cooling. They can be reheated and re-formed if necessary. Most injection molding is accomplished with thermoplastics, and they are seeing a resurgence in various applications in the structural plastics field. Many applications in the automotive field utilize thermoplastics, and the marine industries are also seeing increasing utilization.

The materials selection depends on the design requirements of the component (i.e., the mechanical, thermal, and environmental conditions). Table 6.1 gives a selection of common fibers, and Table 6.2 shows a selection of resins. As for the matrix materials, the choice for aerospace applications (and a few others) is largely dictated by the operating temperature. Figure 6.7 shows a selection of properties of resins that might be considered for various temperatures of operation. Clearly, there is some overlap here, and the temperature regimes are not sharply defined.

TABLE 6.1  Selection of Commonly Used Fibers

| Fiber | Fiber Properties | | Unidirectional Composite Properties | | |
|---|---|---|---|---|---|
| | Diameter ($\mu$m) | Tensile strength | Tensile modulus (Pa) | Tensile strength Mpa (V%) | Tensile Modulus Gpa (V%) |
| *Carbon* | | | | | |
| T-300 (Union Carbide) | 7.0 | 3447 | 230 | 1482(60) | 141(60) |
| Hitex 46 (Hitco) | 6.1 | 5688 | 317 | 2696(64) | 184(64) |
| P-100 (Union Carbide) | 9.6 | 2413 | 760 | — | — |
| *Organic* | | | | | |
| Kevlar 49 | 11.9 | 3169 | 124 | 1380(60) | 76(60) |
| Polybenzoxazole | 10–12 | 3447–4482 | 310-358 | — | — |
| Ceramic | | | | | |
| S-2 Glass (Owens-Coming) | 19 | 4585 | 87 | 1903 | 52 |
| Silicone carbide - Nicalon (Dow) (ceramic grade) | 10–20 | −2758 | 193 | — | — |
| Newtel 480(3M) | 10–12 | 2241 | 193 | — | — |

TABLE 6.2  Selection of Commonly Used Resins

| Resin | Tensile Strength (MPa) | Tensile Modulus (GPa) | $T_r$(K) |
|---|---|---|---|
| *Thermosets* | | | |
| Epoxy(TSMDA) | 103.4 | 4.1 | 463 |
| Bismaleimide | 82.7 | 4.1 | 547 |
| Polyimide | 137.9 | 4.8 | 630 |
| *Thermoplastics* | | | |
| Polyphenylene sulfide | 65.5 | 4.3 | 366(555mp) |
| Polyetherethetherketone | 70.3 | 1.1 | 400 |

| Property | Epoxy | Cyanate | Bismaleimide |
|---|---|---|---|
| Tensile Strength, MN/m$^2$ | 48–90 | 69–90 | 35–90 |
| Tensile Modulus, GN/m$^2$ | 3.1–3.8 | 3.1–3.4 | 3.4–4.1 |
| Tensile Strain at Break, % | 1.5–8 | 2–5 | 1.5–3 |
| $G_{IC}$, J/m$^2$ | 70–210 | 105–210 | 70–105 |
| Specific Gravity | 1.2–1.25 | 1.1–1.35 | 1.2–1.3 |
| Water Absorption, %, Saturated at 100°C | 2–6 | 1.3–2.5 | 4.0–4.5 |
| Dry | 150–240 | 230–260 | 250 - - |
| Water Saturated at 100°C | 100–150 | 150–200 | 200–250 |
| Coefficient of Themal Expansion, ppm/°C | 60–70 | 60–70 | 60–65 |
| TGA Onset, °C | 260–340 | 400–420 | 360–400 |
| Dielectric Constant at 1 MHz | 3.8–4.5 | 2.7–3.2 | 3.4–3.7 |
| Dissipation Factor at 1 MHz | 0.02–0.05 | 0.001–0.005 | 0.003–0.009 |
| Cure Temperature, °C | 150–220 | 177–250 | 220–300 |
| Mold Shrinkage, mm/mm | 0.0006 | 0.004 | 0.007 |

FIGURE 6.7    Properties of resins that might be selected based on operating temperatures.

For operating temperatures of thermosetting resins up to 200°F, the tough-ened epoxy systems are acceptable and are most easily processable. Up to about 250°F, the cyanate esters have acceptable properties and are generally low-viscosity materials that are amenable to resin transfer operations. In the region 200–350T, the bismaleimides are preferable. These classes of resins are additional thermosets and do not generate volatiles during cure. In the region 350–550T it is necessary to use polyimide resins. These are condensation-type thermosetting resins and lose about 10% of their weight and many times their volume as volatiles (usually alcohol and water) during the cure. The management of such large amounts of volatiles increases the complexity of making components with low (less than 2%) void content. Currently, there are no commercially available polymers with reliable long-term integrity when operated for long times above 600T.

### 6.0.3 Composite Manufacturing Technology Overview

The primary subject of this chapter is to introduce the various materials and processes commonly used to manufacture products utilizing composites. After determining the physical properties demanded by the product, the selection of a fiber and a matrix to be used probably occurs next. However, this is closely tied to manufacturing process selection. There are many methods available to combine a fiber and the matrix. The selection of the best process is a major contribution of the manufacturing engineer as part of his or her activities on the design/development team. The task of combining a fiber with a matrix can range from the preparation of a dry fiber preform, and then adding the resin, to preparation of the resin, and then adding the fibers. The range of molds and other tooling available today is quite extensive, as is the increasing technology of process controls and the need for greater automation in the entire process. The following subchapters cover the major processes.

## 6.1 FABRICATION WITH PREPREG BROADGOODS

The drawbacks of the manual layup approach with broadgoods fabrication have been well recognized for a long time. There has been much work and many millions of dollars devoted to attempting to automate this part of the process [6]. Notwithstanding these efforts, the problem has remained stubbornly intractable and real progress is not likely in the near future; though predicting the time scale of innovation is notoriously dangerous.

The basic problem is that sticky cloth is an inherently disorderly medium. Robots and computer vision systems do an exceedingly poor job of recognizing the various shapes that such a material may assume. The tackiness required for successful layup is a further impediment to the efficient implementation of pick-and-place operations for a robotic end effector. Also, the current generation of these devices is nowhere near smart enough to provide the level of pattern recognition, dexterity, or tactile feedback required to genuinely solve the layup problem within the existing technology base. On the other hand, the combination of the human eye and finger is exceedingly well adapted to the following operations:

1. Picking up a piece of floppy, sticky cloth
2. Recognizing that it is slightly wrinkled
3. Smoothing out the wrinkle
4. Placing the pattern shape on a mold surface and providing the tactile feedback to position the pattern on an active geometry around a tight radius, that is, packing it into a delicate radius with the human finger

These types of tasks are difficult or impossible for the current generation of pattern-recognition algorithms, vision systems, and robotic end effectors. For example, without belaboring the point, try writing a foolproof algorithm for "identify the wrinkle." It is something of an exercise in humility to recognize that while we have sent men to the moon, we are nowhere close to solving the latter problem. This is unfortunate, because it is one of those problems that look easy (there are no mathematical formulas) but is very difficult and is therefore of the class on which a great amount of money can be spent with little or no real return.

The manufacturing engineer, therefore, needs to exercise particular caution when contemplating automated layup manufacturing techniques with prepreg broadgoods as a means of cost reduction. In this subchapter we will attempt to provide an overview of the steps required to actually manufacture a composite detail component using technology that exists today. No attempt has been made to cover every possible combination and permutation. The topic of "composite manufacturing" covers a complete section in the New York City Library, and any attempt to cover this much material in a chapter of any reasonable length would be impossible. However, the material covered in the field is characterized more by the breadth and scope of the subject than by its depth, so it is hoped that a discussion of the major topics will be valuable in providing a useful source and some insights for the practicing engineer. Also, the intent has been to provide enough information for the manufacturing engineer to make informed choices of the various manufacturing options and proceed to fabricate an actual component. The major technologies are covered in a self-contained fashion, and the reader is referred to the literature for more detailed discussion [7]. An overview of the basic manufacturing process is shown in Figure 6.8. We will limit our present discussion to the fabrication of

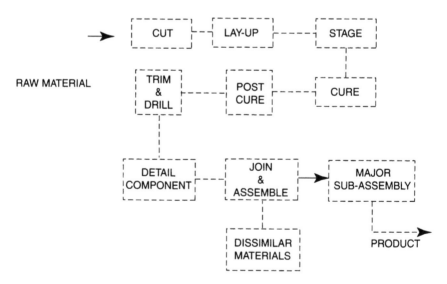

**FIGURE 6.8**  Overview of prepreg broadgoods manufacturing process.

detail components from preimpregnated broadgoods (prepreg cloth), using hand layup and autoclave cure. This is a baseline technology and accounts for a large share of today's production volume. The extension to unidirectional tape is straightforward. The term *prepreg* describes a reinforcement or carrier material that has been impregnated with a liquid thermosetting (or thermoplastic) resin and cured to the B-stage. At this stage, the prepreg is dry or slightly tacky and can be re-formed into a mold. In the prepregging process, the reinforcing material in web form is drawn through a bath of liquid resin. Excess resin is removed as the web leaves the bath to provide a controlled resin content in the reinforcement. The saturated material passes through an oven, where the resin is partially cured, usually to the point where the resin becomes firm but not fully cured. The web is cut into sheet lengths or wound onto rolls to facilitate processing by the fabricator. It must be kept frozen in order to prolong shelf life.

## 6.1.1  Planning

The materials are purchased to conform to an incoming materials specification. It is important to pay attention to developing a good functional specification in order to discriminate against low-quality materials. An example of the criteria addressed in functional specifications is shown in Figure 6.9.

In order to conform to good shop practice, it is essential to develop sufficiently detailed manufacturing planning documentation. This is a list of operations that details the various steps in the fabrication process. An example of the manufacturing process for an airfoil component is shown in Figure 6.10. In developing the fabrication process, the first step is to define the pattern shapes that will be used in the layup sequence. It is desirable to minimize waste material during the cutting operation, and a rough rule of thumb is that offal should not exceed about 10%. Note that if we were to cut circular patterns in a close-packed fashion, only about 90% [exactly $100\pi/(2\sqrt{3})$] of the material could be utilized.

While it is not possible to determine mathematically what the optimum nesting pattern should be, it is possible to use commercially available software to calculate the material utilization for a given pattern and thereby heuristically iterate to an acceptable nest. Since the same pattern may be cut many thousands of times, even minor gains in materials utilization at this stage can translate into significant gains over the production lifetime of a major component. Figure 6.11 shows an acceptable nest for an airfoil component.

Pattern nesting of either broadgoods or tapes may be satisfactorily accomplished using a computer software system that takes into account warp orientation. Commercial systems are available from GOT, CAMSCO Division, and Precision Nesting Company, among others. Some of the more convenient systems allow direct porting of existing computer-aided design data to the nesting and cutting programs. Digitization of the pattern shape may be accomplished by tracing lines on a drawing with a hand-held electronic probe. Alternatively, automatic scanning of drawings by an electro-optical system is another method available for transferring information into the computer.

**1       SCOPE**

a.   This specification establishes requirements for 350 F cure epoxy resin-impregnated BMS 9-8, Type I carbon fiber unidirectional tape and woven fabric forms.

b.   This specification requires qualified products.

**1.1     CONTENTS**

| | EPOXY PREIMPREGNATED CARBON TAPE AND WOVEN FABRIC FOR AIRCRAFT STRUCTURE, 350 F (177 C CURE) | **BMS 8-297D** |
|---|---|---|
| BY | | |
| CK'D | | |
| ENG | **BOEING MATERIAL SPECIFICATION** | PAGE 1 OF 34 |

ORIGINAL ISSUE  11-7-84          CAGE CODE 81205          REVISED  10-15-91

**FIGURE 6.9**   Example of a functional specification for an epoxy-impregnated carbon fiber tape and woven fabric.

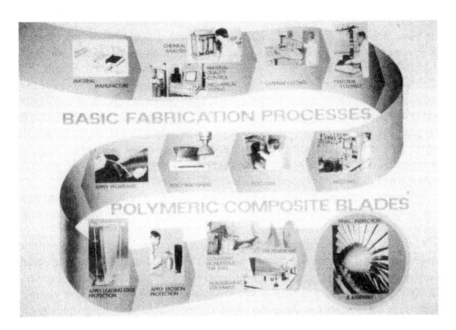

**FIGURE 6.10** Fabrication process for an airfoil component.

**FIGURE 6.11** Pattern nest for an airfoil component made from impregnated broadgoods.

## 6.1.2   Cutting

The basic need in the cutting operation is to obtain a well-defined sharp edge in the minimum amount of time [8,9]. A number of technologies have been developed to do this, including lasers, water jets, reciprocating knives, and die boards using either a clicker press or an incremental (progressive) feed press. Each of these techniques has its strengths and weaknesses, and each can claim specialty applications in which it performs particularly well. In general, the approach that will provide the best trade-off in a particular application depends in large measure on the complexity and the size of the component. For small components where the pattern shapes will conveniently nest on a die board, the clicker or roller-type die press will provide a cost-effective technology that produces a cut of excellent quality.

The manual baseline technology for cutting is simple: a roll of material and a pair of scissors or a Stanley knife. When the manufacturing production rate increases just a little, it is essential to consider more efficient cutting technologies.

Many of the cutting technologies described here provide computer-controlled manipulation of the cutting medium to produce the individual ply shape. The coordinate system sketched in Figure 6.12 shows the convention used for the cutting systems. The ideal prepreg cutting system would produce full-sized patterns, have clean-cut edges, give 100% fiber-cut along the ply periphery, cause no fraying, not alter the chemical composition of the matrix in any way, and leave no scrap material.

It is important to balance the cutting capacity against overall production requirements. The very advanced nesting and cutting technologies that have been developed for the garment industry have capacities far in excess of that which might be required in a small or medium-sized shop. There is no point in having a sophisticated, automated, computerized cutting system that can cut a month's pattern shapes in an afternoon and then sits idle the rest of the time. The textile industry has been very successful in automating the cutting operation—so successful, in fact, that just a few of their cutters in full-time production could supply the total current need of cutting fabric for the entire U.S. advanced composite industry.

While a wide range of methods have been used successfully in industry for the cutting of a variety of prepreg broadgoods and unidirectional tape materials, the following methods cover the most important practical technologies.

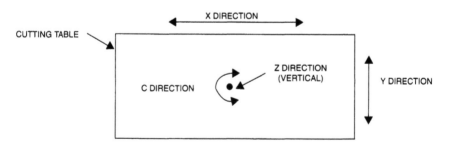

**FIGURE 6.12** Coordinate convention used in systems for cutting prepreg broadgoods.

## Lasers

High-power lasers are currently used in a wide range of industries for cutting and drilling holes in such diverse materials as metal, glass, ceramics, wood, and human flesh. One laser commonly used for cutting applications is the carbon dioxide laser operating at about 10,000 cm$^{-1}$ in the mid-infrared region of the electromagnetic spectrum. Only a few companies (including SAAB) have seriously evaluated the laser as a cutting technology for prepreg graphite material in a production environment. The method has not gained widespread acceptance due to the inability to cut individual plies without curing the edges in the immediate vicinity of the cut. Also, the power densities have to be exceedingly high if an attempt is made to cut multiple stacks of materials.

| Advantages | Disadvantages |
|---|---|
| Very good cut edge, especially on thin gauge metals | Can be used only on a single ply of graphite material |
| Capable of cutting extremely intricate pattern shapes | Noxious fumes when cutting composites |
| May be automated using mirror-driven Technology | Causes curing of the resin immediately adjacent to the cut edge |
| Excellent hole-drilling capabilities in metals | Requires very high power densities and therefore careful isolation of the beam |

## Water Jets

Water jet cutting is a method that uses a thin, high-velocity stream of water as the cutting tool. In order to produce this high-velocity water jet, pressures of 30–60 × 10$^3$ psi are used to force the water through an orifice (0.1–0.5 mm diameter) in a sapphire or other hard material. The resultant water jet is supersonic and therefore extremely noisy. Automated prepreg cutting can be achieved with only a two-axis control system since the water stream is fairly well collimated, making it relatively insensitive to z-axis positioning. A number of commercially available systems come equipped and provide a z axis for added flexibility, if required. The sapphire orifice produces a fully symmetrical water stream because of its circular shape, eliminating any need for a rotational c axis. Since the cutting action is performed by the high-velocity water, there is no wear of the cutting tool. However, the sapphire orifice does require periodic replacement because of gradual water erosion. Epoxy resins absorb moisture readily, so care must be taken with water jet cutting to minimize the amount of water contacting the resin. Water jet cutting has been used in Europe successfully; however, concerns over the moisture-absorption problem have prevented widespread acceptance in the United States. Cutting requires some means of supporting the prepreg material while allowing the water stream to pass through. One simple method uses a support for the prepreg that is made from a thin honeycomb structure, with the cell walls aligned with the direction of the water stream. The walls of the cells support the material adequately while at the same time allow the jet to penetrate the material. As time goes on, the honeycomb material needs to be replaced.

An alternative method of prepreg support [10] using a conveyor belt system with a movable slit in the surface material is shown in Figure 6.13. The four-roller system forms a unit in which the rollers are kept in the same relative position to each other, as shown. Each roller spans the full width of the support surface. The prepreg material lies on the top of the support surface. When the roller unit is moved back and forth in the *x* direction, the slit moves with it, following the jet of water as it traces the ply pattern on the prepreg. The effect is to leave the prepreg material stationary on the support surface while the slit is moved beneath it. In this manner the slit is always maintained directly underneath the jet stream. The water jet head translates in the *y* direction.

Typical water velocity at the nozzle is about 2000 ft/sec. In most advanced applications, the cutting fluid contains an abrasive that allows the cutting of extremely hard materials. In such applications the difficulty is to maintain convergence of the beam. Novel tip geometries can be used to do this—for example, in the system developed at Southwest Research Institute. The advantages and disadvantages of utilizing water jet cutting technology are as follows.

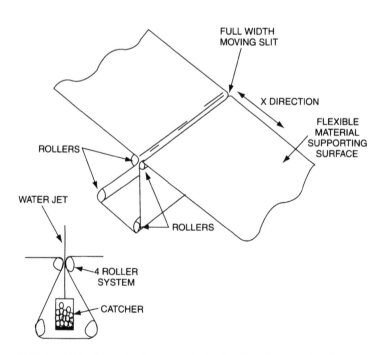

**FIGURE 6.13** Example of conveyor belt with full-width moving slot used to support prepreg broadgoods during abrasive water jet cutting.

| Advantages | Disadvantages |
|---|---|
| Small, clean, and accurate cut | Both a protective top film and backing paper must be left on during cutting |
| Minimal amount of waste | Too slow for high-volume production work |
| Excellent on soft, delicate, or unidirectional material | Unit takes up lots of space |
| Toughness of material virtually irrelevant | Cuts composite material well to a maximum of 10 plies |
| No deterioration of material being cut | Some question as to leaking of the resin in the immediate vicinity of the cut |

Reciprocating knife cutting, an automated computerized cutter and nesting system, is now a fairly popular cutting option. The cutting rate is high (20 in./sec), and the system can handle a broad range of woven broadgoods. A batch of broadgoods materials, typically 54 in. wide, can be thawed and stacked up to about ten layers deep on the power-head table of the cutter. While the cutting operation is in progress on one such table, the cut pattern shapes may be removed and kitted on a second table and the board loaded for the next cutting operation. In this way, one automated cutter can adequately meet the needs of even a large production facility, often with capacity to spare.

The reciprocating knife cutter utilizes a cutting blade mounted on a reciprocating motor. The motor is mounted on a gantry above an X,Y table. The material is then placed over a length of the table and a thin polymeric cover ply is placed on top. Sometimes a vacuum may be applied to hold the plies in place. The gantry-mounted head then cuts specific pattern shapes under the control of a computer.

There are a number of variations on the basic reciprocating knife technology, which can be useful in particular applications. The underlying difficulty with the reciprocating knife is that in some cases it does not yield a clean edge. An acute angle is required relative to the warp or fill direction. In such cases an ultrasonic head may provide an acceptable solution to the problem.

| Advantages | Disadvantages |
|---|---|
| Shapes excellent woven fabric broadgoods, handles many pattern shapes. | Ends of material tend to fray whencutter incidence angle is greater than about 100° or less than 80° |
| High cutting rate may be maintained | Knife replacement fairly frequently |

The simple die board and press is a cheap, reliable, and effective cutting tool. Figure 6.14 shows a commercially available production-type press. It will be noted that die boards can be shifted around on automated roller systems, permitting this function to be semiautomated in high-volume operations requiring efficient materials transport. Also, from the technical point of view, the die board produces excellent-quality cuts even on unidirectional tape.

**FIGURE 6.14** Commercially available production-type press used for die-cutting prepreg broadgoods.

## Roller Presses

The roller press utilizes a hardened steel roller positioned in the head of the press. The material to be cut is pressed between the head of the press and the cutting edge of the steel rule die. The cut material is then unloaded from the die board. New prepreg sheets or tape is then placed on top of the steel rule die to repeat the operation.

| Advantages | Disadvantages |
|---|---|
| Fast—cuts many different-shaped patterns in only a few seconds | Tends to bend steel rule dies |
| Point-to-point cutting action | Pulls material through—accuracy of cuts can be unreliable |

## Incremental Feed Presses

The incremental or progressive feed press incorporates a vertically mounted head on a steel rule die that is fed a section at a time through the press. An electronic base is mounted under the head, and this base transmits pressure to the material between the base and the cutting edge of the die. After the cut is made, the head retracts a few inches, which allows the die board to be fed incrementally a specified distance under the cutting head. The operation is repeated until all the material is cut. Upon completion of the operation, the operator lifts out the patterns, removes the scrap, and the die board shifts automatically to the opposite end of the press, at which point it is reloaded with material. Incremental press cutting with a steel rule die gives clean, accurate cuts without the aid of a top sheet and cuts well without backing paper. The incremental press is a self-contained compact unit that can be easily interfaced with a microprocessor and is one of the fastest methods of cutting multiple patterns. The steel rule die board, when used in conjunction with the press, lends itself readily to the concept of a kit. No extra kitting operations would be necessary. There are several models of excellent-quality presses on the market today. Many are equipped with feed tables and die-board storage retrieval systems and 50–200 tons working pressure.

| Advantages | Disadvantages |
|---|---|
| Clean cuts, fast | Available only in specific tonnages |
| Cuts material well without backing paper | |
| Die board contains a complete kit | |
| No cover sheet needed on material | |
| Point-to-point cutting action | |

For the reasons mentioned in the introduction to this subchapter, there is no fully automated production layup facility in the world. This would be a facility where arbitrary patterns are placed on arbitrary tool surfaces in an active production facility. There are a number of facilities where experimental systems are at various stages of development, and portions of the handling, cutting, and layup tasks are partially automated. One problem is that it is not good enough to have 90% reliability (which is very good given the nature of the considerable challenge), and therefore one still needs a person to watch the robot or to free up the fouled system when a few plies get mangled.

However, in those cases where the design requires unidirectional tape, the geometry is not very active, and the component is large (e.g., wing skins), automated tape laying may be a viable technology. Figure 6.15 shows a tape-laying machine laying up the wing skin for a large aircraft. This is one of a number of such machines that are commercially available.

Finally, it is worth noting that there have been a number of attempts to improve the efficiency of the manual layup operation. Here the most promising approach is that of an optically assisted ply-locating system. This idea was first anticipated by Joe Noyes while he was technical director of the Lear Fan project. Studies conducted by one of the authors about 15 years ago resulted in a functional specification and a request for a quotation for such a system. Such systems are now commercially available and have shown considerable promise as a means of increasing the efficiency of the layup process. The basic mode of operation is as follows:

1. Using an existing kit, the system is taught the location and shape of pattern 1 using a joystick to guide a laser spot around the outer edge of the ply.
2. When the learning step for pattern 1 is complete, the pattern shape is then projected onto the tool surface to ensure that the projected pattern and tool alignment are correct.

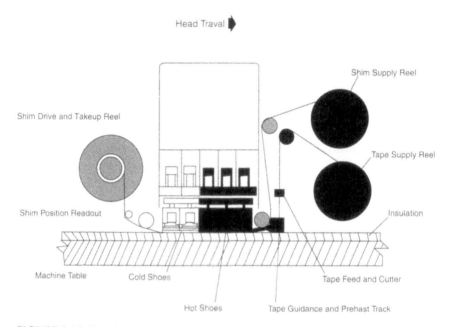

**FIGURE 6.15** Tape-laying machine laying up large aircraft wing skin.

3. When step 2 is complete, a new pattern is placed on the tool surface and the process is repeated until the complete kit has been learned.

This device reduces layup times by providing a direct reference datum to ply placement. Also, there is a considerable advantage in that the planning documentation is paperless and no written instructions need to be followed. Recent trials with optically assisted ply-locating systems indicate that layup efficiency may be increased by approximately 30% using this approach. There are a couple of extensions of the technology that, no doubt, will soon become available, given that the initial technology has now been adopted. The first extension is the coupling of the ply-locating system with a ply-dispensing system. The basic idea here is that when the pattern shape is projected onto the tool surface, an interlock operates that makes the projected pattern physically accessible to the operator. The operator lays up the ply in the usual fashion, using the laser image as a guide. When the pattern layup is complete, the operator prompts the system for release of the next ply. Before an additional ply is made available, a grid is projected onto the tool surface and a vision system (both visible and infrared) confirms that the previous ply has been positioned correctly. Also, common-occurrence inclusions under the ply, or backing paper left in place, can be detected in this way while there is still a possibility for rework (i.e., before the irreversible cure operation has been carried out).

## 6.1.3  Cure

The objective of the cure/postcure operation is to ensure that the laid-up, properly staged, and debulked parts are cured. The parts are converted from a stack of resin-impregnated cloth, in which there is no dimensional stability, to a hard, dimensionally stable near-net-shape component. In this operation it is important to recognize that the process engineer is engaging in an activity that bears on many disciplines. For example, the need to fabricate a component of acceptable dimensional tolerance is certainly an activity familiar to the mechanical engineer. Also, the resin system is in most cases a thermosetting resin, which, by definition, is undergoing a set of chemical reactions that take it from a tacky fluidlike state to a hard, impenetrable, solidlike or glassy state. The control of such reactions is the realm of the chemist or the chemical engineer. On the other hand, such control is actually implemented as the control of a large pressurized oven, and this is often handled by an electrical engineer with a background in control theory. This mix of skills is indeed highly interdisciplinary, and it is interesting to note the differing perspectives that are brought to this part of the processing operation.

An important feature of the curing operation is that it is generally irreversible (i.e., there is no way to reverse the chemical reaction sequence), so after this stage in the process, no rework is possible. If for some reason the cure operation leads to a bad part, then all labor and material costs up to this point are lost. It is important, therefore, that a good physical and chemical understanding of the cure process be achieved, so that cure cycles are designed using an adequate knowledge base. The criticality of the cure cycle depends to a significant degree on the nature of the resin system. For some very mature systems, such as the toughened epoxy resins, the materials, through years of chemical development, have been designed to be exceedingly tolerant. Other systems, such as the polyimides, are notoriously variable and require careful design and control of the cure process.

### Epoxy Resins

An epoxy resin cures by means of an addition reaction. The catalyst, which is usually a Lewis acid, complexes with the oxygen atom in the epoxy ring and thereby activates the carbons in the ring. This activated complex is susceptible to rapid reaction with the nitrogen atom of the hardener. The resulting adduct may undergo a facile equilibration to form an intermediate with both the secondary amine and the alcohol groups. This process then continues with the multifunctional epoxy and amine molecules until a densely cross-linked structure is formed. Notice that there is no elimination of a small molecule such as water, methanol, or HCL, so the cure process itself does not evolve volatiles.

This means that in an epoxy resin, the vacuum schedule can be very straightforward; namely, the vacuum hose is attached to the bag and a full vacuum is applied at the start of the cure schedule and kept on for the complete cycle. In the usual epoxy resin system there are only two minor sources of volatiles. The first is the small amount of air that is dissolved in the resin. As the material heats, this air is released, and if it is not evacuated, it will give rise to voids that will be more or less uniformly distributed throughout the laminate. The second and more serious cause of voids in an addition-type resin system is simply the air that, of necessity, becomes entrapped in the small-volume elements that arise due to the shape of the weave of the material. In a typical system such as an eighthamess satin weave material, small voids are formed when the fiber tow in the fill (or weft) direction goes over the tow running in the warp direction. This air should be removed as far as possible—thus vacuum must be applied.

Insofar as there is a need to determine the cure heatup rate, hold time, and other processing variables, this is best done by an initial investigation using differential scanning calorimetry (DSC). Differential scanning measures the enthalpy change that occurs during the chemical reaction of the resin. The measurements can be made isothermally or, more usually, on a temperature ramp. When an epoxy resin cures, heat is given out (i.e., the reaction is exothermic). The ordinate in a DSC trace is the differential power required to maintain the same temperature in a pan containing the sample relative to a reference pan that contains no sample. In isothermal operation, the resin is heated rapidly to the required temperature and held isothermally. The time evolution of the heat change is measured as the area under the exothermal peak. If the measurement is made at high temperature, the reaction can be brought to completion and this is usually denoted $H^\wedge$. The usual definition of degree of cure of an epoxy is

$$\alpha = H(t)/H\infty$$

(i.e., the degree of cure $\alpha$ is defined as the amount of heat that has been given out relative to the maximum possible amount of heat that could be given out). It is noted without further comment that this widespread definition ignores the fact that the enthalpy is a function of state, so that the quantity $H(t)$ depends on the path taken (i.e., on the heating rate). In any case, one should confirm that the heat rate, ramp, and temperature holds are consistent with the attainment of some high (normally 90%) degree of cure.

From a practical point of view, it will usually be the case that the actual cure cycle will be more dependent on the thermal response characteristics of the autoclave than on the chemical kinetics of the resin system. For this reason, cure cycles are usually considerably longer than would appear warranted on purely kinetic arguments. This whole question is one of considerable complexity and bears on issues of chemical stability, nonlinear kinetics, and structure/property relationships, which are far outside the scope of the present chapter. Suffice it to say that caution should be exercised, and incremental improvement in shortening cycles should be the norm. There has been a good deal of recent progress in designing intelligent cure cycles [11]. Figure 6.16 gives a cure cycle for a number of common epoxy material systems.

## Bismaleimides

The bismaleimides are also addition-type resins, and the cure considerations, apart from the kinetic and maximum hold temperature, are very similar to those for epoxies. A common cure cycle for bismaleimides is shown in Figure 6.17.

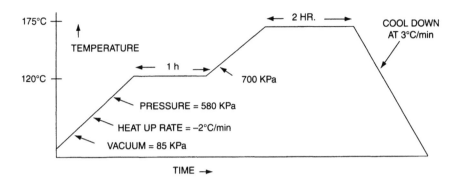

**FIGURE 6.16** Cure cycle for a number of common epoxy material systems.

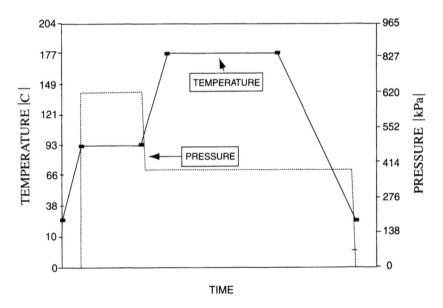

**FIGURE 6.17** Common cure cycle for bismaleimides.

## Polyimides

Polyimides differ fundamentally from either epoxy or bismaleimide systems in that they are condensation resins that emit considerable amounts of volatiles during the cure process. For example, the most widely used commercial polyimide, PMR-15, loses some 10% by weight (and many times the part volume) during cure. This poses considerable challenges in designing cure schedules, bagging, and breather systems to accommodate the large amounts of volatiles that are released.

In this type of condensation system, of which PMR-15 is prototypical, the lone pair of electrons of nitrogen attacks the carbonyl function of either the ester or the anhydride with elimination of a methanol molecule to form a polyamic acid. This polyamic acid will undergo a rapid ring closure to the imide. In this reaction mechanism it is clear that there are a number of competing equilibria among the anhydride, the esters, and the hydrolyzed ester. These equilibria will shift with temperature and pH, so the complete mechanism is one of significant complexity. In any case, the important fact is that in the first part of the cure cycle a polyimide is formed. In the case of PMR-15, this polyimide is end-capped with a norbornene group to form a more or less stable, thermoplastic-like material in the first part of the cycle. Upon raising the temperature further, the norbornene group causes a cross-linking reaction to take place, which gives the final polymer.

In this resin system there are a number of isomers, so a total of some 20 or more distinct chemical compounds are present in the monomer state. There are a number of possible reactions. These include attack by the nucleophile on the back side (Sn2 type) of the methyl group to yield a carboxylic anion and methyl-substituted secondary amin. Also, there is the possibility of nucleophilic attack on the bridging carbonyl of benzophenone, which will yield a Schiff base adduct. In this regard, it is interesting to note that the fully cured resin, if prepared in a void-free state, has a deep magenta red color reminiscent of an azo dye. The important factor to bear in mind from the point of view of part fabrication is that these side reactions can give rise to oil-like materials that can vaporize during cure or postcure or can give rise to chemical structures with inherently low thermal stability. While much research has been done, there is still a good deal of work needed in these areas. Again, it behooves the engineer to err on the side of caution and process materials according to accepted standards.

Figure 6.18 shows a typical cure cycle for a polyimide. In this cure cycle there is an initial hold at 430°F, which allows imidization to complete, after which a ramp is initiated to the high-temperature hold. The vacuum schedule is critical to obtaining void-free laminates with polyimides. On the one hand, it is necessary to engineer a slow boil so that the volatiles will escape through an efficient breather schedule; on the other hand, it is necessary that the rate of volatile evolution not be so vigorous as to displace plies with the laminate. A later subchapter deals with sensors that can help with control of this situation. For most components, which have fairly thin cross sections, the standard polyimides that the processing challenges will be greater, and special procedures, including reverse bag pressurization, should be considered.

## Reverse Bag Pressurization

To understand the basis for the reverse bag pressurization technique, it is necessary to consider the physics of pressure distribution of a bagged part in an autoclave. Figure 6.19 shows a schematic of the situation. The autoclave pressure acts on the bag and causes physical compaction of the laminate. At equilibrium, the force exerted by the bag is exactly compensated by the resistance to further compaction offered by the laminate, breather plies, release plies, and so on. Note that the pressure *under* the bag (i.e., the pressure felt by the resin) is the saturated vapor pressure of the resin

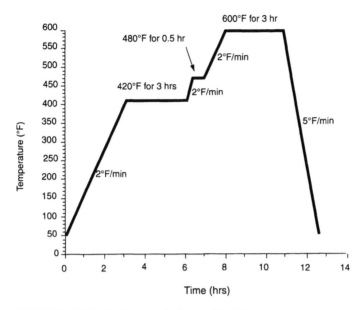

**FIGURE 6.18** Typical cure cycle for a polyimide.

at the temperature. If full vacuum is applied under the bag, volatiles will escape until only those voids that are trapped in interstitial spaces in highly tortuous escape routes remain. It will probably not be possible to remove these voids by application of further vacuum. However, if the resin is still in a fluid state and the vacuum is not removing further volatiles, the void size may be shrunk further by applying pressure under the bag. This increases the hydrostatic pressure on the resin, and any bubble in it will tend to shrink or possibly dissolve in the resin. Of course, it is essential that the hydrostatic pressure applied under the bag be substantially less than the autoclave pressure; otherwise, the bag will blow off. This technique can be used to routinely process thick (1 in.) laminates. Figure 6.20 shows the reverse bag pressurization system developed and employed at Southwest Research Institute.

## 6.1.4 Trim and Drill

A wide variety of methods are available for machining composite components. In general, the presence of a hard fibrous reinforcement leads to excessive wear, and diamond-coated cutting tools are considered the industry norm. Even here the tool life is not notably long.

Conventional machinery such as drilling, grinding, milling, and turning equipment requires contact between the hard surface of the cutting tool and the softer material to be cut. Two major problems arise with cutting composite materials. First, these materials are not ductile, so the picture of a metal as an idealized plastic body that yields in shear either on the shear plane or in the so-called shear zone around the cutting edge will not, even approximately, be met [12,13]. Second, the

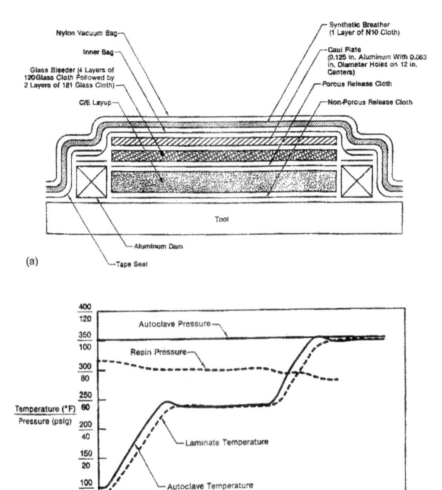

**FIGURE 6.19** (a) Bagged prepreg broadgoods composite layup in an autoclave, and (b) schematic of pressure distribution in an autoclave during typical heat cure cycle.

cut edge invariably shows fiber damage and possibly delamination to a depth that is a function of the pit geometry, material, speed, and material feed rate.

In addition to the traditional mechanical contact methods, a number of nontraditional methods are in routine use in the composite industry. Water jet cutting is a particularly useful technology for cured laminates, and more particularly, cured laminates with honeycomb cores. These are particularly difficult to cut with a good edge. These methods include the following.

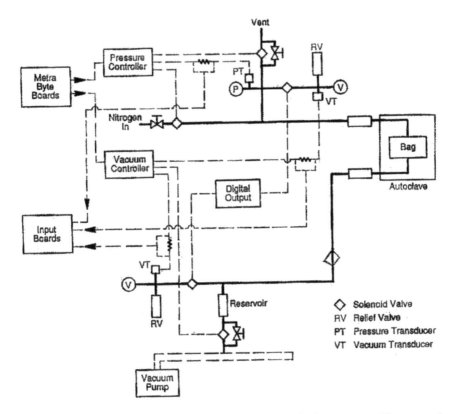

**FIGURE 6.20** Computer-controlled vacuum and pressurization system. (Courtesy of Southwest Research Institute. With permission.)

## Water Jets and Abrasive Water Jets

When applied to cured laminates, water jets and abrasive water jets are used in the same manner as that described previously for prepreg. For laminate cutting, a jet of water, often containing an abrasive medium, is forced through a sapphire orifice as small as 0.1 mm in diameter with water pressure in excess of $50 \times 10^3$ psi. If abrasive water jet cutting is used, including the patented, highly focused jet technology developed at Southwest Research Institute, very hard tough materials can be cut, such as SiC/Al, boron/aluminum, and fiber-reinforced ceramics. For cutting cured graphite/epoxy laminates, an orifice of about 0.754 mm with water pressure of about $50 \times 10^3$ psi has been found to be most satisfactory. An overview of the water jet cutting process is shown in Figure 6.21.

A number of more or less exotic technologies such as ultrasonic, electron beam, and electric discharge machining have all been tried, but none of these has found the widespread acceptance and general applicability of the methods described.

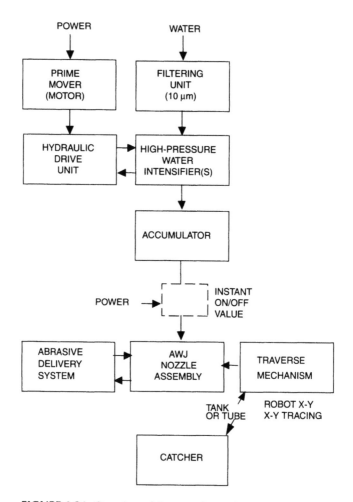

**FIGURE 6.21**  Overview of the water jet cutting process.

## 6.2   FABRICATION WITH CONTINUOUS FIBERS

The goal of the composite designer and the manufacturer is to place the primary load-bearing members (filaments) and the matrix together in the reinforced composite in such a way as to achieve maximum efficiency in the structure. The fact that the individual filaments are primarily loaded unidirectional in tension makes it possible to consider using continuous filaments in the following methods.

### 6.2.1   Filament Winding

For special geometries such as spheres or centrosymmetric parts, filament winding may be a convenient route to fabrication. As the name suggests, this approach consists essentially of winding a single or a small number of fibers onto a rotating mandrel.

By adjusting variables such as the fiber direction and the angle, it is possible to fabricate a wide variety of parts using this technology.

*Filament winding* is a process whereby a rotating mandrel is wound with roving, yarn, or tape in a given angular orientation. The resin-impregnated, filament-wound parts are then cured in an oven or by other means. These parts are often fabricated for high-pressure applications such as gas storage cylinders or rocket motors. Filament winding is easily automated and is relatively inexpensive compared with hand layup or tape-laying technology (see Figure 6.22).

## Winding Methods

From the raw materials standpoint, three methods of winding can be used. Parts can be produced by (1) *wet winding,* in which the roving is fed from the spool, through an impregnating resin bath, and onto the mandrel; (2) *prepreg dry winding,* in which preimpregnated B-staged roving is fed either through a softening oven and onto the mandrel, or directly onto a heated mandrel; or (3) *postimpregnation,* in which dry roving is wound onto the mandrel and the resin is applied to the wound structure by brushing, or by impregnating under vacuum or pressure. This last technique is usually limited to relatively small parts, such as shotgun barrels, since thorough impregnation without the use of pressure or vacuum is difficult.

Currently, wet winding is by far the most common method used. It is lowest in terms of materials cost; and for those producers equipped with plastics-formulating facilities, it offers the benefits of flexibility of resin formulation to meet specific requirements for different parts. Tension of the roving must be altered as the diameter of the part increases if accurate control of the resin–glass content is required. If the winding tension is not altered, resin content varies directly with the diameter.

## Winding Patterns

Two basic patterns can be used in winding, each having a number of variations. Each pattern can be used by itself or combined with the other to provide the desired type of stress distribution in the part.

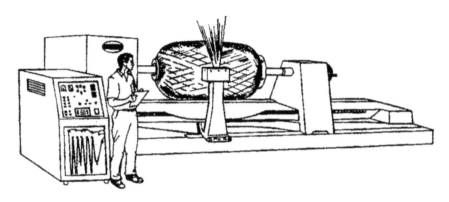

**FIGURE 6.22** Typical filament winding machine.

*Circumferential winding* involves level winding of circumferential filaments. By this method, as shown in Figure 6.23, the impregnated roving, either in single- or multiple-strand band, ribbons, or tape, is laid down on the rotating mandrel at approximately 90° to the axis of rotation. The movement of the carriage that is directing the roving onto the mandrel advances the band of roving a predetermined amount—each revolution depending on the total thickness of wrap desired. These "circs" provide strength only in the hoop direction; longitudinal strength is supplied by applying longitudinal rovings, bands of rovings, or woven fabrics by hand or machine. When such longitudinal reinforcements are applied by hand, usually a unidirectional tape is preferred to provide sufficient pretensioning of the filaments. In the Strickland B process, used by Brunswick, longitudinal rovings are machine applied and can produce open-end structures. Circumferential winding provides maximum strength in the hoop direction, but does not permit winding of slopes over about 20° when wet winding, or 30° when dry winding. Nor does circumferential winding permit effective integral winding-in of end closures.

*Helical winding* is the second most widely used winding pattern (see Figure 6.24). In this technique, the mandrel rotates while an advancing feed places the roving or band of roving on the mandrel. In helical winding, however, the feed advances much more rapidly than in circumferential winding; the result is that the rovings are applied at an angle of anywhere from 25° to 85° to the axis of rotation. In helical winding, no longitudinal filaments need be applied, since the low winding angle provides the desired longitudinal strength as well as the hoop strength. By varying the angle of winding, many different ratios of hoop to longitudinal strengths can be obtained. Generally, in helical winding, single-circuit winding is used: the roving or band of roving makes only one complete helical revolution around the mandrel, from end to end. Young Development has a system for multicircuit winding that permits greater flexibility in angle of wrapping and in the length of the cylinder. According to Young, the optimum helix angle of wrapping to provide a balanced, closed cylindrical structure is 54.75°. The netting analysis used to derive this angle is summarized in Figure 6.25.

All pressure vessels, as well as most other devices that are filament wound, require end closures of one type or another. The most common method has been to design a relatively large collar or flange on the end of a metal closure and wind over it, making the closure integral with the vessel. For winding an integral end

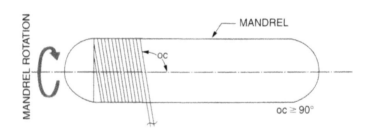

**FIGURE 6.23** Circumferential filament winding.

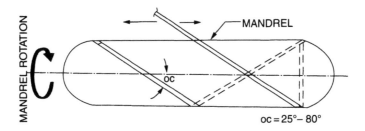

MANDREL ROTATION

$oc = 25° - 80°$

**FIGURE 6.24** Helical filament winding.

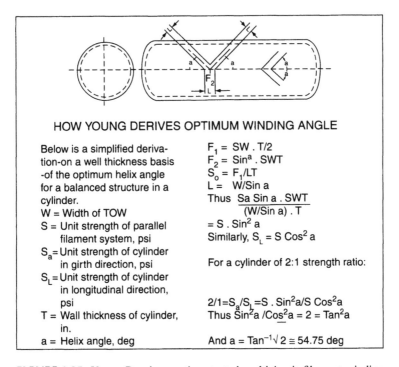

## HOW YOUNG DERIVES OPTIMUM WINDING ANGLE

Below is a simplified deriva-
tion-on a well thickness basis
-of the optimum helix angle
for a balanced structure in a
cylinder.

W = Width of TOW
S = Unit strength of parallel
filament system, psi
$S_a$= Unit strength of cylinder
in girth direction, psi
$S_L$= Unit strength of cylinder
in longitudinal direction,
psi
T = Wall thickness of cylinder,
in.
a = Helix angle, deg

$F_1 = SW \cdot T/2$
$F_2 = Sin^a \cdot SWT$
$S_o = F_1/LT$
$L = W/Sin\ a$
Thus $\dfrac{Sa\ Sin\ a \cdot SWT}{(W/Sin\ a) \cdot T}$
$= S \cdot Sin^2 a$
Similarly, $S_L = S\ Cos^2 a$

For a cylinder of 2:1 strength ratio:

$2/1 = S_a/S_L = S \cdot Sin^2 a/S\ Cos^2 a$
Thus $Sin^2 a\ /Cos^2 a = 2 = Tan^2 a$

And $a = Tan^{-1}\sqrt{2} \cong 54.75$ deg

**FIGURE 6.25** Young Development's patented multicircuit filament winding
angle netting analysis.

closure, a modified elliptical, or ovaloid, configuration is used. Such a design not
only loads the fibers properly in service, it can also be formed using only one angle
of winding; a hemispherical end shape would require several winding angles. The
ellipse is shown in Figure 6.26. Since winding of this type of end requires a low helix
angle, circumferential windings are added to provide the required hoop strength in
the cylinder. Most products can be made on a two-axis machine, or perhaps three
at most. However, state-of-the-art machines today may have six-axis mobility, and
seventh and eighth axes are possible with existing knowhow.

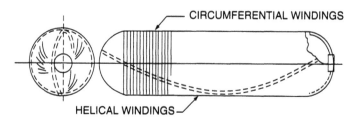

**FIGURE 6.26**  Modified ellipse, or ovaloid, filament winding-in of end closures for a cylindrical vessel.

Speeds depend on the drive motor and the gearing. Generally, the maximum rate for moving the carriage across the mandrel is in the range of 3 ft/sec, with a top spindle rotation of 100 rpm. Larger machines are naturally slower, and a typical production unit might operate 2–3 ft/sec at 30 rpm.

## 6.2.2  Braiding

In the braiding operation, the mandrel is not rotated, but the fiber carriers, which are mounted on a wheel normal to the mandrel axis, rotate around the mandrel axis. The unique feature of the braiding technique is the over-and-under process of adjacent strands of yarn, as shown by the "dancing of the maypole" in  a. The process shown in Figure 6.27b illustrates that by adjusting the braid angle $\theta B$, the hoop and longitudinal strength can be optimized. The greater the braid angle, the greater the hoop strength.

With larger-diameter structures, longitudinal yarn is interwoven at intervals around the circumference. This increases axial and bending strength. Longitudinals may be used on any diameter where additional stiffness is required (see Figure 6.28). Most braiding is accomplished using a *balanced braid;* that is, the widths of yarn lie flat and adjacent so that no gaps or bunching occurs between widths. The braid angle is controlled by adjusting the number of carriers, the speed the carriers travel, and the feed rate of the mandrel through the braider.

Examples of a small braiding machine, usually in the range of up to 36 to 48 carriers, are shown in Figure 6.29. They are often mounted with the wheel parallel to the floor and the mandrel traveling up and down. Figure 6.29a shows the use of a mandrel, while Figure 6.29b shows making a braided sock without using a mandrel. The material in Figure 6.29b will be used as a preform and will require curing in a mold of some type to achieve the desired shape.

For larger diameters and lengths, parts may be braided on the larger machine shown in Figure 6.30. The machine shown has 144 yarn carriers and provisions for 72 longitudinal tows. The wheel diameter is approximately 10 ft and is shown with a traverse mechanism that moves the mandrel back and forth under the direction of a programmable controller and limit switches on the traverse mechanism. The feed and speed can be changed by the operator. The second braiding wheel is used as a yarn reloading station and is easily pushed into position in front of the mandrel

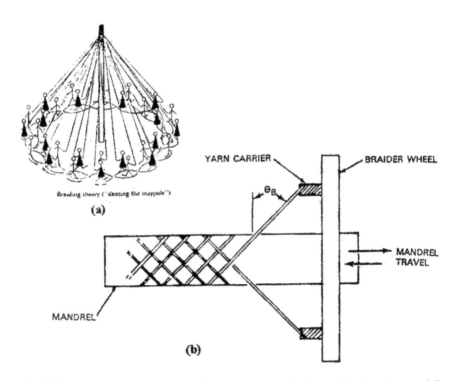

**FIGURE 6.27** Introduction to the braiding process as described by (a) "dancing the maypole"; and (b) in basic machine operation. (Courtesy McDonnell Douglas. With permission.)

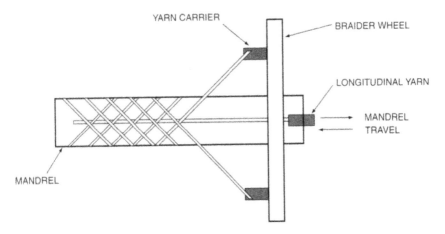

**FIGURE 6.28** Braiding with the addition of longitudinal tows. (Courtesy McDonnell Douglas. With permission.)

by the operator, since the weight is supported by air bearings. Figure 6.31 is a view of the details of the braider mechanism, showing the horn gears that drive the carriers around the circumference of the braiding wheel. The number of carriers and the track in which they travel dictate the diameter of the braider.

## Example of an Automated System

The braided, seamless encasement (which may be round or otherwise) is then impregnated with epoxy resin, cured at elevated temperatures, and the mandrel removed for further use. The proprietary impregnation and cure system used by McDonnell Douglas Corporation at their Titusville, Florida missile production facility is as follows.

1. Braid two layers of 6-end glass yarn over a hollow tubular mandrel, with a thin random fiber mat between the layers (automated system with two braider wheels in tandem, and automatic mandrel transport).
2. Place a premeasured piece of frozen epoxy resin and catalyst on the dry braided tube.
3. Insert the braider-over mandrel with epoxy into a special tubular chamber with a calrod heater in the center and a silicone bladder around the inside wall of the chamber.

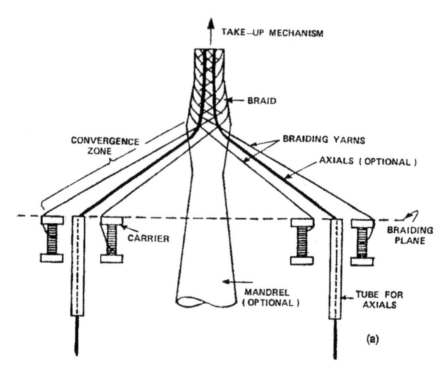

**FIGURE 6.29** Examples of a horizontally mounted small braiding operation: (a) using a mandrel; (b) without using a mandrel. (Courtesy McDonnell Douglas. With permission.)

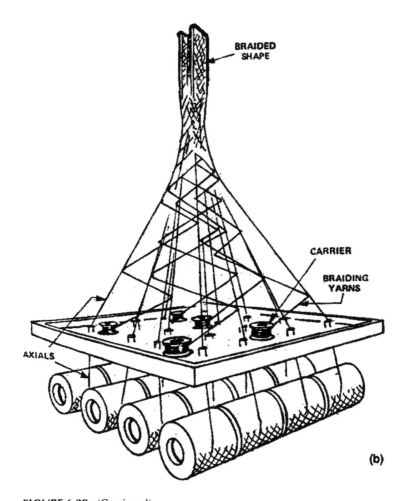

**FIGURE 6.29** (Continued)

4. Start the programmable controller (the system is fully automatic).

   Pull 28 in. vacuum on inside of mandrel and ends of braid.
   Start heating inside of mandrel with calrod (melts the resin).
   Apply 100 psi air pressure on outside of bladder (forces resin into braid).
   Increase temperature up a programmed ramp to 250°F (minimum viscosity).
   Hold temperature for 30 min.
   Turn off heater and flow water through center core to cool mandrel.
   Turn off vacuum and pressure. Eject mandrel and B-staged braid.

5. Remove mandrel and recycle.
6. Cure tube at 300°F in oven to full cure.

Nearly 300,000 rocket tubes (3 in. diameter, 26 in. long) were made with this system. Four people produced 100 tubes per 8-hour shift through this operation (not

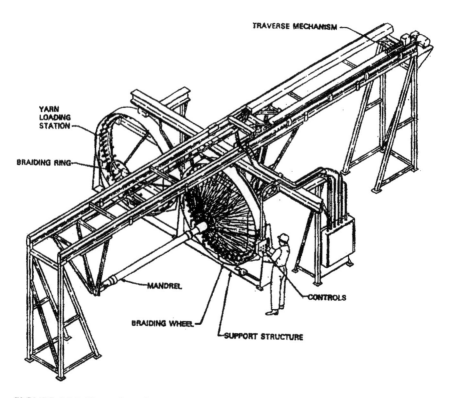

**FIGURE 6.30** Example of a large-diameter production braider setup. (Courtesy McDonnell Douglas. With permission.)

including the additional operations that were required to complete the launch tube assembly).

Glass, carbon, aramid, ceramic, polyester, and polyethylene fibers are being braided—having come a long way from the fabrication of shoe strings, the original use for the process. Figure 6.32 shows tennis rackets braided with glass and graphite prepreg. The original rackets were made from braided graphite only, but were unsuccessful due to the extreme stiffness of the racket. This transmitted the forces to the player's wrist, causing damage to his or her arm from hitting the ball. The addition of glass acted as a damper and made the product very successful.

## 6.2.3  Pultrusion

*Pultrusion* is a technique whereby the fibers are pulled through a heated die, as shown in Figure 6.33. This figure has an H cross section to emphasize the fact that this technology is very useful in the production of beams (I beams or H beams) as well as square or cylindrical beams. In Figure 6.34, the addition of broadgoods to the rovings and an injection system for the resin is shown. In those applications where there is a need to produce bar stock at exceedingly low cost, this is by far the preferred route.

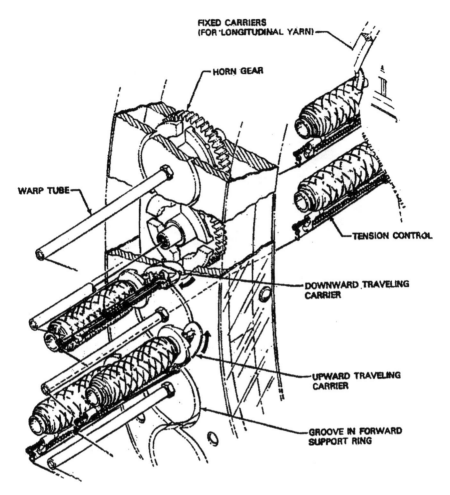

**FIGURE 6.31** Details of the braider wheel mechanism, showing the upward and downward carriers in their continuous figure 8 track; the warp tubes for feeding in longitudinal tows; and the adjustable yarn tension control device. (Courtesy McDonnell Douglas. With permission.)

Such methods can be used to advantage in the production of ribs and stiffeners within the aerospace industry, and in the production of stock materials for the civil engineering, infrastructure, and sporting goods fields.

## Pulforming

*Pulforming* is a process that was developed to produce profiles that do not have constant cross-sectional shape, but do have a constant cross-sectional area at any point along the length of the profile. The materials are pulled from reinforcement creels and impregnated with resin, and in some cases combined with a bulk molding

**FIGURE 6.32**  Braided tennis racket.

compound charge. The material can be preheated using RF energy, as in the pultrusion process.

At this point, the process technology departs from the conventional pultrusion process. Beyond the fiber impregnation area is a horizontal table on which is
mounted a continuous ring of open female molds. As it operates, the table rotates like
a carousel, pulling the wetted fibers through the process. The second mold half, or in
some cases a flexible steel belt, is closed or held against the bottom mold. Since the
mold and the belt are heated, the material within the closed mold is pulled and cured,
accepting the contoured mold profile.

When a two-part mold is used, it opens at the completion of the cycle, moves to the
side, and is redirected to the front of the machine to repeat the process. The finished,
cured product continues its path and moves into a cutoff saw that is synchronized to
cut at the end of each part. To produce curved pulformed parts, a heated steel belt is
used to close the mold. The radius of the part determines the number of molds utilized
in this process. A curved part made in this manner could be an automotive leaf spring.
Other common closed-mold products include hammer handles and axe handles.

## 6.3   FABRICATION WITH CHOPPED FIBERS

In addition to the ratio of fiber to resin in the composite, the orientation, length,
shape, and composition of any of the fibers selected determine the final strength of
the composite—and the direction in which the strength will be the greatest. There
are three types of fiber orientation. In one-dimensional reinforcement, the fibers are
parallel and have maximum strength and modulus in the direction of fiber orientation. In two-dimensional (planar) reinforcements, different strengths are exhibited in

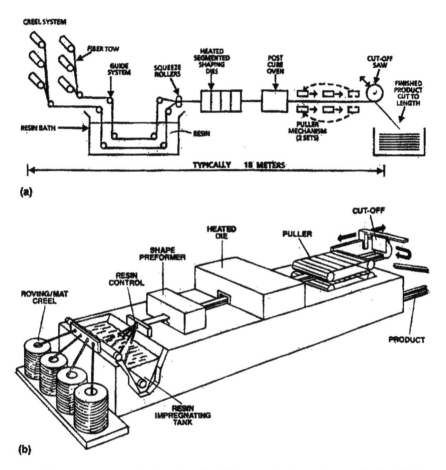

**FIGURE 6.33** Example of pultrusion of an H beam, using dry rovings, and a liquid bath: (a) process schematic; (b) machine set-up.

each direction of fiber orientation, as in broadgoods. The three-dimensional type is more isotropic, but has greatly decreased reinforcing values (all three dimensions are reinforced, but only to about one third of the one-dimensional reinforced value). The mechanical properties in any one direction are proportional to the amount of fiber by volume oriented in that direction. As fiber orientation becomes more random, the mechanical properties in any one direction become lower.

The virgin tensile strength of glass is around 500,000 psi, and the tensile strength of the various matrices is in the range of 10,000–20,000 psi; therefore, the fiber and its orientation is much more important than the matrix selection as far as strength of the composite is concerned. Factors such as cost, elongation, strength at temperature, and general workability in the manufacturing process may become more important in matrix selection.

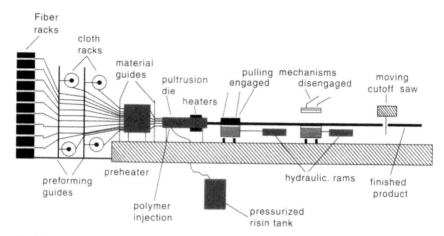

**FIGURE 6.34** Example of pultrusion with fiber rovings and broadgoods. Resin is injected under pressure after preforming and preheating the material, and before final cure in the pultrusion-forming die.

### 6.3.1  Sprayup

In sprayup, the roving is fed into a specially designed chopper gun, which chops the roving into approximately 1-in. lengths and simultaneously sprays a predetermined amount of resin on the fibers and into the mold. Sprayup is best suited to low to medium production, but has a greater production rate (faster mold turnover), produces more-uniform parts with skilled operators, and can utilize more-complex molds than hand layup of broadgoods. Factory installations can be highly mechanized, while portable equipment allows in-field repair, on-site fabrication, and product maintenance.

In this process, reinforcement fibers (usually glass) are simultaneously deposited in a mold by specialized spraying equipment. Hand or automatic spray guns/dispensing devices, either airless (hydraulic) or air-atomization types, are in common use. A marriage of both, called air-assisted, augments airless with external air to shape and improve the pattern. This utilizes the best features of both types and is rapidly becoming the standard. Glass roving passes through a chopper, is chopped to predetermined length, and is projected to merge with an atomized resin-catalyst stream. The stream precoats the chop, and both are simultaneously deposited on the mold surface. Special equipment controls fluid volumes and pressures, catalyst-to-resin ratio, fiber volume, and chop length. The deposited composite is rolled with a hand roller to remove air and to lay down fiber ends. Composite deposition is dependent on operator skill. Multiple passes can build up nearly any required thickness. Typical glass fiber-reinforced uses are boats, automobile components, electrical consoles, aircraft parts, helmets, and tanks.

Special automatic units (vertical, horizontal, rotary—or universal, articulating robots) greatly enhance quality and production rates. Panels in various widths made in endless lengths, or seamless necked tanks, can be made completely by machine. Waist and shoulder pivot, elbow extension, and wrist pitch-yaw-rotate movements can be combined with a transverse axis to provide a total of seven axes. Products produced by robots include snowmobile hoods and car fenders. Custom automatic mechanical (nonrobotic) package installations can include conveyors, exhaust systems, resin supply systems, pumping stations, and a traversing carriage with spray gun and chopper. Controls allow automatic gel-coat and sprayup with automatic rollout.

A vacuum bag or a film of polyvinyl acetate, polyethylene, or nylon may be used to cover the sprayup (see Figure 6.35). The film is sealed at the edges, vacuum applied, and the part is cured—either at room temperature or in an oven—depending on the matrix used. Atmospheric pressure forces out entrapped air, improves resin distribution, lays down fiber ends, and glazes the surface. Physical properties are enhanced, and surfaces away from the mold are improved. The most commonly used matrix is polyester, although others can be applied if required.

## 6.3.2  Compression Molding

*Compression molding* in general is the process of inserting a measured amount of molding material into one half of a mold (or die) and closing the mold under pressure in a press. The die is heated, the part formed, and the mold opened for part ejection. We will discuss one of the more common methods of making composite parts with this method.

### Sheet Molding Compounds

The basic process for making sheet molding compounds (SMCs) was developed and refined at the Dow Corning Research Labs in Ohio. A thin plastic film is continuously unrolled onto a wide (36–48 in.) belt conveyor. The film passes under a series of chop guns, where the glass roving is cut to 1/4–2 in. lengths, and deposited dry onto the moving film. As this dry, random fiber mat progresses down the moving belt, liquid polyester resin is applied to the fiber mat. A second plastic film is automatically placed over the wetted mat, and at the next workstation the material enters a series of

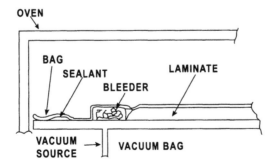

FIGURE 6.35 Example of vacuum cure system for layup or sprayup. Polyvinyl acetate (or other film) is placed over the wet part, and joints are sealed with plastic.

corrugated rollers that knead the material, mixing the resin into the random fibers. It is then rolled up and partially cured. The material is frozen and shipped to the fabricator as SMC. It is then unfrozen as needed for production and cut into strips or other shapes needed to make the particular part. The pot life of the SMC at this point is quite reasonable, and in most cases it does not have to be refrozen to halt the cure. At ambient temperature, the material at this stage feels like thick leather or rubber sheets and can be handled with some amount of automation. In practice, the sheets are then stacked onto half of the heated mold, to achieve the desired thickness (after removing the film backing, which prevents the SMC from sticking to itself), and the mold is closed in a press. Although the shape of the precut SMC need not be exact prior to compression molding, the total weight must be correct, and the various pieces of SMC are placed in the mold, where tests have shown they are needed to achieve the correct thickness and fiber location.

### 6.3.3  Transfer Molding

There are at least two types of transfer molding. One system utilizes a fiber preform in a closed mold, and a resin is pumped in under pressure. The other method heats and forces a fiber matrix into the mold. Both are cured in the mold and then ejected.

**Resin Transfer Molding**

In classical resin transfer molding (RTM), the fiber reinforcement is constructed as a somewhat rigid preform assembly. One popular technique is to make a form from a porous material that approximates one side of the finished part, such as common screenwire. A chop gun, as described earlier under sprayup techniques, is used to build up a thickness of fibers on the wire premold, making a preform. The binder may be a version of the final matrix to be used in the production part, or more often an inert binder that holds the preform in the approximate shape desired. Overcure of the preform will speed up the drying process, permit handling, and allow reuse of the wire form. The fiber preform is then inserted into the lower half of the mold (die), and the mold is closed and locked. Application of vacuum to the cavity is followed by injection of the heated resin matrix into the mold, where it is held under pressure until the resin is set up by the application of additional heat, or the appropriate time has passed.

Applications of this technique include using a dry filament-wound or braided preform, or by laying broadgoods into the cavity prior to injecting the resin. The resin content tends to be somewhat high in many applications, and is totally dependent on the fiber density of the preform. However, it does permit placing the fibers in the product at the point where they are most needed.

**Other Types of Transfer Molding**

Preforms of SMC or bulk molding compound (BMC) are placed under a hydraulic ram in a cylindrical pot above the mold cavity. The material is heated to the point of softness, and the warm material is forced through a small gate into the mold cavity, which has been preheated to 430–600°F. Increased shear at the gate during transfer molding induces random fiber orientation, and the resulting components have more isotropic properties.

This process has been used for molding thermosets such as phenolics for some time. More recently, the same process has been applied to polyimides. Both SMC and BMC forms of PMR-15 polyimide can be used in transfer molding because prior imidization removes all volatiles. Materials must be volatile-free for transfer molding of this type, since the mold is a closed system with limited venting capacity. With sufficiently large runners and gates, transfer molding is done successfully with relative long-fiber SMC (to 1/2 in.). Strips are cut and rolled up, imidized in an oven, preheated, and inserted into the transfer pot. Successful transfer molding may offer economical shorter molding cycles than compression molding, ranging from 8 to 20 min.

## 6.4 REFERENCES

1. Suppliers of Advanced Composites Materials Association (SACMA) Personnel Needs in the Advanced Composites Materials Industry: An Assessment, 1988; for a market assessment, see Peter D. Hilton and Peter W. Kohf, *Spectrum: Chemical Industry Overview Portfolio,* Arthur D. Little Decision Resources, 1987.
2. Watts, A. A., ed., *Commercial Opportunities for Advanced Composites,* ASTM Special Technical Publication 704, American Society for Testing Materials, Philadelphia, 1980.
3. Economy, J., High-strength composites, in *Biotechnology and Materials Science,* Good, Mary L., ed., American Chemical Society, Washington, DC, 1988.
4. Maguire, J. F., P. Paul, and M. Sablic, Computer-Aided Molecular Design, Synthesis, and Magnetic Processing of Polymer Composites, Final Report 06-9730, Southwest Research Institute, San Antonio, TX, 1995.
5. Lee, Stewart M., ed., *International Encyclopedia of Composites,* VCH, New York, 1990.
6. Klein, Allen J., Automated tape laying, *Advanced Composites,* pp. 44–52 (January 1989).
7. Strong, A. B., *Fundamentals of Composites Manufacturing,* Society of Manufacturing Engineers, Dearborn, MI, 1989.
8. Leonard, LaVerne, Composites cutting comes of age, *Advanced Composites,* pp. 43–46 (September 1986).
9. Cook, R. J., Waterjets on the cutting edge of machining, *SAMPE Int. Symp.,* 31:1835 (1936).
10. Hall, Terence F. W., Manufacturing automation/polymer composites, in *International Encyclopedia of Composites,* Vol. 3, VCH, New York, 1910, pp. 133–142.
11. Maguire, J. F., Peggy L. Talley, Sanjeev Venkatisan, Mark Wyatt, and Tom Rose, Cure Sensing Systems for the Efficient and Cost-Effective Manufacture of Parts Fabricated in Composite Materials, Final Report 06-3658, Southwest Research Institute, San Antonio, TX, 1995.
12. Lee, E. H., and B. W. Shaffer, *J. Appl. Mech.,* 18:4 (1951).
13. Everstine, G. C., and T. G. Rogers, *J. Composite Materials,* 5:94 (1971).

# 7 Finishing

*Frank Altmayer*

with

*Jack M. Walker*

and

*Robert E. Persson*

## 7.0  INTRODUCTION TO FINISHING

Metals are usually found in nature as ores in the oxide form, their lowest energy level state. When humans refine a metal, nature begins at once to return it to its lowest energy level. Today we understand the mechanism of corrosion, and the treatments and protective finishes required to make it feasible to use steel and other metallic structures in environments where they would not otherwise have an economic life.

This chapter starts with an explanation of the corrosion mechanism and the various types of corrosion. This is followed by a discussion on the cleaning and surface preparation processes, which are necessary to prepare the metal for its ultimate finish. Electroplating, the most widely used protective and decorative finish, is introduced in considerable detail. The various processes are explained, and the process steps outlined for them. The final section covers the other decorative and protective finishes, including paint, powder coating, metal spraying, and porcelainizing.

## 7.1  CORROSION

Robert E. Person, senior corrosion engineer, EG&G Florida, Cape Canaveral, Florida, and Jack M. Walker, consultant, manufacturing engineering, Merritt Island, Florida

### 7.1.1  Introduction to Corrosion

Metals are usually found in nature as ores in the oxide form, their lowest energy-level state. When humans refine a metal, nature begins at once to return it to its lowest energy level. *Corrosion* is defined as the adverse reaction of a refined metal with its environment. The rate of corrosion on any specific metal varies because of temperature, humidity, and chemicals in the surrounding environment. Variations of intensity of these factors have great impact on the rate of corrosion. Rules of thumb indicate that a temperature increase of 10°C doubles the chemical activity if a suitable electrolyte is present. An increase in humidity or conductivity increases the rate of corrosion accordingly. When the environment contains several factors that aid corrosion, the corrosion rate is usually greater than the resultant of the forces would indicate. For example, Figure 7.1 illustrates corrosion rate differences resulting from the two environmental variables of oxygen concentration and temperature.

The contamination of expensive chemicals and the corrosion of valuable structures and equipment are going on about us every day. It has been estimated that the direct cost to industry and home approaches $10 billion annually in the United States. When the indirect costs are added, this becomes $15 billion. The corrosive effects on *all* metal products are a very large concern to everyone in the manufacturing industries. Several types of corrosion, however—those associated with specific areas such as high temperature, nuclear activity, and the like—are not discussed in this chapter. Most of the problems that arise in the manufacturing industries are of an electrochemical nature. We need to have a clear understanding of the processes by

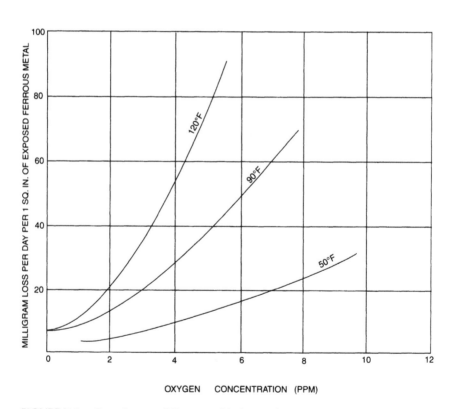

**FIGURE 7.1**    Corrosion rate differences with changes in temperature and oxygen content.

which a useful item is reduced to a collection of rusty or corroded scrap if we are to be effective in controlling corrosion. Intelligent selections of metals and protective coating systems during the design and manufacturing phases are an important part of the manufacturing engineer's task. Continuing maintenance of facilities, equipment, and products is greatly reduced by correct design and initial manufacture.

## 7.1.2 Principles of Corrosion

Corrosion is almost always a detrimental reaction, but there are some exceptions. One that serves humanity is the wet-cell automobile storage battery. To understand the principles of corrosion, we will examine the operation of this familiar object in some detail.

### The Automobile Storage Battery

A *wet-cell storage battery* is a box holding a number of lead plates. Half of the plates are made of metallic lead, and the others of lead oxide. If we were to hook a wire to two of these plates and insert a light bulb in a circuit, as shown in Figure 7.2, nothing would happen without the addition of an electrolyte. This is a mixture of battery acid and water. Battery acid is a 30% solution of sulfuric acid. The electrolyte conducts current inside the battery in addition to taking part in a chemical change.

When sulfuric acid is mixed with water, it undergoes a change that makes it possible to function as an electrolyte. A particle of acid is split into ions. These are charged chemical particles capable of conducting an electrical current. Each particle of acid splits into hydrogen ions, which carry positive electrical charges, and sulfate ions, which carry negative electrical charges. In dilute solutions of sulfuric acid, all of the acid present undergoes this change. Equal amounts of positive and negative charges are developed. The ability of an electrolyte to conduct current depends directly on how many ions are available. Pure water has relatively few ions and is a poor conductor of electricity. It has a pH of 7.0, which means that 1 part in 10 million

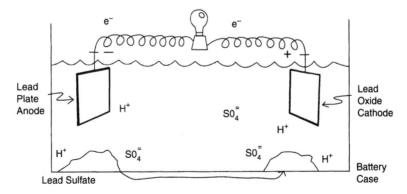

**FIGURE 7.2**  Wet-cell storage battery with light bulb in circuit.

parts is present as hydrogen ions and hydroxyl ions. The absence of available ions to carry electricity in pure water accounts for the addition of sulfuric acid to make the automobile storage battery electrolyte.

Everything in nature, including metals, has some tendency to dissolve in water and hence in dilute acid solutions. In the case of our battery, we have lead in two forms. Of these two, metallic lead has the greater tendency to dissolve in battery acid solution and goes into solution in the form of lead ions. These are *positively* charged electrical particles. Since opposite charges attract, the positive lead ions are attracted to the *negative* sulfate ions and produce the chemical compound lead sulfate. A deposit of sulfate builds up on the lead plates of a storage battery as it discharges. The lead in the battery plate was *neutral* electrically, and small particles go into solution as positive lead ions. There is then a surplus of negative electricity (electrons) left on the plate. These electrons flow through the wire circuit to light the electric bulb in the external circuit. Electrons are negative particles of electricity and subsequently flow to the lead oxide plate, where further chemical changes take place. Upon receipt of electrons, the lead oxide is converted to lead ions and oxide ions. The lead ions combine with passing sulfate ions to form lead sulfate. In Figure 7.2, we show two piles of lead sulfate in the bottom of the battery. In actual practice, the lead sulfate is deposited on the plates. This action can be reversed by driving an outside source of electrical current in the opposite direction and restoring lead, lead oxide, and sulfuric acid as ions. The lead plate is the positively charged anode, and the lead oxide plate is the negatively charged cathode.

Sulfuric acid is not unique in its ability to undergo a change upon solution in water to form ions. Indeed, all soluble salts, alkaline or caustic compounds, and mineral acids share this ability to split into active ions. Other mineral acids, such as hydrochloric, nitric, phosphoric, chromic, and so forth, liberate the same hydrogen ion in solution with water as sulfuric acid. The hydrogen ion is the essence of acidity. Sulfuric acid rather than the other popular mineral acids is used in storage batteries simply because of its low cost and low volatility.

This same phenomenon explains the reduction of steel and other metals into useless rust and corrosion products. Corrosion is an electrochemical change identical in fundamental principle to the transformation of lead and lead oxide to lead sulfate in a storage battery.

### 7.1.3  Surface Corrosion

We can now set up an electrolytic cell or *corrosion battery* to describe the type of rusting suffered by steel in normal atmospheric or marine environments. This corrosion battery exists whenever metal is exposed to a combination of oxygen, water, and ions. This combination, the "big three" of corrosion, is not hard to find. All the *oxygen* we will ever need is available in the atmosphere, and all exposed metal is constantly in contact with it. The atmosphere, except for desert areas, is usually *moisture* laden. Either moisture vapor or dew droplets are in contact with the iron surface. Smog, salt spray, marine locales, industrial soot and fumes, splash and spillage, soil contamination, and other sources can provide minerals, salts, acids, or alkaline materials that will dissolve in available moisture to produce *ions*.

## Corrosion on a Steel Plate

Figure 7.3 shows the tiny corrosion battery, which will produce rust when current flows. We have shown a greatly magnified portion of the iron surface. Adjacent areas, not visible to the eye, act as anode and cathode. On and just above the surface is moisture loaded with ions, to serve as an electrolyte. Gaseous oxygen is dissolved in the electrolyte and is freely available to the surface of the metal. Note that in this case, the electrolyte provides the external circuit. The internal circuit is provided by the metal part itself, since it is highly conductive from point to point. In Figure 7.4, the following process is operating in our corrosion battery:

1. At the anode, an atom of metallic iron is converted to *iron ions* and immediately reacts with water to form rust.
2. *Current flow* is set up between the anode and cathode.
3. The small currents generated here are dissipated as *heat* through the adjacent body of metal.
4. At the cathode, receipt of the current results in transformation of oxygen to *caustic ions*. This results in a highly alkaline solution on the surface, as the caustic ion dissolves in the moisture on the surface.
5. At and above the surface, current is carried by the dissolved ions.
6. As the iron anode is eaten away, a growing deposit of rust forms.

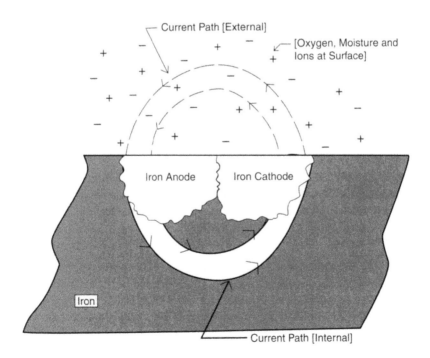

**FIGURE 7.3** Typical corrosion battery.

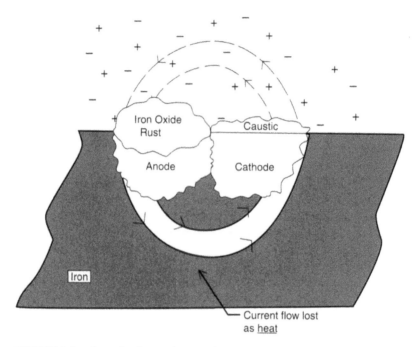

**FIGURE 7.4**   Corrosion battery in operation.

7. This tiny battery is duplicated millions of times over a large iron surface, so that eventually the eye sees what appears to be a uniform layer of rust building up.
8. Note that the iron cathode is not attacked. However, it performed its function in providing the locale at which oxygen could undergo its necessary change to a caustic.
9. This process will continue until (a) all the iron is converted to rust, or (b) something happens to weaken or break the circuit.

Edges and other areas subjected to mechanical work (shearing, drilling, sawing) are prone to be anodes and will corrode preferentially. Particular care must be paid to protection of edge areas or other sharply prominent parts of a metallic structure.

## Crevice Corrosion

The presence of crevices or pockets in the design of a steel structure presents a special corrosion problem and one that is extremely difficult to combat successfully. Crevices have a tendency to corrode at a far greater rate than the adjacent flat metal areas.

We may use a simplified version of the corrosion battery to examine crevice corrosion, as shown in Figure 7.5. The essentials of the corrosion process—oxygen, water, and ions—must be present. Crevices certainly hold soluble salts and liquid solutions of these salts, so more than enough water and ions are present to form an

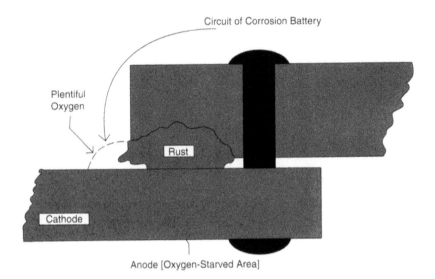

**FIGURE 7.5** Crevice corrosion with two areas of metal.

electrolyte. The remaining ingredient involved is oxygen. As we have observed in the iron corrosion battery and in the galvanic corrosion battery, the change of oxygen to caustic is associated with the cathode area (the area protected). If the two areas of metal are involved in a corrosion battery, such as in Figure 7.5, the area exposed to the most oxygen will become the cathode, thereby forcing the other area (the oxygen-starved area) to act as an anode and be corroded. This is the situation that exists in crevice corrosion. The metal in the crevice is an oxygen-starved area and is therefore subject to corrosion and acts as an anode.

It is known from field experience that it is very difficult to stop this type of corrosion. Application of a coating over the crevice is usually not satisfactory for two reasons:

1. Trapped salts and rust laden with corrosion products and water will promote early blistering beneath the coating, leading to premature film failure.
2. The sharp edges of the crevice are likely to be inadequately covered with the coating; therefore, edge rusting can start promptly. Film rupture along the edges, followed by progressive adhesion loss, will result in unsatisfactory coating performance.

The most practical way to eliminate crevice corrosion is to avoid structural design details that create crevice conditions (lap welds, lap sections riveted together, etc.). In existing structures, crevices should be filled with appropriate mastic or fillers, and proper welding techniques should be observed. Weld spatter, pockets, cracks, and other crevice-type spots in the area to be coated should be eliminated.

## Pitting Corrosion

Pitting corrosion takes a heavy toll on machinery and equipment due to the concentration of the electrical cycle on a very small anodic area. When protective films break down or pinholing occurs, local corrosion or pitting may follow. An anode will be formed at the point where the film break occurs and, being in contact with the external electrolyte, will thus bear the full onslaught of any attack by a surrounding cathodic area. (See Figure 7.6.) Generally, the smaller the area that is in contact with the electrolyte, the more rapid and severe the pitting will be.

## Strong Acid Corrosion

A corrosion battery of great industrial importance and frequent occurrence is that formed by the action of *strongly acidic* materials on iron or steel. (See Figure 7.7.)

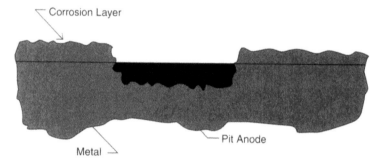

**FIGURE 7.6**    Pitting corrosion caused by a breakdown in the protective coating, or pinholing in the coating.

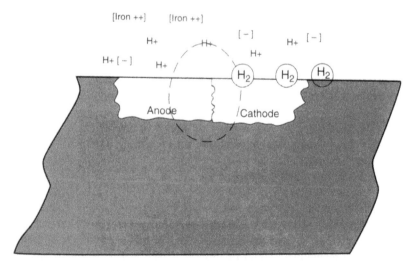

**FIGURE 7.7**    Corrosion battery caused by exposure to a very strong acid solution.

Briefly, a vigorous reaction occurs to dissolve the metal into solution at the anode, and transforms hydrogen ion ($H^+$) into hydrogen gas bubbles ($H_2$) at the cathode. The reaction involved here is a vigorous one and will continue until all of the available metal is consumed, or until all available acid is used up.

The dissolved metal in this battery will not form rust (at least not initially), but will remain dissolved in the acid electrolyte as an iron salt as long as sufficient moisture is present and as long as an excess of acid is present. When the available acid is consumed (for example, after an acid splash on steel), and the moisture evaporates somewhat, copious rust will then form and be deposited over the corroded areas of the metal surface.

## 7.1.4   Internal Corrosion

The corrosion phenomena discussed in Subsections 7.1.2 and 7.1.3 generally apply to all metals. However, some metal alloys have additional corrosion problems. When the alloy consists of two or more metal elements that are widely separated on the galvanic scale, as discussed in Subsection 7.1.5, the corrosion battery may develop at the boundary layer between grains, or within the grain.

As an example, the initial strength of the heat-treatable aluminum alloys is improved by the addition of alloying elements such as copper, magnesium, zinc, and silicon. In the heat-treatment process, the initial heating is designed to put the soluble element or elements in solid solution with the aluminum. This is followed by rapid quenching, which momentarily "freezes" the grain structure. In some cases, this is followed by a second controlled heating at a slightly elevated temperature. The precipitation of the alloying constituents into the boundary layer between grains essentially "locks" the grains in position to each other. This increases the strength and reduces the formability.

### Intergranular Corrosion

The aluminum alloys that are given high strength by heat-treating contain considerable amounts of copper or zinc. If those containing copper are not cooled rapidly enough in the heat-treating process to keep the copper uniformly distributed, there is a concentration of copper at the grain boundaries, setting up a potential difference there. Corrosion occurs in damp air, eating into the boundaries and causing intergranular corrosion and resultant brittleness, with lowered fatigue resistance from the notches thus formed. Precise control of the heating and quenching temperatures and times is a minimum requirement. (See Figure 7.8.)

### Stress Corrosion

The aluminum alloys that are rather high in zinc or magnesium are subject to *stress corrosion*. This behavior is not confined to aluminum alloys but is quite prevalent in many alloys that are constantly loaded and simultaneously subjected to a corrosive environment. Even slow general corrosion may be accelerated by stress, but accelerated local attack may occur either at grain boundaries or in the grain. Some magnesium alloys display stress corrosion, and in some corrodents, even steel shows such attack. In an oversimplification, one may look at this behavior as due to the tensile stress tending to pull the grains apart more and more as corrosion occurs between them. (See Figure 7.9.)

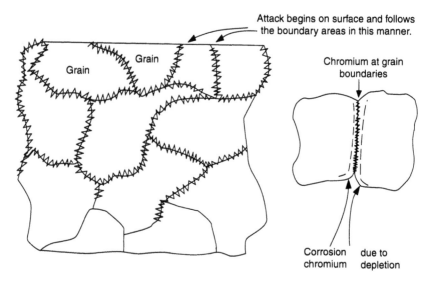

**FIGURE 7.8**   Corrosion battery showing intergranular corrosion of aluminum alloy due to copper migration to the grain boundary layers as a result of improper heat treatment.

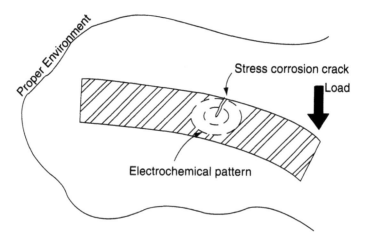

**FIGURE 7.9**   Stress corrosion crack in metal due to material being constantly loaded and simultaneously subjected to a corrosive environment.

## Corrosion Fatigue

All corrodible alloys are also subject to *corrosion fatigue*. Corrosion pits or intergranular penetrations naturally act as stress raisers. Corrosion, simultaneous with repeated stress, produces far more rapid damage than these stress raisers would inflict were corrosion not going on after they were formed. The intermittent release of stress, or its reversal, permits the pits and fissures to close a bit, and if corrosion products are trapped therein, they may act like chisels to help pry the metal apart. Steel, which in the absence of corrosion has an *endurance limit* (a stress whose repetition it will withstand indefinitely), has no such definite limit under corrosion fatigue. Sometimes the endurable repeated stress for even a reasonable life before replacement is necessary in only some 10% of the repeated stress it would withstand indefinitely in a noncorrosive environment.

To defeat intercrystalline corrosion, stress corrosion, and corrosion fatigue, the attackable metal must be kept out of contact with the corrodent, by a complete and impervious coating. Subchapters 7.3 and 7.4 describe some of these coatings and their application processes.

## 7.1.5   Galvanic Corrosion

Situations often arise where two or more *different* metals are electrically connected under conditions permitting the formation of a corrosion battery. A situation then exists where one metal will be corroded preferentially in relation to the other metal to which it is physically connected. This is termed *galvanic corrosion*. Three ingredients are required—an electrolyte, a material to act as an anode, and another to act as a cathode—in addition to the metals. The electrolyte is the medium in which ionization occurs. The electrons flow from the anode to the cathode through a metal path. The loss of metal is always at the anode. Figure 7.10 illustrates the electrical and chemical

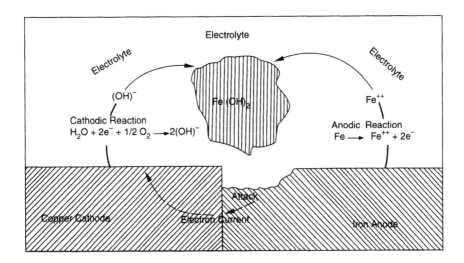

**FIGURE 7.10** Example of galvanic corrosion between iron and copper.

interplay of electrochemical corrosion of two different metals. While this is similar to the corrosion battery discussed previously, it is usually faster acting and more severe.

Anodic reactions are always *oxidation* reactions, which tend to destroy the anode metal by causing it to go into solution as ions or to revert to a combined state as an oxide. Cathodic reactions are always *reduction* reactions and usually do not affect the cathode material. The electrons that are produced by the anodic reaction flow through the metal and are used in the cathodic reaction. The disposition of the reaction products is often decisive in controlling the rate of corrosion. Sometimes they form insoluble compounds that may cover the metal surface and effectively reduce the rate of further corrosion. At other times the reaction products may go into solution or be evolved as a gas, and do not inhibit further reaction. Galvanic corrosion is an extremely important corrosion process and one that is frequently encountered. An understanding of it will be helpful in rounding out our knowledge of corrosion processes. The principles of galvanic corrosion may actually be utilized to advantage in the cathodic protection of surfaces by using sacrificial metal anodes or inorganic protective coatings.

## The Galvanic Scale

Corrosion occurs at the rate that an electrical current, the *corrosion driving current*, can get through the corroding system. The driving current level is determined by the existing electrode potentials. *Electrode potential* is the tendency of a metal to give up electrons. This can be determined for any metal by measuring the potential between the specimen (metal) half-cell and the standard (hydrogen) half-cell. Tabulating the potential differences between the standard (hydrogen) half-cell and other elements opens an extremely important window to view one part of the corrosion spectrum. Such a tabulation, known as the *electromotive series*, is illustrated in Figure 7.11.

Utilizing the electromotive series, an engineer can determine the electrical potential between any two elements. This electrical potential is the algebraic difference between the single electrode potentials of the two metals. For example:

Zinc and copper
    $+0.76 - 0.34 = +1.10$ V potential
Iron and copper
    $+0.44 - 0.34 = +0.78$ V potential
Silver and copper
    $-0.8 - 0.34 = -0.46$ V potential

The electromotive series forms the basis of several possibilities for controlling and decreasing corrosion rates. It provides the data required to calculate the magnitude of the electrical driving force in a galvanic couple. The electrical driving force of an iron and copper couple may be thought of as promoting the following activities:

Oxidation of the iron (the anodic reaction)
Flow of electrons through the solid iron and copper couple

| LITHIUM | – Li, Li+ | +3.02 volts* |
|---------|-----------|--------------|
| POTASSIUM | – K, K+ | +2.92 |
| SODIUM | – Na, Na+ | +2.71 |
| MAGNESIUM | – Mg, Mg++ | +2.34 |
| ALUMINIUM | – Al, Al+++ | +1.67 |
| ZINC | – Zn, Zn++ | +0.76 |
| CHROMIUM | – Cr, Cr++ | +0.71 |
| IRON | – Fe, Fe++ | +0.44 |
| CADMIUM | – Cd, Cd++ | +0.40 |
| COBALT | – Co, Co++ | +0.28 |
| NICKEL | – Ni, Ni++ | +0.25 |
| TIN | – Sn; Sn++ | +0.14 |
| LEAD | – Pb, Pb++ | +0.13 |
| HYDROGEN | – $H_2$, H+ | 0.00 |
| BISMUTH | – Bi, Bi++ | −0.23 |
| COPPER | – Cu, Cu++ | −0.34 |
| MERCURY | – Hg, Hg++ | −0.80 |
| SILVER | – Ag, Ag++ | −0.80 |
| PLATINUM | – Pt, Pt++ | −1.2 |
| GOLD | – Au, Au+ | −1.7 |

*Oxidation reaction voltages
+indicates valences

**FIGURE 7.11**  Electromotive series.

Cathodic reaction on the copper (reduction), where the electrons are used
Current (ionic) of $Fe^{2+}$ and $(OH)^-$ in the electrolyte

It is possible to arrange many metals and alloys in a series, known as the *galvanic scale*, which describes their relative tendency to corrode. Figure 7.12 is a listing of some of the industrially important metals and alloys, including the ones most frequently encountered. Bearing in mind that a metal located higher on the scale will corrode preferentially and thereby protect a metal lower on the scale from corrosion attack, the example shown in Figure 7.12 may be set up.

## Tendency to Corrode

When copper and zinc are connected, the zinc will dissolve or be corroded preferentially, thus protecting the copper. Since the metal attacked is defined as the anode, the zinc will then serve as an anodic area. The copper will be the cathodic area, or the metal protected. The function of each will be identical to that found in the corrosion battery set up for the rusting of iron. The intensity with which two metals will react in this preferential manner may be measured by the distance between the two metals in the galvanic scale. Thus, between brass and copper there would be only a weak tendency for the brass to corrode; whereas zinc or magnesium would dissolve

| | |
|---|---|
| Magnesium | Copper |
| Zinc | Aluminium Bronze |
| Alclad 3S | Composition G Bronze |
| Aluminium 3S | 90/10 Copper–Nickel |
| Aluminium 61S | 70/30 Copper–Nickel—Low Iron |
| Aluminium 63S | 70/30 Copper–Nickel—High Iron |
| Aluminium 52 | Nickel |
| Low Steel | Iconel (nickel–chromium alloy) |
| Alloy Steel | Silver |
| Cast Iron | Type 410 (Passive) |
| Type 410 (Active) | Type 430 (Passive) |
| Type 430 (Active) | Type 304 (Passive) |
| Type 304 (Active) | Type 316 (Passive) |
| Ni-Resist (corrosion-resisting, nickel cast iron) | Monel (nickel–copper alloy) |
| Muniz Metal | Hastelloy (Alloy C) |
| Yellow Brass | Titanium |
| Red Brass | Gold |
| | Platinum |

**FIGURE 7.12**  Galvanic series.

very readily to protect silver. It may be noted in Figure 7.12 that magnesium is at the top of the scale and will have a tendency to corrode in preference to any other metal shown on the galvanic scale. Conversely, platinum, which is extremely inert, never corrodes preferentially.

## Rate of Corrosion

While the tendency to corrode depends on the *kinds* of metal coupled together, the *rate* at which the corroding anode is attacked depends on the relative area of anodes and cathodes hooked together. Thus, in Figure 7.13, if we couple a small magnesium anode to a large area of steel (as in protection of a ship's hull), the anode area (being small compared to the cathode area) will corrode very readily. This is due to the entire impact of the galvanic current being concentrated on a small area of active metal.

Conversely, if the cathode area is small compared to the anode area, the corrosion of the anode will be relatively slow, since the demand on the anode is spread thin and any local spot loses little metal. The areas of each metal involved are those in electrical contact and not just the areas of metal in physical contact. The area of metals in electrical contact will be determined by those areas in contact with an external conductive circuit (electrolyte).

## Rivets as Fasteners

The use of rivets of one metal to fasten together plates of a different metal is an excellent example of the possible effects of galvanic corrosion. (See Figure 7.14.) When

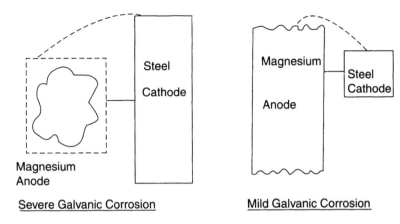

**FIGURE 7.13** The rate of galvanic corrosion is dependent on the relative area of the anode and cathode.

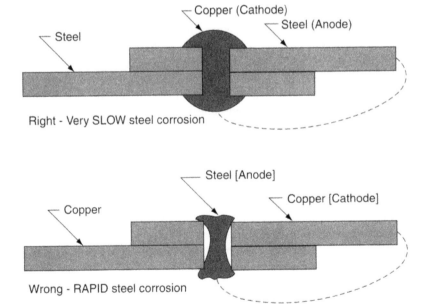

**FIGURE 7.14** Effects of galvanic corrosion shown in rivets.

steel plates (anodes) are fastened with copper rivets (cathodes), only a very slow corrosion of the steel occurs, since the galvanic corrosive effect is spread out over a large area of steel. On the other hand, if copper plates (cathodes) are fastened with steel rivets (anodes), rapid rusting of the rivets will occur. The small area of rivets will be attacked by all the galvanic current generated by the large copper plates and could corrode rapidly.

## 7.1.6  Methods of Corrosion Control

The tendency of metals to corrode is a natural occurrence that must be recognized as inevitable. The corrosion engineer's job is to control these destructive effects at minimum cost. There are five principal methods in use:

1. Altering the environment
2. Using corrosion-resistant materials of construction
3. Cathodic protection
4. Overdesign of structures
5. Placing a barrier between the material and its environment

Each of these methods has certain characteristic advantages and disadvantages and certain areas of use where it is most economical. Since an industrial plant is a composite of many and various types of corrosion areas, no single method should be classified as a universal cure-all. Each situation must be studied individually and a decision reached based on such factors as available downtime, the possibility of equipment obsolescence, operating temperatures and cycles, appearance, environment, and so forth. For each separate corrosion problem, it is necessary to weigh these individual factors and to pick the corrosion tool that provides the most economical means of protection.

Often this choice must be made in the design stage of a plant, so it is important that the corrosion or maintenance engineer be closely consulted during this period. If this practice is not followed, the architects and the engineers may find that they have eliminated various economical weapons of corrosion control from the arsenal.

### Altering the Environment

Altering the environment usually involves controlling accidental discharge of corrosion vapors, or the addition of inhibitors to liquids in a closed system. The use of chemical inhibitors is normally restricted to water supply or water circulating systems, steam and condensate lines. and brine systems. Since they are strictly in immersion solution, their usefulness in the maintenance field is definitely limited. In addition, proper caution must be exercised in the selection of quantities and types of chemicals. Improper selection or maintenance of inhibitor systems can often accelerate rather than retard corrosion. However, if used properly within their limited scope, they provide a simple and relatively low-cost solution to corrosion control.

### Corrosion-Resistant Materials

Largely because of their price and structural qualities, steel and iron are the most widely used metals in industrial construction. Unfortunately, as has been previously discussed, these materials tend to corrode or revert to their oxides more rapidly than other types of metal. Therefore, the corrosion or design engineer may turn to the more inert metals or alloys to retard the corrosion process. In extremely severe exposures, this may be the only feasible answer. High-temperature operations combined with highly corrosive chemicals may produce a situation too severe for any other type of structural material or protection. In such cases, the relatively high initial cost of these metals is easily justified by their long life span.

Among the most common metals alloyed with steel or iron are chromium, copper, nickel, and molybdenum. Of the metals used in their natural state, aluminum is the most reasonably priced and widely used, while such rare metals as titanium and tantalum are employed only under the most severe conditions. The decision to use this type of protection versus other means of control will depend to a great extent on the severity of exposure and the ultimate cost of the alternative methods. In the majority of plant maintenance, exposures are only mildly corrosive and the use of alloys and rare metals as construction materials is not economical.

In addition to alloys, many corrosion-resistant plastic materials are now available. Thermoplastic materials such as polyvinyl chloride and polyethylene are used in pipe and fume ducts. Glass-reinforced epoxy and polyester compounds are employed in process piping with a higher temperature limit, and can also be used in tanks or reaction vessels.

## Cathodic Protection

A third tool available to the corrosion engineer is cathode protection. We have discussed how two metals can be coupled together to produce galvanic corrosion. While we are often concerned with reducing the corrosion of the active metal, it is possible to take advantage of the fact that the cathode metal is protected while the anode corrodes. By deliberately coupling two dissimilar metals together, we can prevent corrosion of the less active (cathode) at the expense of the other metal.

Therefore, to protect a steel surface, we would choose a more active metal, that is, one higher on the galvanic scale. Magnesium is commonly used for this purpose. When it is coupled electrically to steel, a magnified corrosion battery is set up in which the magnesium, because of its greater activity, becomes the anode and the steel the cathode. In so doing, the magnesium anode corrodes preferentially, leaving the steel cathode intact.

The same results can be achieved by providing the current from an external source applied to the metal we wish to protect. On such systems, generators, rectifiers, or batteries may be used as the DC source and, to prevent rapid disintegration of the anode, usually an inert metal will be selected. This electrical current and voltage must be of the correct amount—too little or too much will worsen the corrosion problem.

In the protection of marine equipment, hot-water tanks, and underground and underwater pipe lines, cathodic protection has found its greatest use. Simplicity is certainly one of its prime values, and its effectiveness in the presence of a good electrolyte is unquestionable. In dry or damp areas, its usefulness is limited, and it is replaced or complemented by protective coatings or other methods.

## Overdesign

Overdesign of structures refers to the common use of heavier structural members or thicker plates in anticipation of corrosion losses. This method is often used unwittingly, and excessive plate thickness is frequently specified as a matter of habit or custom, where lighter weights could be used if corrosion were prevented. There is now a trend toward the use of lighter structural members when suitable protective methods are employed.

The principal disadvantage of overdesign, or built-in corrosion allowance, is that neither the exact length of life nor the replacement cost of a corroding material can be predicted. The cost and the effectiveness of other methods can be determined much more accurately.

## Barriers

Barriers are corrosion-resistant materials that can be used to isolate a material of construction, such as steel or concrete, from a corrosive environment. Examples are acid-proof brick or tile in conjunction with suitable cements, grouts, and bedding compounds; plastic sheeting; troweled-on resinous membranes; and spray-applied protective coatings. These barriers are the principal weapons in the arsenal of the corrosion engineer. Although each has its own area of use, careful analysis of an individual corrosion problem is necessary in choosing the most effective and most economical system.

Not only is it important to select the proper barrier for a given situation, of equal importance is the proposed surface preparation and the application technique. A variety of equipment and procedures is available for these purposes, and the ultimate performance of any coating or lining hinges on the correct decision and the proper follow-up during the application. In the following sections, these points will be thoroughly investigated. Faying surfaces of two metals presents one of the most difficult areas for protective coatings, and is even worse when friction is involved. Use of inert barrier sheets may be most effective—even when there may be welding performed later. Another difficult task is in the proper coatings of steel springs, where friction is inherent in their function. A zinc-rich primer with an ethyl silicate carrier may be effective.

## Corrosion Maintenance Plan

The importance of good design and proper protective coating systems cannot be over-emphasized. However, an element sometimes overlooked is the maintenance plan. This can start with inspections to identify the type and location of corrosion, and to establish the beginning of a database. Problems can then be classified as to their severity, as very urgent, or to be corrected next month, or next year, as an example. This permits the preparation of a comprehensive maintenance plan, where all items subject to corrosion can be scheduled for maintenance at a specific interval—and eventually on a specific date.

## 7.1.7 Bibliography

Baumeister, Theodore, *Marks Handbook for Mechanical Engineers*, McGraw-Hill, New York, 1979.

Bendix Field Engineering Corporation, *Corrosion Control*, NASA report MG-305, under NASA contract NAS5–9870, 1968.

Gillett, H. W., *The Behavior of Engineering Metals*, John Wiley, New York, 1951.

The National Association of Corrosion Engineers (NACE), P.O. Box 218340, Houston, TX 77218–8340, publishes several books and articles devoted to corrosion, including:

Munger, C. G., *Corrosion Prevention by Protective Coatings*

Van Delinder, L. S., *Corrosion Basics—An Introduction*

The Steel Structures Painting Council (SSPC), 4516 Henry Street, Pittsburgh, PA 15213, also publishes books and other information, including the following:

*Steel Structures Painting Manual, Vol. 1, Good Painting Practice*

*Steel Structures Painting Manual, Vol. 2, Systems and Specifications*

## 7.2 CLEANING AND PREPARING METALS FOR FINISHING

Frank Altmayer, president, Scientific Control Labs, Inc., Chicago, Illinois

### 7.2.1 Preparing Metals for Plating

Before any part can be electroplated, it must be properly prepared. In fact, the secret to being a successful electroplater or finisher is knowing how to clean and prepare a substrate for coating. An improperly prepared substrate typically results in poor adhesion of the plated metal, resulting in peeling or blistering. When prepared properly, the plated metallic layer will have about the same adhesion to the substrate as the individual metal atoms within the substrate have to themselves. There are some exceptions to this, notably plated metals on plastics, and metals that have a perpetual layer of tenacious oxide, such as aluminum, magnesium, titanium, and stainless steel.

Preparing a part for plating is typically performed in a series of tanks and rinses referred to as the *preplate cycle*. The preplate cycle is typically customized to the type of substrate that is to be plated. It may involve degreasing the part, then cleaning and acid pickling it, or it may be far more complicated. The process steps that are common to most parts are discussed below.

### 7.2.2 Precleaning for Plating

Most metal parts are covered with grease or oil during their manufacture, by operations such as stamping, drilling, forging, buffing, and polishing. Heat-treated steel parts may have a "scale" of iron oxides (rust) that would interfere with the appearance and adhesion of the plated coating. Other ferrous-based parts may have some rust due to unprotected storage. This "soil" must be removed by the metal finisher to avoid contamination of the processing solutions and to obtain adhesion of the coating to the plated part. *Precleaning* refers to the processing a metal finisher may perform on parts before they are routed through the plating line. Precleaning is typically accomplished in one or a combination of the following procedures.

## Vapor Degreasing

A vapor degreaser typically consists of a stainless steel tank with a compartment at the bottom for boiling one of several solvents, such as trichloroethylene, perchloroethylene, methylene chloride, Freon. or 111—Trichloroethane. The boiling solvent creates a vapor zone within the walls of the tank. The vapors are condensed using cooling coils near the top of the tank. There are several variations of how to vapor degrease, but the basic principle involves hanging the greasy parts in the vapor zone on stainless steel hooks or wire. Larger degreasers for small parts are automated, and the parts enter in steel trays and are routed through the equipment automatically. Since the parts entering the degreaser are cooler than the vapor, the solvent condenses on the parts and flushes off the oil/grease. The parts emerge from the degreaser in a relatively dry state, although some solvent may be trapped in pockets or drilled holes. The solvent/oil mixture returns to the boiling chamber, where the solvent is reboiled, and the oil/grease remains in the mix. Eventually, the oil/solvent mix must be removed, replaced, and disposed of. (See Figure 7.15.)

Since vapor degreasing equipment utilizes organic solvents, and since many of these solvents are either currently banned, to be banned, or possibly banned in the future, many companies are eliminating vapor degreasers and substituting parts-washing systems that use aqueous cleaners. These systems are designed to

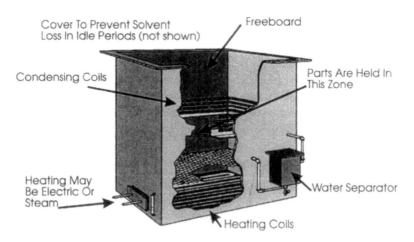

**FIGURE 7.15** Cutaway of typical vapor degreaser used for removing organic soils.

remove oil and grease from a specific type of substrate. Parts washers designed to degrease multiple types of substrates invariably cause problems with one of those substrates.

Vapor degreasing using presently exempted solvents and modern systems designed to emit extremely low amounts of solvent are available today and may allow vapor degreasing to continue in the future. For removing waxes and polishing/ buffing compounds, vapor degreasers currently have little or no competition.

### Pickling/Descaling/Blasting/Shot Peening

Heavily rusted or scaled parts need to be processed through an operation that removes the heavy oxide from the surface of the part. There are a number of ways to accomplish this, including blasting with Dry Ice, nutshells, sand, or other grit; pickling in strong solutions of acids; or descaling in an alkaline-descaling solution consisting of concentrated sodium hydroxide and potassium permanganate at high temperature. Pickling and descaling are normally performed off-line, because they are time-consuming operations.

Steel and stainless steel parts that have been subjected to heat-treating processes or are designed to carry high tensile loads are typically shot-peened prior to finishing to relieve surface stresses that can enhance hydrogen pickup and result in hydrogen embrittlement. The peening process also imparts a slight compressive stress into the surface of the part, enhancing fatigue resistance. Shot-peening is conducted using equipment designed to impact the surface of the parts with media that may consist of steel, stainless steel, or other composition of shot. Impact force, medium geometry, and composition are all important parameters to control.

### Soak Cleaning

A soak-cleaning tank typically consists of a steel tank containing a cleaning solution consisting of strong alkalies and various other ingredients and detergents mixed with water at temperatures from 160 to 200°F. The parts to be cleaned are racked or put in plating barrels and are immersed in the cleaning solution; the oil and grease are either emulsified or converted to soaps through saponification. The parts emerge from the soak cleaner coated with hot cleaner, which then is either rinsed off before further processing or is dragged into the next process. Soak cleaners are normally incorporated into the plating process line and are usually the first tank the parts go into, so they may also be considered part of the preplate operations described below. (See Figure 7.16.)

## 7.2.3   Preplate Operations

Precleaning removes the bulk of the oils, greases, rust, scale, or other soils present on the parts. In the old days, when most plating solutions contained large concentrations of cyanide, precleaning was often all that was required prior to plating. Modern metal-finishing operations have replaced many cyanide solutions with non-cyanide chemistries and have reduced cyanide concentrations in those processes

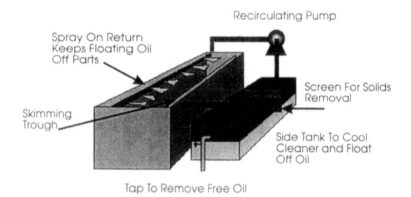

Recirculating Pump

Spray On Return
Keeps Floating Oil
Off Parts

Screen For Solids
Removal

Skimming
Trough

Side Tank To Cool
Cleaner and Float
Off Oil

Tap To Remove Free Oil

**FIGURE 7.16**  Typical soak-cleaning tank with continuous oil removal.

where cyanide remains, so precleaning is not enough to prepare the parts for plating.
The preplate operations must be tailored to the type of metal processed and the conditions of the surface of the parts. The object of the preplate process is to remove the
last traces of surface soils and to remove all oxides from the surface. The following
preplate processing steps are typical.

## Cleaning

Degreased parts still require additional cleaning to remove traces of soils left
behind by the precleaning step. This is normally accomplished by an electrocleaning operation. Electrocleaning systems consist of a heated steel tank that contains
a solution similar to the soak cleaner. The tank is equipped with a rectifier and
steel or stainless steel electrodes hanging from bus bars on either side of the tank.
Cleaning is normally the next process step after degreasing or soak cleaning. If the
parts were vapor degreased, they are either racked on plating racks or scooped/
shoveled into plating barrels for cleaning. The racks or barrels are immersed into
the electrocleaner and a DC current is passed through the parts. The current decomposes the water into two gases, oxygen and hydrogen. It is these gases, which are
discharged in finely divided bubbles, that do the cleaning. If the parts to be cleaned
are negatively charged during this process, then hydrogen bubbles (commonly
referred to as "direct" cleaning) are generated on the parts. If the parts are positively charged (commonly referred to as "reverse" cleaning), then oxygen bubbles
perform the cleaning task. In each case the opposing electrode generates the other
gas. The choice of direct versus reverse cleaning depends on the type of metal to be
cleaned and its tendency to react with oxygen to form an oxide, or the tendency for
direct cleaning to deposit smuts. Both direct and reverse cleaning are sometimes
performed either in sequence, with a periodic reverse rectifier, or through use of a
rectifier and a reversing switch. Following electrocleaning, the parts are rinsed in
water. (See Figure 7.17.)

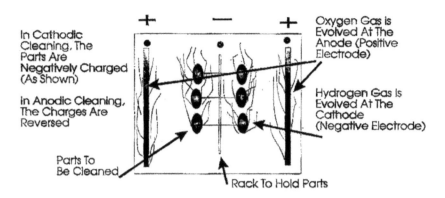

In Cathodic Cleaning, The Parts Are Negatively Charged (As Shown)

In Anodic Cleaning, The Charges Are Reversed

Parts To Be Cleaned

Oxygen Gas Is Evolved At The Anode (Positive Electrode)

Hydrogen Gas Is Evolved At The Cathode (Negative Electrode)

Rack To Hold Parts

**FIGURE 7.17** Illustration of the electrocleaning process.

## Acid Dip/Pickeling

Parts that have been electrocleaned still have a thin alkaline film remaining, even after prolonged rinsing. This film must be removed for adequate adhesion to take place in the plating tank. Additionally, the parts may have a thin oxide film, either formed during electrocleaning or formed by exposure of the clean metal to air. This oxide must also be removed in order to obtain adequate adhesion of the plating. Lastly, some metals contain alloying elements that interfere with good adhesion. An example is lead added to brass to enhance the machining properties of the brass. The lead in the brass forms oxides that are not removed by the acids that are normally used before plating, such as sulfuric and hydrochloric. A special acid must be used to remove lead oxide from the surface of such brasses. The acid dip must therefore be of a chemistry that will neutralize alkali and remove all surface oxides present on the part to be plated. After the acid dip, the parts are rinsed in water.

## Special Dips

Some parts are made of metals that re-form oxides as soon as the metal is exposed to air. An example is aluminum. Aluminum parts can be cleaned and acid dipped before plating and the plating will still not adhere, because the aluminum forms an oxide by reacting with the air as soon as the part is removed from the acid, rinsed, and exposed to air. A special dip is therefore needed to prevent this from happening. The cleaned and acid-dipped aluminum is dipped into a solution of sodium hydroxide and zinc oxide (often other ingredients are added) and water. In this solution, a controlled galvanic reaction occurs, where some of the aluminum dissolves and, at the same time, some of the zinc coats the aluminum with a very thin film. The part that leaves this dip (called a zincate) is now coated with zinc, so there is no aluminum surface to react with the air.

## Strikes

Some parts are made out of metals that react galvanically with certain plating solutions. An example is a zinc die casting or an aluminum part that has been dipped into a zincate. If we want to plate the zinc (or zinc-coated) part, most plating solutions will chemically or galvanically attack the zinc. We therefore must use a specially designed solution called a *strike* to apply a thin protective coating of metal that will not react with the plating solutions. For zinc, or zincated aluminum, such a strike typically is a cyanide copper strike solution (described later). Some metals cannot be adherently plated without first applying a thin strike deposit from a specialized strike plating solution. An example is stainless steel, which has a rapidly forming oxide that must be simultaneously removed in a special nickel strike solution, while a thin film of nickel is deposited over the stainless steel to prevent the re-formation of the oxide. The nickel strike solution is purposely formulated to yield a thin deposit while generating a large amount of hydrogen that reacts with the oxide on the stainless steel.

## Postplating Processes

Some plated parts are further processed to yield additional corrosion protection or to change the color of the deposit. Examples of such further treatment include the application of waxes or lacquers to enhance tarnish resistance and chromate conversion coatings following zinc, cadmium, or other plated deposits to yield chromate films that range in color from transparent to olive drab green. Brass plating is often treated with various chemical solutions to turn the brass to different colors ranging from green to black (even red is possible). All such subsequent treatments typically involve dipping the rack or plating barrel in one or more chemical solutions in various tanks and then rinsing off those solutions. Such solutions often contain ingredients such as nitric acid, sodium dichromate, selenium, arsenic, antimony, or other hazardous ingredients. The processing tanks and associated rinses may be incorporated into the plating line, or the operation may be carried out off-line.

## 7.2.4   Typical Cleaning Process Sequences for Plating

The following are typical processing sequences for commonly plated metals (vapor degreasing is not included, but may be required).

## Leaded Brass

| Process Step | Temperature | Time (sec) |
|---|---|---|
| Cathodic clean | 140 | 10–30 |
| Rinse | | |
| Reverse clean | 140 | 10 |
| Rinse | | |
| Fluoboric acid | 70 | 5–20 |
| Rinse | | |

| Process Step | Temperature | Time (sec) |
|---|---|---|
| Copper strike | 140 | 180–300 |
| Rinse | | |
| Copper plate | 140 | As needed |
| Rinse/dry | | |

*Note:* For nonleaded brass and copper or copper alloys, substitute acid salts or sulfuric acid for fluoboric acid.

## Zinc Alloy Die Castings

| Process Step | Temperature | Time (sec) |
|---|---|---|
| Soak clean | 140 | 120 |
| Anodic clean | 140 | 5–10 |
| Rinse | | |
| Dilute acid | 70 | To first gassing |
| Rinse | | |
| Copper strike | 140 | 180–300 |
| Rinse | | |
| Plate | | |

*Note:* A high pH nickel plate may be an alternate to the cyanide copper strike, but may pose some operational problems.

## Case-Hardened/High-Carbon Steel

| Process Step | Temperature | Time (sec) |
|---|---|---|
| Soak clean | 180 | 120 |
| Anodic clean | 180 | 120 |
| Rinse | | |
| Hydrochloric acid | 70 | 30 |
| Rinse | | |
| Cathodic clean | 180 | 120 |
| Rinse | | |
| Hydrochloric acid | 70 | 30 |
| Rinse | | |
| Anodic etch (25% Sulfuric) | 70 | 60 |
| Rinse | | |
| Plate | | |

*Note:* A Woods nickel strike or sulfamate nickel strike may be substituted for anodic etch.

## Aluminum or Magnesium Alloys

| Process Step | Temperature | Time (sec) |
|---|---|---|
| Soak clean | 140 | 30 |
| Rinse | | |

| | | |
|---|---|---|
| Nitric acid | 70 | 20 |
| Rinse | | |
| Zincate | 70 | 10 |
| Rinse | | |
| Copper strike | 70 | 180–300 |
| Rinse | | |
| Plate | | |

*Notes:* Add bifluoride salts to nitric acid if the aluminum alloy contains silicon. If the parts are magnesium, substitute a solution of 10% ammonium bifluoride in 20% phosphoric acid for the nitric acid, and substitute a pyrophosphate zincate for the normal zincate. Pyrophosphate zincate contains 1–1.6 oz/gal zinc sulfate plus 10–12 oz/gal sodium pyrophosphate. The bath operates at 170–180°F, and the immersion coating forms in 3–5 min.

Zincate may also be applied in "double step." The first zinc coating produced is dissolved in nitric acid. The part is then rinsed and zincate coated a second time. For unexplained reasons, this two-step process often enhances adhesion.

## 400 Series Stainless Steel, Inconel, Hastelloys

| Process Step | Temperature | Time (sec) |
|---|---|---|
| Soak clean | 180 | 60–120 |
| Rinse | | |
| Cathodic clean | 180 | 60–120 |
| Rinse | | |
| Hydrochloric acid | 70 | 30 |
| Rinse | | |
| Anodic in Woods nickel strike | 70 | 10 |
| Cathodic in Woods nickel strike | 70 | 30 |
| Rinse | | |
| Plate | | |

## 300 Series Stainless Steel, Monel, Tungsten Steel

| Process Step | Temperature | Time (sec) |
|---|---|---|
| Soak clean | 180 | 60–120 |
| Rinse | | |
| Cathodic clean | 180 | 60–120 |
| Rinse | | |
| Hydrochloric acid | 70 | 30 |
| Rinse | | |
| Anodic in sulfuric acid 50% | 70 | 10 |
| Rinse | | |
| Cathodic in Woods nickel strike | 70 | 30 |
| Rinse | | |
| Plate | | |

## Beryllium and Tellurium Copper Alloys

| Process Step | Temperature | Time (sec) |
|---|---|---|
| Brite dip | 70 | 10–20 |
| Rinse | | |
| Hydrochloric acid 50% | 70 | 10–20 |
| Rinse | | |
| Ammonium persulfate 8–32 oz/gal | 70 | 15–45 |
| Rinse | | |
| Cyanide copper strike | 140 | 180–300 |
| Rinse | | |
| Plate | | |

*Note:* Brite dip contains 2 gal sulfuric acid plus 1 gal nitric acid plus 1/2 fluid oz hydrochloric acid, no water. Optional substitute: 20–30% sulfuric acid at 160–180°F.

## Bronze Alloys Containing Silicon or Aluminum

| Process Step | Temperature | Time (sec) |
|---|---|---|
| Cathodic clean | 180 | 60–120 |
| Rinse | | |
| Hydrochloric acid 15% | 70 | 30 |
| Rinse | | |
| Nitric acid 75% containing 10 oz/gal ammonium bifluoride | 70 | 2–3 |
| Rinse | | |
| Copper strike | 140 | 180–300 |
| Rinse | | |
| Plate | | |

## Titanium Alloys

| Process Step | Temperature | Time (sec) |
|---|---|---|
| Blast clean | | |
| Hydrochloric acid 20% | 70 | 30 |
| Rinse | | |
| Electroless nickel | | |
| Plate | 200 | 300–600 |
| Rinse | | |
| Diffuse | 1000 | 1800 |
| Sulfuric acid 20% | 70 | 30 |
| Rinse | | |
| Plate | | |

## 7.2.5  Cleaning and Preparing Metals for Painting

While the same cleaning processes detailed for electroplating could theoretically be employed to prepare metals for painting, anodizing, and other metal-finishing processes, the type of finishing and the base metal to be finished will usually dictate some variation in the cleaning procedure or method.

To yield maximum performance of painted films on metallic substrates, we must overcome surface problems detrimental to good adhesion and corrosion resistance. If a metal part is simply cleaned and painted, the smooth surface of the cleaned metal does not allow any anchoring points for the paint to hold on to, should it become chipped or scratched. If such a painted metal part is chipped or scratched in service, the resulting corrosion between the paint film and the metal substrate can lift the paint off in large sheets. This is called *creepage* and leads to rapid failure of a painted part. We must therefore utilize a preparation cycle that not only cleans the surface of the base metal, but also creates a surface that is conducive to enhanced adhesion, even after the paint becomes damaged (scratched or chipped). Such enhancement is obtained when a preparation cycle first cleans the surface and then deposits a porous film that will allow the paint to seep into the pores and anchor itself onto the surface. The mechanism used to produce the porous surface will depend on the metal substrate.

### Steel Substrates

The most common method of cleaning and preparing steel substrates for painting is by employing a five-stage washer. This equipment first cleans off surface oils and greases using alkaline cleaning. Following a rinse, the steel surface is either sprayed or immersed in an acidic chemical solution that converts the surface of the steel into a crystalline material. Depending on the chemicals employed, the crystals can be thin and composed mostly of iron phosphate, or moderately heavy and composed mostly of zinc phosphate. Following a rinse, the crystalline surface is sealed using a dilute chromate solution. The sealing process neutralizes residual acidity and imparts a small amount of hexavalent chromium into the crystals, which enhances the adhesion of the paint to the crystals.

### Aluminum and Magnesium Substrates

There are numerous methods for preparing aluminum and magnesium substrates for painting (after a mildly alkaline cleaning to remove surface oils), including use of phosphoric-based primers, application of phosphates, and anodizing. The most common method of preparing aluminum and magnesium surfaces for painting is to clean the surface with mildly alkaline cleaners, followed by a preparation cycle that results in the conversion of the surface to a chromate film.

For aluminum substrates, the preparation cycle following alkaline cleaning may employ an alkaline etching step, and will usually employ chemicals for desmutting the aluminum prior to application of the chromate. The chromate film may be of either the chromate–phosphate or the chromate–oxide type. The chromate–phosphate film can be applied by spray or immersion, yielding an iridescent green to gray film. The chromate–oxide film can also be created by immersion or spray and is considered to be superior to the chromate–phosphate film in corrosion-resistance performance. (See Figure 7.18.) Each of the chromate films may be sealed by immersion in a dilute chromate solution.

Magnesium substrates are typically vapor degreased, cleaned in a mildly alkaline cleaner, acid pickled, and either anodized or immersed in chromate conversion solutions (most of which have been developed by Dow Chemical specifically for use on magnesium).

### Copper and Copper Alloys

Copper alloys can be painted after alkaline cleaning, followed by acid pickling or bright dip, followed by application of a chromate conversion coating.

### Lead and Lead Alloys

Lead and lead alloys require alkaline cleaning (or vapor degreasing), followed by pickling in dilute fluoboric acid. Following rinsing, the parts can be painted directly.

### Stainless Steels and Nickel Alloys

Stainless steels and nickel alloys can be painted after alkaline cleaning (or vapor degreasing). Roughening the surface by grit blasting or sanding will promote adhesion. On stainless steels, passivating treatments or application of a commercial black-oxide conversion coating will also enhance paint adhesion (corrosion is not an issue, since these alloys are corrosion resistant without the paint).

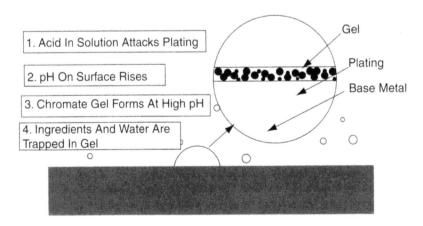

**FIGURE 7.18** Formation of chromate film.

## Zinc Alloys

Zinc alloy parts are typically vapor degreased, followed by alkaline cleaning, rinsing, immersion in a mild acid, rinsing, and then immersion in a chromate conversion solution. Alternately, the zinc part can be converted to a crystalline phosphate film using a five-stage parts washer (see steel; the alkaline cleaner would need to be a milder version).

Galvanized parts can be alkaline cleaned and then primed with paints formulated with phosphoric acid. Alternatively, the galvanized parts can be processed for phosphating.

## 7.2.6  Cleaning for Other Purposes

While it is difficult to provide specific cleaning guidance for purposes other than those already covered, in general, use of any of the above methods prior to resistance welding, soldering, brazing, or other metal-working operations will result in satisfactory work. Metals that are highly reactive, such as aluminum and magnesium, should be cleaned in mildly alkaline solutions, followed by application of chromate films, unless the operation is carried out immediately after cleaning.

## 7.3  ELECTROPLATING

Frank Altmayer, president, Scientific Control Labs, Inc., Chicago, Illinois

## 7.3.1  Introduction to Electroplating

This subchapter on electroplating contains information on the design of products for cost-effective, high-quality plating; the advantages and disadvantages of using various substrates; selection of the proper plating materials and processes; and the use of electroless plating. The various processes and chemistries are definitized, with sufficient detail to permit understanding and application of the many choices. However, consulting with the plater during design of the product and process is highly recommended.

## 7.3.2  Factors in Product Design That Affect Electroplating

### Choosing the Right Substrate

By choosing the right substrate to produce a part, the design engineer can lower final product cost and help protect the environment by reducing the generation of pollutants and wastes. In general, the fewer processing steps that a substrate requires to achieve the final appearance, the less expensive it will be to process. Also, by avoiding certain metals that require strong or unusual chemicals/acids, the design engineer can help generate less waste (also lowering costs).

The designer must be aware of the difficulties imposed on the electroplater by intricate designs and use of metal alloys in the manufacturing of parts that are difficult to plate. The electroplating process does not coat a part uniformly, due to concentrations of electricity that occur at corners and sharp edges. The plater can alleviate this to some extent through use of conforming anodes and shielding. However, this normally drives up costs and increases pollution loading. Exotic alloys create problems for electroplaters in properly cleaning the parts prior to plating (see Subchapter 7.2). The designer should minimize or avoid use of alloys such as stainless steels, inconel, hastalloy, titanium, and combinations of metals, as these are extremely difficult to process for electroplating. If such metals or combinations of metals must be used, the designer should locate and consult with the plater *before* completing his or her design in order to minimize costs and difficulties.

The following are typically plated metals and alloys, problems associated with processing them, and the possible reasons for specifying them.

## Zinc/Zinc Die Castings

Zinc die casting is an inexpensive method of manufacturing a part from a low-cost-basis metal. The die castings are deflashed, polished, and buffed prior to cleaning and plating, adding significant labor costs to the product. The die-casting process must be carefully designed (especially the gating) and controlled (temperature and pressure) to minimize air entrapment, which results in excessive casting porosity. Castings with excessive surface porosity cannot be successfully plated (they will evidence blistering and peeling). Zinc die castings are typically plated first with a cyanide copper strike. This initial plated layer promotes adhesion of subsequent plated metals. Noncyanide solutions for initially plating zinc die castings are available, but are only in the trial phase at this time. If a presently die-casted zinc part can be manufactured using steel, the overall cost of the finished product will normally be reduced due to elimination of manual labor involved in polishing/buffing and increased ease of cleaning and plating.

## Aluminum/Aluminum Die Castings

Numerous aluminum alloys can be used in manufacturing. All aluminum alloys are highly reactive when cleaned, forming surface oxides immediately upon contact with air. These surface oxides inhibit adhesion of electroplated metals. Therefore, aluminum alloys are typically cleaned and processed in a manner that will form an initial immersion coating that does not form oxides upon contact with the air. Two such processes are the zincate process and a proprietary process called the Alstan (trademark of Atotech USA) process. These processes form thin layers of zinc (zincate) or tin (Alstan) on the surface of the cleaned aluminum. The electroplated coating is applied over the zinc or tin immersion coating. There are other methods of obtaining adhesion of electroplate on aluminum, including anodizing the aluminum prior to plating, but the immersion coatings are most commonly used and are the least expensive to apply. Problems with aluminum die castings are similar to those discussed above for zinc die castings. If a part can be manufactured from steel instead of aluminum die cast, significant savings can be realized in finishing the part.

## Stainless Steels

All stainless steels have a tendency to form surface oxides upon contact with the air, and therefore pose similar problems to those discussed above for aluminum. The method of processing to obtain adhesion, however, differs in that stainless steel parts are cleaned and first plated in either a sulfamate nickel strike solution or a Woods nickel strike solution. These solutions are specially formulated to create high volumes of hydrogen gas, which removes surface oxides while simultaneously depositing a thin layer of nickel, which can then be plated with other metals. Because stainless steel has much lower conductivity than most other plated metals, the designed part should allow for a larger number of electrical contact points.

## Common Steel

Common, mild, low-carbon steel is the most easily cleaned and electroplated substrate. However, these steels can be subjected to machining, stamping, drilling, and heat-treating, which can significantly alter the ease of plating. Manufacturing operations that use lubricants and corrosion-inhibiting fluids should use only those types of fluids compatible with the electroplating process. Especially to be avoided are any lubricants/fluids that contain silicones. Corrosion-inhibiting products can also create problems for the plater. Products such as calcium stearate or calcium sulfonate are not easily removed and lead to adhesion problems. Heat-treating steel causes the formation of heat-treat scale, which must be removed through mechanical means to prevent adhesion problems after plating. Heat-treating can also alter the structure of the steel, making it much more difficult to prepare for plating, especially case-hardening. Case-hardened alloys must be cleaned in a double cycle or plated with a Woods nickel or sulfamate nickel strike in order to obtain adhesion. Case-hardened steels and steels hardened above Rockwell C 40 should be stress-relieved at 400°F for 1 hr prior to processing for plating. Avoid use of leaded steels, as such alloys cause considerable adhesion problems for a plater. Leaded steel alloys must be pickled in acid containing fluoride salts to remove surface oxides.

## Cast Iron

Cast iron can contain enough graphite (carbon) to be impossible to plate in a *cyanide* zinc plating solution (due to low hydrogen overvoltage potential on carbon). Such parts will require a copper deposit first, or an *acid* zinc plating process can be used. Cast iron also contains silicon as an alloying element. These alloys will require acid pickling in acid containing fluoride salts or hydrofluoric acid.

## Copper/Copper Alloys

Copper and copper alloys develop an adherent tarnish over prolonged storage that is extremely difficult to remove. Tarnished parts are typically bright dipped to remove the oxide/tarnish present on these alloys. The bright-dipping process produces a significant amount of toxic nitric oxide fumes, and the solution itself must be periodically disposed of, increasing the cost of plating such parts. Parts manufactured from copper/copper alloys should be coated with a tarnish-preventative material such as a chromate or organic antitarnish product. Heat-treated copper alloys should be carefully treated to create as little oxide scale as possible. Tellurium (0.5%) is added to copper to increase machinability. Beryllium (0.5–2.5%) is added to yield high hardness upon heat treatment. Avoid manufacturing parts from tellurium or beryllium containing copper alloys if at all possible, as these alloys are extremely difficult to clean and prepare for plating. Brass often has lead added to enhance machining properties. Leaded brass must be processed through fluoboric acid in the electroplating line, to obtain adhesion. Bronze alloys often contain silicon or aluminum. These alloys require the plater to use a nitric acid dip containing ammonium bifluoride, to promote adhesion of plated deposits.

## Titanium

Parts manufactured from titanium alloys require an electroless nickel deposit that is diffused at 800°F for 1 hr, in order to obtain adhesion. Few platers are equipped to prepare and plate onto titanium parts, so a premium can be expected on the price for plating them.

## Nickel Silver

Nickel silver actually contains no silver (copper 55–66%, nickel 15–30%, balance zinc). Some alloys of nickel silver contain high concentrations of lead. Such alloys must be pickled in acid containing fluorides to remove lead oxide from the surface. Special treatments, such as cathodic charging in sulfuric acid or a Woods nickel strike, are also commonly required.

## Powder Metallurgy Products

Avoid the plating of parts fabricated by the powder metallurgy process, as such parts yield high numbers of rejects due to adhesion problems originating from high levels of porosity, which traps processing solutions. If powder metallurgy must be used, the parts should be vacuum impregnated prior to processing for plating.

## Design Parameters That Affect Plating Uniformity

A part that is to be electroplated should be designed to allow for uniform plating thickness over the part geometry, low liquid retention upon withdrawal from the processing tanks, ease of racking, good electrical contact between the part and the rack, and ease of handling. Following are some guidelines in this regard.

## Part Geometry

Electricity concentrates along sharp edges, ribs, corners, and other points. Conversely, recesses, deep troughs, slots, and other depressed areas are deficient in electric current. The amount (thickness) of plating obtained is directly proportional to the amount of electricity that a specific area of the part obtains during the plating process, and thickness is normally directly related to corrosion resistance. One can therefore expect excess thickness at high current densities and thickness deficiency (poor corrosion resistance) at low current densities. Sharp edges and points should be reduced as much as practical. Gently curving surfaces, grooves, and serrations yield more uniform plating. Edges should be beveled/rounded to a radius of at least 1/64 in. (0.5 mm), 1/32 in. (1 mm) being preferred. The inside/outside edges of flat-bottomed grooves should be rounded off, and their depth should be limited to 33% of their width. Avoid V-shaped grooves. Other indentations should also be limited to a depth of 33% of their width. The depth of *blind holes* (holes that do not go all the way through the part) should also be limited to 33% of their width. Avoid blind holes with very small diameters [less than 1/4 in. (6 mm)]. Apply countersinks to drilled and threaded holes. The height of fins and other projections should be reduced as much as possible, with rounding at the base and tips by a minimum 1/16 in. (1.5 mm) radius.

Avoid manufacturing a part from different types of metals or metals with distinctly different treatments. For example, if a steel stamping is made from mild steel with a case-hardened steel rod attached, during processing, the mild steel will need to be treated with harsh chemicals to be able to plate the rod, and severe etching of the mild steel or poor plating on the rod will result.

## Influence of Manufacturing Processes on Electroplating

Certain methods of manufacturing a product can cause trouble for a plater, which translates to higher costs, rejects, and waste generation.

## Welding

Welding should be performed using material that matches the basis metal as closely as possible. The weld must be pore free. Avoid lap welding, unless the lap can be completely pore free. Pores in welds and porous laps will trap processing liquids, contaminating the process solutions and yielding adhesion and appearance problems. Parts that are welded at high temperatures can develop a scale that will require blasting or pickling to remove.

Weld spatter must be avoided, as these spots will have a reduced amount of corrosion resistance. Spatter should be removed by grinding/sanding.

## Brazing

Brazing yields the same basic problems as welding, except that the creation of dissimilar metals cannot be avoided. In such cases, the plater must be informed as to what was used to braze the components together, so that he or she can adjust the preparatory cycle accordingly or make other modifications to the cleaning of the part.

## Soldering

When possible, soldering should be performed *after* plating. If soldering must be performed before plating, the operation should be carefully controlled in terms of temperature and fluxing, to yield as pore free a joint as possible. Avoid use of silver solder, if possible, as this requires extra preparatory steps by the plater. Remove excessive amounts of flux from the parts before sending them to the plater.

## Drawing

The drawing operation utilizes lubricants that can be either easy or difficult for the plater to remove. Consult the plater you intend to use to determine which lubricants cause problems. Drawing at the wrong speed, with poorly maintained equipment, or without adequate lubrication can create surface fissures that trap plating chemicals or can force lubricant deep into surface defects, yielding blistering.

## Annealing

Annealing at the wrong temperature or in the wrong atmosphere can leave oxides on the surface that are very difficult to remove.

## Case-Hardening

Case-hardening yields a high-carbon surface that yields large amounts of smut upon cleaning and pickling. Careful control of the carbon content of the case to the minimum that will still yield the desired case will reduce plating problems.

## Shot Peening

Use of shot media that leave little residue on the surface of the part is very important to the plater. Glass beads leave residual glass on the surface, which can be very difficult to remove. Cast iron can leave graphite residues on the surface, which can also be difficult to remove. Steel, ceramic, or stainless steel shot usually results in a surface that is easier for the plater to prepare for plating.

## Choosing the Correct Plated Coating

The choice of what metals are to be plated onto a part is usually made by the design engineering team at the manufacturing site. Plated and chemically applied coatings are typically applied to enhance one of the following properties:

Corrosion resistance
Appearance
Abrasion resistance

Intrinsic value
Solderability
Rubber bonding
Wire bondability
Electrical contact resistance
Reflectivity (UV, visible, infrared)
Diffusion barrier
Lubricity
High temperature resistance
Susceptibility to hydrogen stress cracking

The designer will usually find that there is not much choice as to which combination of coatings is the best for his or her application, but usually there are *some* choices. The plating that will provide the best compromise of cost versus the above benefits will usually be the one specified for the part. The designer must utilize knowledge of the galvanic properties of the metal combinations that he or she will be creating, the corrosion characteristics, and any of the other properties mentioned above. Often, a combination of metallic coatings will be applied to achieve a combination of the benefits available.

The designer can utilize any of the numerous military, ASTM, or corporate specifications that are presently available for most any type of part contemplated (contact the U.S. Government Printing Office for copies of specifications and an index). If necessary, the designer will create a unique specification for a part, which will detail the type of plating, the thickness range that each plated layer is to have, special properties (hardness, solderability, etc.) the plating is to have, and the subsequent coatings, if any, that are to be applied after plating.

If the designer is not careful in properly specifying the plating to be performed, there is a great likelihood of the parts failing to meet expectations. For example, the designer cannot simply specify "zinc plating" if he or she is looking for highly conductive zinc plating. There are three process chemistries that can yield zinc plating, and only the cyanide system yields highly conductive zinc. If the plating comes from the acid or alkaline system, it may not have sufficient conductivity for the application. It is always best to consult a knowledgeable person in the field to review a specification before proceeding to produce the part.

The engineer must develop a knowledge of the types of process chemistries that a plated metal/alloy can be produced in, and the properties of that deposit versus those obtained from other chemistries.

Table 7.1 is a summary of plated metal/alloys, the process chemistries most commonly available, features of those chemistries, and the properties most commonly sought from those deposits. There are numerous other metals and alloys that can be plated to obtain specific benefits. These include precious metals other than gold and silver (rhodium, palladium, ruthenium, platinum), and some uncommon "common" metals and alloys such as bismuth, iron, and Alballoy (copper–tin–zinc). There are also composite coatings that can be plated. For example, one can plate a nickel–cobalt alloy containing finely dispersed particles of silicon carbide to enhance

TABLE 7.1  Summary of Plated/Metal Alloys, the Process Chemistry, Features of Those Chemistries, and Beneficial Properties of the Deposits

| Deposit | Available chemistries | Beneficial property of chemistry | Beneficial property of deposit | Common use |
|---|---|---|---|---|
| Zinc | Cyanide | High throw | Pure high conductivity | Hardware Fasteners |
| | Acid | Leveling | Bright appearance | Same |
| | Alkaline noncyanide | High throw | Fine grain | Same |
| Zinc–cobalt | Acid | Leveling | High corrosion resistance | Same |
| Zinc–cobalt | Alkaline | Uniform alloy composition on complex shapes | Same | Same |
| Zinc–nickel | Acid | Leveling | Same | Same |
| Zinc–nickel | Alkaline | Less corrosion on complex shapes | Same | Same |
| Zinc–tin | Alkaline | — | Solderability High corrosion resistance | Same |
| Cadmium | Cyanide | Pure deposit | Solderability Corrosion resistance No mold growth Lubricity | Same |
| Cadmium | Acid | Avoids cyanide | Appearance | Hardware Fasteners Automotive |
| Copper | Acid | Fine grained Can plate onto zinc | Pure | Cables Pennies Carburizing-stop-off Automative |
| Copper | Alkaline noncyanide | Avoids cyanide | — Die castings | Hardware Same zinc |
| Copper | Acid-sulfate | Leveling or high throw | Appearance | Printed circuit boards |
| Copper | Pyrophosphate | High throw | — | Printed circuit boards |
| Brass | Cyanide | High throw | Apperance Hardware Vulcanizing | Furniture |

(Continued)

TABLE 7.1  (Continued)

| Deposit | Available chemistries | Beneficial property of chemistry | Beneficial property of deposit | Common use |
|---------|----------------------|----------------------------------|--------------------------------|------------|
| Bronze | Cyanide | High throw Lubricity | Appearance Bearings | Hardware |
| Nickel | Acid | Leveling Bright | Decorative Hardness | Consumer items Electroforming Electronics Jewelry Tools Automative Plated plastics Printed circuit boards |
| Chromium | Acid-hexavalent | — | Appearance Hardness | Consumer items Wear surfaces |
| Gold | Cyanide | High throw | Pure | Jewelry Electronics |
| Gold | Acid | No Cyanide | Hardness | Jewelry Electronics |
| Gold | Neutral | No Cyanide Low corrosivity | Alloys | Jewelry Electronics |
| Gold | Sulfite | No Cyanide | Pure | Jewelry Electronics |
| Silver | Cyanide | High throw | Pure | Jewelry Consumer items Electronics |
| Silver | Non–cyanide | No cyanide | Pure | Same |
| Tin | Alkaline | High throw | High solderability | Electronics Consumer items |
| Tin | Acid | Bright | appearance | Consumer items |
| Tin-lead | Acid (fluoboric) | High throw | Solderability | Electronics Printed circuit boards |
| Tin-lead | Acid (sulfonic) | Uniform alloy composition over complex shapes. | Solderability | Electronics Printed circuit boards |

abrasion resistance. This, or some other composite, may one day be a substitute for chromium plating.

## Hydrogen Stress Cracking

More than 11 major theories on the mechanism involved in hydrogen stress cracking—commonly referred to as hydrogen embrittlement—have evolved, yet no single mechanism explains the complete characteristics of this phenomenon.

The following background information should be helpful to the designer:

1. Steels are susceptible to hydrogen stress cracking in varying degrees depending on the composition, microstructure, and type/amount of defects.
2. Embrittlement occurs at all strength levels, but is most prevalent in steels with strength levels above 200,000 psi.
3. Under tensile load, hydrogen stress cracking is manifested mainly as a loss of ductility. There is no influence on the yield point or the plastic properties up to the point where local necking starts. Hydrogen apparently prevents the local necking from continuing to the normal value of hydrogen-free steel.
4. In static loading with notched specimens, hydrogen causes delayed failures at loads as low as 20% of the nominal tensile strength.
5. No macroscopic plastic flow occurs during static loading.
6. Unnotched tensile specimens exhibit hydrogen stress cracking at higher yield strengths than notched specimens.
7. Hydrogen stress cracking effects are diminished or disappear at low temperatures (around −100°C) and high strain rates.
8. Embrittlement increases with increased hydrogen content.
9. For a given hydrogen content, embrittlement increases with increased stress concentration.
10. The surface hardness of affected metals is not altered.
11. While it is most critical to the proper functioning of high-strength steels, hydrogen stress cracking also manifests itself in other metals/alloys:
    Copper: Pure copper alloys are not subject to the problem, while alloys that contain oxygen have been reported to be hydrogen embrittled.
    High-temperature alloys: Nickel, cobalt-based, and austenitic iron–based exotic alloys show no apparent susceptibility to the problem, while body-centered cubic alloys containing cobalt or titanium are susceptible.
    Stainless steels: Austenitic alloys (200–300 series) are not susceptible, while martensitic (400 series) alloys are. Ferritic (400 series) alloys exhibit varying degrees of susceptibility depending on the amount of work hardening (work hardening appears to reduce the effects).
    Steels: Alloys with Rockwell hardnesses above C 35 and tensile strengths above 140,000 psi are susceptible to the problem. Additionally, reports

of low-strength, plain carbon steel also being embrittled by hydrogen
are in the literature. Under the right conditions, almost any steel alloy
can exhibit susceptibility.

## Mechanism

Hydrogen embrittlement involves most or all of the following conditions:

1. Stress
2. Adsorption
3. Dissociation
4. Dislocations
5. Steel microstructure/composition
6. Other variables such as strain rate and temperature

The preliminary step is the adsorption of the hydrogen onto the iron surface. The adsorption of hydrogen does not occur unless the iron surface is chemically clean. This is why plain carbon steel cylinders can be used to store high-pressure hydrogen without catastrophic failure.

Once the steel contains hydrogen gas, it must dissociate to nascent hydrogen for embrittlement to proceed. If the hydrogen is absorbed during acid pickling or plating, it is already dissociated to the nascent state. Experimenters have shown that large increases of crack growth occurred when a steel specimen was exposed to partially ionized hydrogen gas versus nonionized.

Under stress, the hydrogen within the steel diffuses to regions of maximum triaxial stress. If microcracks are present, maximum triaxial stress occurs just beneath the crack tip.

Microcracks present may be filled with hydrogen gas that can dissociate, enter the iron-lattice structure, and also diffuse to regions of triaxial stress. If no microcracks are present, hydrogen gas can be transported by dislocations that can pile up at inclusions and second phases forming microcracks. Competing actions therefore deliver hydrogen to areas where the damage can occur: diffusion and mobile dislocations.

By adsorption, hydrogen ions are supplied to areas of high triaxial stress. In the region of high triaxial stress, the formation of dislocations is accelerated. Numerous pileups occur just below the crack tip, forming voids filled with hydrogen gas that can instantly dissociate and diffuse to dislocations outside the void, adding to the growth of the voids.

As the stress intensity decreases due to void growth, decohesion (cleavage) takes over, resulting in catastrophic failure.

Inclusions are not necessary to the mechanism, because dislocation pileups assisted by hydrogen can substitute for microvoids. Steels of low strength are not as susceptible, because these steels are too ductile for the cleavage mechanism to take over from microvoid coalescence. The reversibility of the effect can be explained by the fact that if nascent hydrogen is given enough energy to form gas and diffuse out of the steel before dislocation pileups and diffusion of nascent hydrogen to areas of stress can occur, then embrittlement will not result. Since diffusion plays a major role, low temperatures reduce the effect.

## Finishing Processes Influencing Hydrogen Embrittlement

Any aqueous process that contains hydrogen ions or yields hydrogen on the surface of a susceptible alloy can result in embrittlement. These include:

Acid pickling
Cathodic cleaning
Plating

Plating processes that are not 100% efficient can result in hydrogen embrittlement. Zinc electroplating from cyanide baths results in the most embrittlement, with cyanide cadmium next. This is probably because cyanide plating baths do an excellent job of cleaning the surface of the steel, and the mechanism requires a relatively clean surface.

Chromating usually does not increase embrittlement unless the plated deposit is too thin to accept a chromate film adequately.

## Minimizing Hydrogen Stress Cracking

Although it is not totally preventable, hydrogen embrittlement can be minimized by use of the following techniques:

1. Plate at the highest current density allowable, to improve current efficiency.
2. Use two steps for plated deposits approaching 0.0005 in. or more, with a bake after 0.0002 in. followed by a second plate to the final thickness and a second bake.
3. Stress-relieve all highly stressed steels and any steels with a hardness greater than Rockwell C 40 prior to processing.
4. Bake hydrogen out of the steel after prolonged acid pickling.
5. Avoid cathodic electrocleaning.
6. Use inhibited acids.
7. Substitute mechanical cleaning for chemical cleaning.
8. Keep brightener content in plating baths low.
9. Bake parts as soon as possible after processing, usually within an hour and within 15 min for alloys above Rockwell C 40.
10. Observe the following bake times:
    0.0002 in. or less: 4 hr
    0.0002–0.0003 in.: 8 hr
    0.0003 in. or more: 24 hr

The bake temperature should be 375 + 25—0°F. Baking times must be at temperature: Do not start from cold.

11. Minimize the amount of descaling required by controlling the heat-treatment process.
12. Remove surface defects (tumble, deburr, shot peen) prior to processing to reduce the number of areas of stress.

### 7.3.3  The Electroplating Process

*Electroplating* is a process for coating a metallic or nonmetallic substrate with a metallic coating through the use of a combination of electricity and a chemical solution that includes ions of the metal in the coating.

To conduct the process, we first need to purchase some *hardware*. Simple electroplating hardware consists of:

A rack or barrel to hold parts
A tank
Electric cables or copper bus
A rectifier
Filtration equipment (may be optional)
Agitation equipment (may be optional)
Ventilation equipment (may be optional)
Plating solution
Other processing tanks for cleaning, rinsing, and acid pickling

### Plating Methods

Plating can be performed using any of three main methods of part handling, each requiring different hardware.

### Rack Plating

Rack plating is sometimes referred to as still plating and is used whenever the parts are too large, delicate, or complicated to be barrel plated. Rack plating is much more expensive than barrel plating because of the labor involved in putting the parts on the rack and taking them off after they are processed. Rack plating is performed by hanging the parts to be plated on racks, which are typically plastic-coated copper or aluminum rods, with stiff wires that hold the parts in place, protruding at various intervals. Racks come in numerous designs and are most often constructed by outside vendors and sold to the plater. Some small parts are racked simply by twisting a thin copper wire around them. The wire, with perhaps 20–50 pieces hanging on it, is then handled as a rack. During plating, the part of the rack that makes electrical contact with the part being plated is also plated, so after several cycles these contacts have a lot of metal buildup. The racks are then sent through a stripping solution that removes the excess metal, or the plater physically removes the excess metal using pliers or a hammer. Chemical rack strippers are usually strong solutions of cyanide or acid and can be difficult to waste-treat.

Noncyanide and regenerative strippers are available for some processes, but are expensive to use or strip very slowly. (See Figure 7.19.)

## Barrel Plating

Barrel plating is the most efficient and least costly method. Plating barrels of varied designs are purchased from manufacturers of such products. The basic barrel consists of a hexagonal cylinder, closed at both ends, with perforated walls made of polypropylene. A door, which is held in place with plastic-coated clips, is installed in one wall of the barrel to allow entry and exit of the load. Electrical contact between the saddle on the tank and the parts inside the barrel is made by a copper or bronze rod attached to the barrel, which sits in the electrified saddle. The rod has a cable attached, and this cable is routed inside the plating barrel through the end of the barrel. Sometimes two cables are used, one entering each end of the barrel. The barrel has a hole in the center that allows the cable to enter. The end of the cable, inside the barrel, has a stainless steel ball attached, called a *dangler*. This dangler makes contact with the parts inside the barrel by gravity. There are other methods of making electrical contact inside the barrel, including rods and button contacts, but the dangler is the most commonly used. Parts to be plated are scooped or shoveled into the barrel. The load is often weighed to make certain that the parts are uniformly plated. As a general rule, the barrel is never filled beyond one half of its total volume. As plating proceeds, a motor mounted either on the plating tank or on the barrel turns the barrel at 3–5 rpm, through either a drive belt or a set of gears mounted on the barrel. If the barrel is not rotated during plating, the top of the load will be plated and the bottom part will remain bare. See Figure 7.20 for typical portable horizontal barrels. Other designs that allow automatic loading and unloading, such as oblique and horizontal oscillating barrels, are also commonly employed (see Figure 7.21).

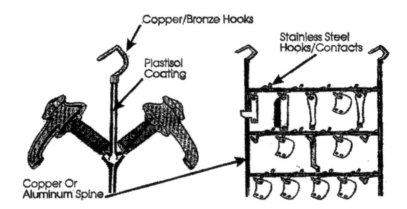

Illustrations Courtesy of Belke Mfg. Chicago IL

**FIGURE 7.19** Parts being prepared for rack plating.

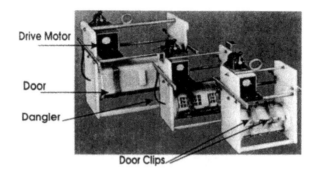

Drive Motor

Door

Dangler

Door Clips

**FIGURE 7.20**  Small plating barrel.

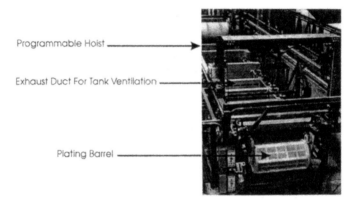

Programmable Hoist

Exhaust Duct For Tank Ventilation

Plating Barrel

**FIGURE 7.21**  Automated horizontal barrel plating line.

Barrel plating can be performed by manually transferring barrels from one station to the next, or through use of automated processing lines.

## Continuous/Reel-to-Reel Method

The continuous or reel-to-reel method of plating is highly efficient and competes effectively against all other methods when the parts are small, uniform, of simple geometry, and amenable to being stamped from a thin strip of metal. This method is used to electrogalvanize (zinc plate) steel strip that is used to stamp automobile bodies and to plate brass or copper strip for stamping electrical connectors for tele-communications. In this method of plating, the parts to be plated consist of long strips of metal that are rolled up on a wheel. The wheel is mounted on the equipment and the strip goes through a sequence of rollers directing it through various process-ing tanks, including the plating tank. The strip may partially dip into the plating tank, or it may be completely immersed. Electrical contact is made through metal brushes, rollers, or by a principle called bipolarity, which does not actually contact the strip.

The strip may travel at speeds ranging from 50 to 1000 or more feet per minute. At the other end of the continuous strip plating line, a second wheel takes up the processed strip. (See Figure 7.22.)

Rack and barrel plating are typically performed by manual and automated methods. In manual operations, platers transfer the racks or barrels from one processing tank to another by hand or with electric hoists. In automated operations, racks or barrels are transferred using programmable robots. Continuous strip plating operations are always automated, as shown in Figure 7.23.

## Brush Plating and Other Methods

Numerous other techniques are also used to electroplate substrates. One of these is called *brush plating*. To perform brush plating, the plating solution is formulated with ingredients that create a paste or thick liquid. A "brush" consisting of an anode

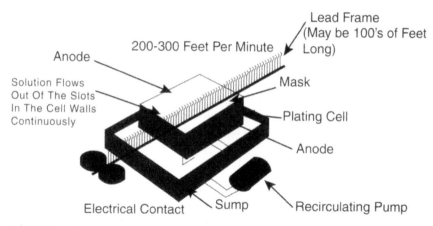

**FIGURE 7.22** Continuous reel-to-reel plating.

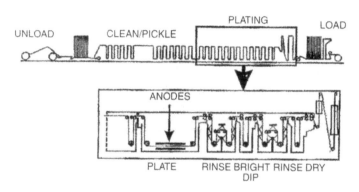

**FIGURE 7.23** Continuous strip plating schematic.

covered with either a sponge or layers of absorbent material is used to hold the plating paste onto the substrate to be plated. The part to be plated is connected to the negative electrode of the rectifier, while the brush is connected to the positive electrode. The area to be plated is stroked with a back-and-forth motion (the area must be cleaned prior to plating). (See Figure 7.24.)

Brush plating is an ideal method for covering selected areas of large parts that must be repaired, or plating large parts without the need for a large tank and large volumes of plating solution. For example, an entire sports car has been gold plated using this method. To immerse even a small fender into a conventional gold plating solution would require a plating tank containing hundreds of gallons of plating solution at over $500 per gallon, while with brush plating, the car was plated using only a few gallons of solution.

Another method of plating is called *out-of-tank plating*. This method is commonly used to plate deep cavities that are difficult to plate in conventional tanks, due to solution-flow problems. In out-of-tank plating, the plating solution is pumped into the cavity and flows continuously out of the cavity, returning to a holding tank, which may hold hundreds of gallons of solution (which keeps the chemical constituents in balance). Plating is performed with an internal anode.

## Hardware

The hardware required to perform plating is typically purchased from a company that specializes in producing this equipment, although some platers produce their own hardware. Let us take a closer look at the hardware used for plating.

## Tank

The plating tank must resist chemical attack from the plating solution. Tanks containing cyanide plating solutions are often made of bare steel. Tanks for other plating

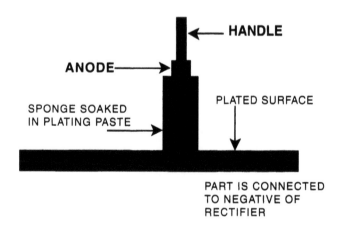

FIGURE 7.24 Schematic of brush plating method.

solutions are typically made of steel with PVC lining, polypropylene, polyvinyl chloride, or polyethylene. Plating tanks should not have any wall perforations below the liquid line, to prevent accidental discharge of contents. A rack plating tank typically has three copper bus bars mounted on top of the tank. One bus is in the center and is used to hang the parts in the plating solution. The other two bus bars are located near the walls of the tank and are used to hang anodes or "baskets" for anodes.

A barrel plating tank has the same anode bus, but there is no center bus. Instead, the tank typically has four "saddles" made of copper or bronze, mounted to the lips of the tank, so that the barrel contact rods can sit firmly in the saddles. At least one of the four saddles has a cable or copper bus attached to it for contact with the rectifier.

## Rectifier

The rectifier for plating converts AC current to DC. The rectifier is typically installed near the plating tank, but it may be located in another room. In either case, cable or a bus bar is used to connect from the positive terminal of the rectifier to the anode bus bar on the tank. The negative terminal of the rectifier is connected by cables or copper bus to the saddle of barrel plating tanks or to the cathode bus of rack tanks. Rectifiers generate heat as a by-product. This heat must be removed either by using a fan (air cooled) or by circulating cooling water through the rectifier (water cooled). The water used to cool the rectifiers can be routed to other plating operations such as rinsing. (See Figure 7.25.)

## Filter

Some plating solutions require continuous filtration; others do not. A general rule is that alkaline solutions can usually operate satisfactorily without a filter, whereas

**Bus Bars**

**FIGURE 7.25** Bank of air-cooled rectifiers located above process tanks.

acidic solutions need filtration to avoid particulates suspended in the solution from being incorporated into the coating, yielding roughness.

## Agitation

Most plating processes require some form of solution agitation to deliver the brightest, densest, most uniform deposit, and to plate at higher current densities without burning. A common method of agitation is to move the cathode rod back and forth in the solution using a motor attached to the cathode rod. This is called *cathode rod agitation*. A second common method is to install an air sparger in the bottom of the plating tank and use low-pressure air bubbles to perform the agitation. A less common technique is to use a prop mixer. Plating tanks for barrel plating are not agitated, because the rotation of the barrel provides sufficient solution movement.

## Ventilation Equipment

Various metal-finishing process solutions can emit vapors, fumes, and mists. To maintain a safe working environment, such emissions are captured using ventilation equipment. There are numerous methods of ventilating a process tank, including side-draft, push–pull, pull–pull, and four-sided ventilation. Highly sophisticated systems may even totally enclose a process tank during use. Depending on the nature of the emissions, the ventilation system may be connected to a system for removing the emissions prior to discharge of the air to the outside of the building. These systems may employ water wash scrubbing, mesh pad impingement systems, or combinations of both, as shown in Figure 7.26.

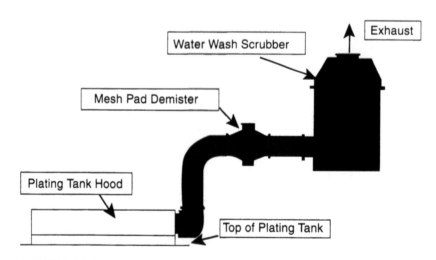

**FIGURE 7.26**  Ventilation of a plating tank.

## Plating Lines/Departments

Except for vapor degreasing, which is normally performed off-line, plating operations are normally incorporated into a sequence of tanks, called a *line*. A plating line may be designed to produce a single coating or a number of coatings. The process line usually contains tanks lined up in sequential order. Automated lines may or may not contain tanks in sequential order.

A zinc plating line, as shown in Figure 7.27, may therefore consist of 13 tanks, each containing a chemical processing solution or rinse water: soak clean, electroclean, rinse, acid, rinse, zinc plate, rinse, bright dip, rinse, chromate, rinse, hot-water rinse, dry. If the line is for barrel plating, each tank may have one or more stations, that is, places to put a barrel. A six-station zinc plating tank can plate six barrel loads of parts at one time. To economize, some shops may have one cleaning line that services several plating lines. There also are tanks for rack stripping, stripping rejects, purifying contaminated solutions, or holding solutions that are used only sporadically. The entire lineup of tanks and lines creates the shop layout, with parts entering the plating department from one direction, traveling through the process lines, and then leaving the plating department.

## Electroplating Process Summary

The plating step may be a single plate or a series of different deposits. If a series of deposits is to be applied, there usually is a rinse and an acid dip between the different plating steps. For example, a zinc die casting may be plated with a cyanide copper strike, followed by a cyanide copper plate, followed by a semibright nickel plate, followed by a bright nickel plate, followed by a thin deposit of chromium. No rinsing or acid dipping may be required between the cyanide copper strike and the cyanide copper plate, or between the semibright nickel and the bright nickel, because the chemistries of the sequential baths are similar. However, there will be a rinse and an acid dip between the cyanide copper plate and the semibright nickel plate, because one solution is alkaline and the other is acidic. There will be a rinse between the bright nickel plate and the chromium plate, to prevent contamination

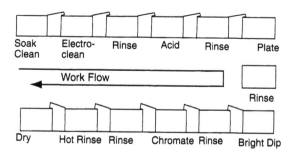

**FIGURE 7.27**  Zinc plating line showing typical process sequences.

of the chromium plating solution with nickel. There may or may not be an acid dip, since both solutions are acidic. After all plating has been performed, the parts are rinsed and dried before packaging them for shipment.

The plating solution normally contains water and a number of ingredients that determine if the coating produced is dense, bright, hard, a certain color, or has a number of other desirable properties that can be obtained through chemistry. It also contains an ingredient that forms ions of the metal to be plated when added to water. For example, one ingredient in a Watts nickel plating solution is nickel sulfate. When nickel sulfate is added to water, it dissolves and forms nickel ions and sulfate ions, just as when one adds salt to water, it dissolves to form sodium and chloride ions. The dissolved nickel ions can be converted to nickel metal by passing a direct current through the plating solution using a rectifier (which converts AC current to DC), the anode as the electrode with positive polarity, and the part to be coated as the electrode with negative polarity. The conversion of the nickel ions to nickel metal will occur on the surface of the negative electrode, where excess electrons (which make the electrode negatively charged) reduce the nickel from ions to metal. While nickel ions are converted to metal at the cathode (negative electrode), at the anode, nickel metal is converted from the metal back to the ions. Ideally, for each nickel atom plated out at the cathode, a new ion is formed at the anode to replace the one plated out. This is the case in most plating solutions, but with some solutions the anode is not converted to metal ions.

### 7.3.4  Typical Plating Solution Chemistries

We will now briefly discuss the ingredients of the most commonly used plating solutions.

### Zinc

Zinc is the most commonly plated metal, normally applied over ferrous substrates for the purpose of enhancing the corrosion resistance. Zinc can be plated from a number of different chemistries, but the three most common are cyanide, alkaline noncyanide, and acid chloride.

### Cyanide Zinc

Cyanide baths are favored when high thicknesses are required or when parts are to be plated and then deformed. Still widely used, cyanide zinc plating solutions contain

Zinc cyanide: 2–8 oz/gal
Sodium cyanide: 1–6 oz/gal
Sodium hydroxide: 10–14 oz/gal
Organic brightener: as required
pH: 14+

### Operating Conditions

Temperature: 70–90°F
Anodes: zinc/steel baskets
Plating current density: 10–50 ASF
Agitation: cathode rod/none
Filtration: not required
Color of solution: straw colored, sometimes has floating oily layer
Odor of solution: aldehyde

## Alkaline Noncyanide

Alkaline noncyanide baths can be substituted for cyanide baths without the need for major equipment modifications. The deposits tend to become brittle as the thickness increases, and some parts that have been heat-treated yield poor adhesion.

### Ingredients

Zinc oxide: 1–2 oz/gal
Sodium hydroxide: 10–15 oz/gal
Sodium carbonate: 0–3 oz/gal
Additives: as required
pH: 14+

### Operating Conditions

Temperature: 70–90°F
Anodes: zinc/steel baskets
Plating current density: 5–45 ASF
Agitation: cathode rod/none
Filtration: not required
Color of solution: pale yellow
Odor of solution: sharp odor

## Acid Chloride

Acid chloride baths yield the brightest deposit. The process requires excellent cleaning and corrosion-resistant equipment, however, and thick deposits tend to be brittle.

### Ingredients

Zinc chloride: 4–12 oz/gal
Potassium chloride: 14–20 oz/gal
Ammonium chloride: 3–5 oz/gal
Boric acid: 4–5 oz/gal
pH: 4.5–6.0

**Operating Conditions**

Temperature: 70–115°F
Anodes: zinc/titanium baskets
Plating current density: 10–150 ASF
Agitation: air or cathode rod
Filtration: required
Color of solution: pale yellow
Odor of solution: none; sharp with air agitation

## Cadmium

The vast majority of cadmium plating is performed from the cyanide-based chemistry. The sulfate chemistry has made small inroads, but often does not adequately cover heat-treated steel parts that have high surface hardness. The sulfate process also requires a much higher degree of cleaning, or adhesion becomes marginal.

### Cyanide Cadmium Solutions

#### Ingredients

Cadmium oxide: 3–5 oz/gal
Sodium cyanide: 10–20 oz/gal
Sodium carbonate: 3–14 oz/gal
Sodium hydroxide: 2–8 oz/gal
Brighteners/additives: none
pH: 14+

#### Operating Conditions

Temperature: 70–90°F
Anodes: cadmium/steel baskets
Plating current density: 5–90 ASF
Agitation: cathode rod/none
Filtration: not required
Color of solution: pale yellow
Odor of solution: aldehyde

### Acid Sulfate Solutions

#### Ingredients

Cadmium chloride: 1–1.5 oz/gal
Ammonium sulfate: 10–15 oz/gal
Ammonium chloride: 1.5–3 oz/gal
Brighteners/additives: none
pH: 5–6

### Operating Conditions

Temperature: 70–100°F
Anodes: cadmium, titanium baskets
Plating current density: 2–15 ASF
Agitation: cathode rod
Filtration: required
Color of solution: pale yellow
Odor of solution: sharp

## Copper

Copper is plated from three popular chemistries: cyanide, acid sulfate, and pyrophosphate. Recently, patented alkaline noncyanide copper plating processes have been developed and marketed by at least three companies, but these processes are troublesome and expensive to operate.

## Cyanide Copper Process

There are several cyanide copper plating processes, but they can be divided into two basic chemistries: a strike bath and a plate or high-speed bath.

### Ingredients

| Strike plate | | |
|---|---|---|
| Copper cyanide | 1.5–2.5 oz/gal | 4–6 oz/gal |
| Sodium cyanide | 3–4 oz/gal | 8–12 oz/gal |
| Sodium carbonate | 3–14 oz/gal | 3–14 oz/gal |
| Sodium hydroxide | 0–1.5 oz/gal | 2–4 oz/gal |
| Brightener/additives | None | As required |
| pH | 10–10.5 | 12–14 |

*Note:* Potassium salts are most often used in high-speed cyanide copper plating solutions, at approximately the same concentrations.

### Operating Conditions

Temperature: 140–160°F
Anodes: oxygen-free, high-conductivity copper
Plating current density: 10–100 ASF
Agitation: cathode rod
Filtration: required
Color of solution: pale yellow
Odor of solution: pungent

## Acid Sulfate Processes

Two main acid sulfate chemistries are used in electroplating copper. One is termed *conventional* and is often used as an underlayer for plated plastic, or in applications where a high degree of leveling (smoothing of scratches) is desired. The second process is called a *high-throw bath* and is used mostly by printed wiring board manufacturers because of its ability to produce uniform thicknesses on the outside of a circuit board and on the inside of tiny holes drilled into the board.

### Ingredients

|                       | Conventional | HighThrow   |
| --------------------- | ------------ | ----------- |
| Copper sulfate        | 26–33        | 10–14 oz/gal |
| Sulfuric acid         | 6–12         | 20–30 oz/gal |
| Chloride              | 40–80        | 40–80 ppm   |
| Brighteners/additives | Yes          | Yes         |
| pH                    | <1           | <1          |

### Operating Conditions

Temperature: room
Anodes: copper containing 0.02–0.06% phosphorus, bagged
Plating current density: 20–200 ASF
Agitation: air
Filtration: continuous
Color of solution: deep cobalt blue
Odor of solution: no specific smell, inhaled mist may yield sharp odor/burning
    of nose

## Pyrophosphate Copper Solutions

Pyrophosphate copper plating solutions are used almost exclusively by printed circuit board manufacturers. Their major benefits include low copper concentration and the ability to deposit an even thickness over complex geometries, such as the top of a circuit board versus the inside of a drilled hole. The majority of these solutions have been replaced with bright throw acid sulfate systems because the pyro baths have much more difficult chemistries to analyze and control. The baths may be found in some job shops as substitutes for cyanide copper strike baths on zinc die castings or for copper-striking zincated aluminum.

### Ingredients

Copper pyrophosphate: 10–12 oz/gal
Potassium pyrophosphate: 40–45 oz/gal
Potassium nitrate: 1–1.5 oz/gal
Ammonia: 0.15–0.5 oz/gal
Additives: variable
pH: 8–9

### Operating Conditions

Temperature: 125–135°F
Anodes: oxygen free, high conductivity/titanium baskets
Plating current density: 10–90 ASF
Agitation: air
Filtration: yes
Color of solution: iridescent blue/purple
Odor of solution: no specific

## Brass

Currently, there are no commercially viable processes for plating brass and alloys of copper and zinc, other than from a cyanide-based chemistry. Brass is applied mostly for decorative purposes, wherein it is subsequently stained to yield an antique or colored finish. Brass is also applied to enhance adhesion of rubber to steel. Brass tarnishes readily, so most often it is finished off with a coat of lacquer.

### Ingredients

Copper cyanide: 4–8 oz/gal
Zinc cyanide: 1–2 oz/gal
Sodium cyanide: 2–4 oz/gal
Sodium carbonate: 3–14 oz/gal
pH: 10–11.5

### Operating Conditions

Temperature: 125–135°F
Anodes: brass of same alloy composition as plated
Plating current density: 5–15 ASF
Agitation: cathode rod
Filtration: yes
Color of solution: pale yellow
Odor of solution: no specific odor

## Bronze

Bronze (80% copper, 20% tin) can be plated only from a cyanide-based chemistry. The plating equipment is identical to that for copper or brass plating.

### Ingredients

Copper cyanide: 4.6 oz/gal
Potassium cyanide: 10.3 oz/gal
Potassium stannate: 5.6 oz/gal
Potassium hydroxide: 1.7 oz/gal

Sodium potassium tartrate: 6 oz/gal
pH: 12+

### Operating Conditions

Temperature: 150–160°F
Anodes: copper/carburized steel, graphite, or stainless steel
Plating current density: 20–100 ASF
Agitation: cathode rod
Filtration: yes
Color of solution: pale yellow
Odor of solution: pungent

## Zinc Alloys

Alloys of zinc have been the major focus for a good substitute for cadmium plating.
If an alloy of zinc contains a small amount of a more noble metal, such as nickel, tin,
cobalt, or iron, the zinc retains its cathodic relationship with steel, but the alloying
metal reduces the activity of the coating so that it corrodes sacrificially at a slower
rate, thereby enhancing corrosion protection over plain zinc. Numerous zinc alloy
processes are being touted as the best cadmium alternative, including zinc–nickel,
zinc–cobalt, zinc–tin, and zinc–iron. Of these, zinc–nickel appears to be a favorite at
this time, while some zinc–cobalt installations have been made. The others are either
too expensive or do not produce a pleasing-enough appearance to be applicable for
anything other than as a paint undercoat for automobile body panels. The equipment
for plating zinc alloys is identical to that used for nickel plating (see below).

## Zinc–Nickel

Zinc–nickel alloys can be plated from both alkaline and acidic chemistries, with the
alkaline process the most often favored.

## Alkaline Zinc–Nickel

### Ingredients

Zinc: 1–2 oz/gal
Sodium hydroxide: 12–17 oz/gal
Nickel: 0.1–0.2 oz/gal
Additives: as required
pH: 14+

### Operating Conditions

Temperature: 70–90°F
Anodes: zinc and steel
Plating current density: 10–45 ASF

Agitation: cathode rod
Filtration: yes

## Zinc–Cobalt

### Ingredients

Zinc chloride: 10–12 oz/gal
Potassium chloride: 26–33 oz/gal
Cobalt: 0.25–0.50 oz/gal
Boric acid: 3–4 oz/gal
pH: 5–6

### Operating Conditions

Temperature: 65–95°F
Anodes: zinc, bags
Plating current density: 1–40 ASF
Agitation: air
Filtration: yes

## Nickel

Nickel is most often plated from the Watts chemistry, although there are numerous other formulations, including a specialized Woods nickel strike that is used to obtain adhesion on stainless steels. The Watts bath is used to obtain bright or semibright deposits for decorative applications. In decorative applications where deposit appearance and corrosion resistance are highly important, as on the exterior of an automobile, two or more layers of nickel from Watts baths are applied. The most common such layered nickel plating is referred to as *duplex nickel* and consists of two layers of nickel. The first layer, called *semibright nickel,* contains no sulfur-bearing brighteners, and the second layer is a fully bright nickel deposit containing a controlled amount of sulfur-bearing brightener.

The duplex nickel is normally topped off with a thin coating of chromium plating. The bright nickel layer corrodes in favor of the semibright layer, protecting it galvanically and delaying the onset of corrosion of the base metal.

Another often-used nickel plating formulation is the sulfamate-based chemistry. It is used in electroforming or other applications where a nickel deposit containing no or low internal stress is desired.

Nickel is also used to plate composite deposits, where the plated nickel contains finely dispersed diamond dust or other abrasives such as silicon carbide. Such composite coatings are used to create long-lasting cutting tools.

## Watts Nickel

Watts nickel plating chemistry can contain a variety of additives to control pitting and yield leveling, and can produce brightness from a medium (semibright) to a full mirror-bright deposit. Additives are normally patented products sold by suppliers along with the plating chemicals.

### Ingredients

Nickel sulfate: 30–45 oz/gal
Nickel chloride: 4–12 oz/gal
Boric acid: 4–6 oz/gal
Additives: as required
pH: 3–5

### Operating Conditions

Temperature: 125–135°F
Anodes: nickel or nickel containing 0.02% sulfur or others, bagged
Plating current density: 25–100 ASF
Agitation: cathode rod or air
Filtration: yes
Color of solution: deep green
Odor of solution: no specific odor

## Woods Nickel Strike

Woods nickel strike is purposely designed to generate high volumes of hydrogen gas while depositing only a thin layer of nickel, even at the highest current densities. The major use is to obtain adherent thin nickel deposits that can then be plated with other metals.

### Ingredients

Nickel chloride: 30 oz/gal
Hydrochloric acid: 16 fl oz/gal
pH: <0.1

### Operating Conditions

Temperature: 70–90°F
Anodes: nickel
Plating current density: 100–300 ASF
Agitation: none
Filtration: none
Color of solution: dark green
Odor of solution: sharp hydrochloric fumes

## Sulfamate Nickel

The sulfamate nickel chemistry is used mainly for electroforming purposes, although some electronic applications requiring a low-stress nickel underplate for gold overplates also use this bath. The equipment for sulfamate plating is identical to that used for Watts baths, with the exception that the sulfamate process will typically have a purification compartment attached to the tank, incorporated into the tank, or alongside the tank. The purification compartment is about one fifth the size of the plating tank, and the solution is recirculated through the compartment, using the filtration system. In the purification compartment, electrolytic nickel anodes and dummy electrodes plate out metallic contaminants, and the polarization occurring at the anodes decomposes some of the sulfamate ions into stress-reducing compounds.

### Ingredients

Nickel sulfamate: 60–70 oz/gal
Magnesium chloride: 1–2 oz/gal
Boric acid: 6–7 oz/gal
Additives: as recommended
pH: 3–5

### Operating Conditions

Temperature: 125–135°F
Anodes: sulfur depolarized nickel
Plating current density: 20–140 ASF
Agitation: air
Filtration: yes
Color of solution: deep green
Odor of solution: no specific odor

## Sulfamate Nickel Strike

Sulfamate nickel strike is an alternative strike solution that can be used to activate stainless steel.

### Ingredients

Nickel sulfamate: 320 g/L
Sulfamic acid: 150 g/L

### Operating Conditions

Temperature: 50°C
Anodes: electrolytic nickel
Plating current density: 50 ASF
Agitation: none
Filtration: yes
Color of solution: deep green
Odor of solution: no specific odor

## Chromium

Chromium plating generally falls into two categories: decorative and hard. Both categories can be plated from the same chemistries based on hexavalent chromium, while decorative chromium can also be plated from one of several trivalent chemistries. The terms *decorative* and *hard* are confusing and really mean "thin" and "thick." All chromium plates have a hardness in the same range (900–1100 Vickers). Decorative chromium is a very thin layer of chromium applied over a substrate that has been bright-nickel-plated. The appearance of decorative chromium is, to a large extent, due to the appearance of the nickel. The chromium is so thin (3–20 millionths of an inch) that it is essentially transparent. Decorative chromium plating equipment is identical to that for hard chromium. An exception is the trivalent decorative chromium equipment, which typically has no exhaust system and requires continuous filtration.

Hard chromium should be called engineering chromium, because it is usually applied when a hard, wear-resistant metallic coating is required on a part that is subject to abrasive forces during service. A typical example is the chromium applied to hydraulic shafts for heavy equipment, on the piston rings of internal combustion engines, and on the shafts of landing gear for aircraft. A typical chromium plating tank is constructed of steel with a PVC lining. It is equipped with heating elements and an exhaust system to remove the chromic acid fumes from the workers' breathing zone.

## Hexavalent Chemistries

The hexavalent chromium plating chemistries fall into two categories: conventional and mixed catalyst. The conventional is a simple chemistry that anyone can mix up and use. The mixed-catalyst chemistries are patented processes that have a few advantages, including faster plating, fewer problems caused by current interruption, and fewer problems plating onto passive nickel deposits. They also tend to be more difficult to control and expensive to operate. In hard-chromium applications, mixed-catalyst baths also tend to etch steel in areas where plating is not intended, making masking more critical.

### Conventional Chemistry Ingredients

Chromium trioxide: 30–35 oz/gal
Sulfate: 0.3–0.35 oz/gal

### Operating Conditions

Temperature: 125–135°F (for hard plating applications, temperatures may be 140°F)
Anodes: lead
Plating current density: decorative 100–150 ASF, hard 150–250 ASF
Agitation: none (hard may use some air agitation)
Filtration: no

Color of solution: deep, dark red-brown
Odor of solution: no specific odor

### Mixed-Catalyst Ingredients

Chromium trioxide: 30–45 oz/gal
Sulfate: 0.15–0.18 oz/gal
Fluoride: 0.13 oz/gal
*Note:* Fluoride may be present as one or more of a variety of fluoride-containing compounds.
Operating conditions are the same as for the conventional chemistry.

## Trivalent Chemistries

Trivalent chemistries were developed in response to concerns about the detrimental effects of hexavalent chromium on the environment and on workers' health. Hexavalent chromium is a powerful oxidizer that readily attacks human tissues and has been linked in some studies to lung cancer. Trivalent chromium has a much lower toxicity level, is not an oxidizer, and to date has not been linked with cancer. Platers have been slow to accept trivalent chromium chemistries as substitutes for hexavalent chemistries because the former tend to plate deposits that are noticeably darker or not consistently of the same color. Since trivalent baths are used only for decorative applications, this is a major drawback, but solution manufacturers have made great progress toward solving these problems. A major benefit from trivalent processes is that these baths contain very low concentrations of chromium (about one fifth as much), and the chromium can be waste-treated without a reduction step, so waste treatment and sludge disposal costs are reduced. Equipment generally consists of a rubber- or plastic-lined steel or plastic tank, an air agitation system, a heating and cooling system, and a filtration system.

### Ingredients (Atotech Process)

TC additive: 52 oz/gal
Chromium: 2.7 oz/gal
TC stabilizer: 8% vol
TC-SA: 1.2% vol
TC regulator: 1 ml/liter
pH: 3.2

### Operating Conditions

Temperature: 70–90°F
Anodes: graphite (Atotech) (lead in membrane cell Enthone-OMI)
Plating current density: 90–200 ASF
Agitation: air
Filtration: yes
Color of solution: deep blue-green
Odor of solution: no specific odor

## Gold

Gold can be plated from three major chemistries: alkaline cyanide, neutral, and acid. All three chemistries utilize gold from potassium gold cyanide salts. A proprietary gold plating process that does not utilize potassium gold cyanide is on the market, but is expensive to operate and limited in alloying capability. Most gold plate is an alloy of gold and some other metal or combination of metals such as nickel, cobalt, copper, and silver. Gold can be plated in any commercial karat desired. The neutral and acid gold plating chemistries utilize chelating agents to perform the tasks normally performed by cyanide: control of metallic impurities and alloying elements. While these chelates could cause waste-treatment problems, they rarely enter the wastewater treatment system in high concentrations, since most gold plating operations have meticulous recovery systems to use as little rinse water as possible and recover the plating chemicals.

### Alkaline Cyanide Chemistry

Alkaline cyanide chemistry is most often used to apply a thin film of gold over bright nickel in decorative applications such as jewelry. The plating tank is a typical layout with a plastic or lined steel tank, filtration, cathode rod agitation (optional), and heating elements.

#### Ingredients

Potassium gold cyanide: 0.1–0.5 oz/gal
Potassium cyanide: 1–1.5 oz/gal
Potassium carbonate: 3–30 oz/gal
Additives: none
pH: 10

#### Operating Conditions

Temperature: 125–150°F
Anodes: gold or stainless steel or platinized titanium
Plating current density: 1–35 ASF
Agitation: cathode rod
Filtration: yes
Color of solution: dark yellow
Odor of solution: no specific odor

### Neutral Gold Plating Chemistry

Neutral gold plating chemistry is favored for barrel plating applications of high-purity gold.

#### Ingredients

Potassium gold cyanide: 1–1.5 oz/gal
Monopotassium phosphate: 10–12 oz/gal

Potassium citrate: 8–10 oz/gal
pH: 6–6.5

### Operating Conditions

Temperature: 125–135°F
Anodes: platinized titanium
Plating current density: 1–3 ASF
Agitation: cathode rod/recirculation pump
Filtration: yes
Color of solution: pale yellow/clear
Odor of solution: no specific odor

## Acid Gold Plating

Acid gold baths can produce a variety of gold deposits, including the hardest, most wear-resistant ones. They are favored for plating of printed circuit board connectors and in the semiconductor industry.

### Ingredients

Potassium gold cyanide: 0.5–1.0 oz/gal
Citric acid: 4–6 oz/gal
Ammonium citrate: 4–6 oz/gal
pH: 3–5

### Operating Conditions

Temperature: 90–140°F
Anodes: platinized titanium or platinized niobium
Plating current density: 1–5 ASF
Agitation: cathode rod
Filtration: yes
Color of solution: range from clear to purple
Odor of solution: no specific odor

## Silver

While noncyanide silver plating chemistries based on sulfites or succinimides have been available for some time, almost all silver plating is presently being performed in the cyanide chemistry. The noncyanide baths are far more expensive to install and operate and cannot tolerate contamination to the same degree as can the cyanide process. The plating equipment for cyanide silver plating is typically a lined steel or plastic tank equipped with a filter and cathode rod agitation. Silver can also be barrel plated.

### Ingredients (for Rack or Barrel Plating)

Silver cyanide: 1–4 oz/gal
Potassium cyanide: 2–4 oz/gal
Sodium carbonate: 3–14 oz/gal
Potassium hydroxide: 1–2 oz/gal
Potassium nitrate: 0–2 oz/gal
pH: 12–14

### Operating Conditions

Temperature: 70–90°F
Anodes: silver
Plating current density: 1–40 ASF
Agitation: cathode rod
Filtration: yes
Color of solution: dark brown/black
Odor of solution: organic

## Tin

Tin can be plated from more than four major chemistries: the alkaline stannate process, the fluoborate, the proprietary halogen and sulfonate processes, and the sulfuric acid–based process. The alkaline stannate, fluoborate, and sulfate chemistries are most often encountered in job shops. The alkaline stannate process typically consists of a heated steel tank, while the sulfuric acid–based process uses a plastic- or PVC-lined steel tank and has filtration. The alkaline process produces a matte, pure tin deposit that has excellent solderability, while the sulfate process produces matte or bright deposits with lesser or marginal solderability but superior appearance. The alkaline process is a bit more difficult to operate than the sulfate process. The alkaline bath is favored for barrel plating applications, although the fluoborate bath can also be used.

## Alkaline Stannate Process

### Ingredients

| Rack barrel | | |
|---|---|---|
| Potassium stannate | 13.3 | 26.6 |
| Potassium hydroxide | 2 | 3 |
| Potassium carbonate | 3–14 | 3–14 |
| Additives | None | None |
| pH | >14 | >14 |

*Note:* 1/4 to 1/2 oz/gal of cyanide is sometimes added to reduce the effects of metallic contaminants.

### Operating Conditions

Temperature: 150–180°F
Anodes: tin
Plating current density: 1–100 ASF
Agitation: cathode rod
Filtration: no
Color of solution: pale yellow/clear
Odor of solution: no specific odor

## Fluoborate-Based Chemistry (for Rack and Barrel Plating)

### Ingredients

Tin (from concentrate): 4–6 oz/gal
Fluoboric acid: 25–35 oz/gal
Boric acid: 3–5 oz/gal
Additives: as recommended, but required
pH: <0.1

### Operating Conditions

Temperature: 90–120°F
Anodes: tin
Plating current density: 1–80 ASF
Agitation: cathode rod
Filtration: yes
Color of solution: pale yellow
Odor of solution: no specific

## Sulfate-Based Chemistry

### Ingredients

Stannous sulfate: 2–6 oz/gal
Sulfuric acid: 1–3 oz/gal
Additives: as recommended, but required
pH: <1

### Operating Conditions

Temperature: 55–85°F
Anodes: tin
Plating current density: 1–25 ASF
Agitation: cathode rod
Filtration: yes
Color of solution: pale yellow
Odor of solution: sweet

## Tin–Lead

Tin–Lead is applied to electronic components that require high solderability. The plating hardware is a typical plating setup with continuous filtration optional. Two basic chemistries are used in plating the alloy: the fluoboric and the (proprietary) sulfuric acid–based chemistries. These baths are most commonly found in printed circuit board manufacturing shops and job shops specializing in plating for electronics. The electronics industry uses a high-throw formulation to allow for plating inside drilled holes. Others use a conventional bath that yields the best solderability. The fluoborate bath is made by mixing liquid fluoborate concentrates with water.

### Fluoboric Bath Ingredients

| | Conventional | high throw |
|---|---|---|
| Tin (from concentrate) | 7–8 oz/gal | 1.6–2.7 oz/gal |
| Lead (from concentrate) | 3–4 oz/gal | 1.1–1.9 oz/gal |
| Fluoboric acid | 13–20 oz/gal | 47–67 oz/gal |
| Boric acid | 3–5 oz/gal | 3–5 oz/gal |
| pH | <0.1 | <0.1 |

### Operating Conditions

Temperature: 70–90°F
Anodes: tin–lead alloy
Plating current density: 15–25 ASF
Agitation: cathode rod
Filtration: yes
Color of solution: pale yellow
Odor of solution: sweet

## 7.3.5  Electroless Plating Processes

One major drawback to electroplating is the nonuniform coating produced, because electric current tends to concentrate on sharp edges, corners, and points. Electroless plating processes are used when it is necessary to obtain a very uniform coating on complex geometries, because these processes do not depend on electricity delivered from a rectifier. As the name implies, the coating is produced without an outside source of current. The reducing electrons are provided chemically.

Some electroless deposits are also more corrosion resistant than their electroplated counterparts. The following are two of the most often applied electroless processes.

## Electroless Nickel

Electroless nickel is applied to numerous complex electronic and industrial components for a high degree of wear resistance and corrosion protection. An example is the sliding plates that mold the hamburger patties served in fast-food restaurants. Two plates slide against each other to form the cavity that is used to injection-mold the patty. The plates slide against each other at lightning-fast speeds. Electroless nickel is also used in the plating of plastics, to provide the first metallic layer on the plastic to yield conductivity for subsequent deposits.

The electroless plating process normally consists of two plating tanks and a nitric acid storage tank. Each plating tank contains heating elements, an air sparger, and a recirculating filter. The solution will eventually deposit nickel on everything it contacts, so periodically the tank walls and associated equipment must be stripped with nitric acid (thus the nitric storage tank). The plating solution has a finite life (8–14 turnovers), after which it must be waste-treated or disposed of through a commercial disposal firm. The electroless nickel plating solution contains strong chelating agents that interfere with a conventional wastewater treatment system, so they must be treated separately using electrowinning, proprietary treatment methods, or special chemical treatments. The rinse water from electroless nickel operations is usually segregated and treated separate from other rinse water.

Electroless nickel is typically plated from one of two basic chemistries, yielding either a nickel–phosphorus alloy (most common) or a nickel–boron alloy. Each alloy can be plated from a number of different solutions to yield varying alloy compositions. Following are two such chemistries.

## Nickel–Phosphorus Alloys

| | | |
|---|---|---|
| Nickel sulfate | 21 g/L | 11.8 g/L |
| Acetic acid | 9.3 g/L | |
| Lactic acid | 27 g/L | |
| Molybdic acid | 0.009 g/L | |
| Propionic acid | 2.2 g/L | |
| Lead acetate | 0.001g/L | |
| 1,3 Diisopropyl thiourea | 0.004 g/L | |
| Sodium hypophosphite | 24 g/L | 22.3 g/L |
| pH | 4.6 | 5.5 |
| Temperature | 95°C | 95°C |

## Nickel–Boron Alloys

| | |
|---|---|
| Nickel chloride | 0 g/L |
| Sodium hydroxide | 40 g/L |
| Ethylene diamine | 86 ml/L |
| Sodium borohydride | 0.6 g/L |
| Thallium nitrate | 0.007 g/L |
| Sodium gluconate | 5 g/L |
| Diethyl amine borane | 1.0 g/L |
| Lead acetate | 0.02 g/L |
| pH | 13–14 |
| Temperature | 90°C |

## Electroless Copper

The major use for electroless copper is in the manufacture of printed wiring boards. The electroless copper is used to apply a thin coating of copper over the top side and into the drilled holes of the boards. The drilled holes are initially nonmetallic, since the boards are made of epoxy–fiberglass. With the electroless copper, the holes become conductive for further plating.

Equipment for electroless copper plating usually consists of a polypropylene or PVC tank and filter. Some baths operate at room temperature, so heating is not required; others require heating. The rinse water and spent electroless copper often contain chelating or complexing agents, so waste treatment is difficult.

Two typical compositions for an electroless copper solution are:

| | | |
|---|---|---|
| Copper sulfate | 13.8 g/L | 5 g/L |
| Rochelle salts | 69.2 g/L | 25 g/L |
| Sodium hydroxide | 20 g/L | 7 g/L |
| MBT | 0.012 g/L | |
| Formaldehyde | 38 ml/L | 10 ml/L |
| Temperature | 50°C | 25°C |

## Other Electroless Processes

There are numerous other electroless plating solutions in the literature, although they are rarely used. The following solutions may be encountered.

## Electroless Cobalt

Cobalt sulfate: 30 g/L
Ammonium chloride: 84 g/L
Sodium hypophosphite: 20 g/L
pH: 10
Temperature: 95°C

## Electroless Silver

Silver cyanide: 1.34 g/L
Sodium cyanide: 1.49 g/L
Sodium hydroxide: 0.75 g/L
Dimethyl amine borane: 2 g/L
Temperature: 55°C

## Electroless Gold

Potassium gold cyanide: 5.8 g/L
Potassium cyanide: 13 g/L
Potassium hydroxide: 11g/L
Potassium borohydride: 21.6 g/L
Temperature: 75°C

## 7.3.6 Typical Postplating Operations

After plating, parts are often further treated in various chemical solutions to enhance the appearance or corrosion/tarnish resistance of the plated coating. Examples of such operations include chromate films on zinc, cadmium, and copper plates, and stains produced on copper or copper alloy deposits. Application of lacquers, waxes, and other organic topcoats are also popular postplating methods for improving shelf and service life of parts. Chromate conversion coatings are so popular that almost every zinc and cadmium plating line has chromating tanks and rinses built into it, and most all zinc and cadmium plated parts have some form of chromate film on top of the metal deposit.

### Chromating

A chromate is a very thin complex film created by converting a small amount of the top surface of the plated metal into the film, thus the term *conversion coating*. The chromate film is formed by immersing the plated deposit into an acidic solution containing a variety of chemicals depending on the color and corrosion resistance to be obtained. If, for example, we want the zinc to turn bright, reflective, and with a hint of blue (typically called a blue-bright dip), the solution will contain nitric acid and potassium ferricyanide, along with some trivalent chromium compounds. If we want a yellow iridescence, hexavalent chromium in the form of sodium dichromate may be added. If we want olive drab green, even more dichromate along with sulfates may be added. If we want a black coating, a small amount of silver nitrate is added (finely divided silver particles create the black color).

In each case, the mechanism for forming the chromate film is the same: the plated metal is attacked by the acid in the chromate dip, releasing hydrogen as a by-product of the attack. As the hydrogen is released from the metal surface, the pH of the solution near the metal surface rises high enough to deposit a film of metal hydroxides and other trapped ingredients from the solution. The film at first is a delicate gel, but quickly hardens into a thin coating only a few millionths of an inch thick (or less).

The chromate film protects the plated metal from corrosion by acting as a barrier layer against corrosive atmospheres. In general, the more hexavalent chromium is trapped in the film, the higher will be the coloration, and the better the corrosion resistance.

Since the chromate functions by attacking and dissolving some of the plated metal, eventually the solution becomes so contaminated with plated metal that it stops producing acceptable coatings. At this point the chromate becomes spent and must be waste-treated.

Rinses following chromating operations are not recoverable (since recovery would only hasten the demise of the chromating solution) and therefore must be routed to a wastewater treatment system before discharge to sewer.

There are several variations on the chromating process, including some films that are applied with reverse current and others that are applied and then leached to remove the coloration. The most commonly applied chromates are the blue brights, followed by the yellow and then the black. The military favors the olive drab for its color and high corrosion resistance.

A significant amount of research is being made into substitutes for chromates that do not contain hexavalent chromium, although not much is on the market now.

## Bright Dipping

Some operations utilize a bright dip after zinc or cadmium plating. This dip can be used by itself or before chromating. It is simply a very dilute solution of nitric acid (0.25−0.50% vol). The bright dip removes a thin organic film from the surface of freshly plated zinc or cadmium, rendering the deposit far brighter than before. This is not a chromate, nor is it a true conversion coating.

## Other Postplating Operations

Numerous other postplating operations can be performed, including application of lacquer, wax, dyes, or stains. Invariably, such operations involve additional tanks, equipment, and chemical solutions.

### 7.3.7   Bibliography

The following was a source of information included in this subchapter and is recommended for additional reading.

*Intensive Training Course in Electroplating,* Illustrated Lecture Series, American Electroplaters and Surface Finishers Society, Orlando, FL.

## 7.4   COATINGS

Jack M. Walker, consultant, manufacturing engineering, Merritt Island, Florida

### 7.4.1   Introduction to Coatings

In keeping with the premise that manufacturing engineers working today cannot possibly be familiar with the total factory operation and all manufacturing processes, this subchapter on coatings is an introduction to the principles of coatings. The modern protective coatings are an outgrowth of the paint that started out as an artist's material. From the early history of humanity we find that it was the artist who gathered the materials and developed the methods for making paint. Since early types of paint were made by artists, paint and varnish making was in itself an art for many centuries.

Decoration seems to have been the original purpose of paint, but in time its power to protect the vulnerable surfaces of manufactured objects became of equal importance. With the industrial development that took place in the eighteenth and nineteenth centuries, paint began to emerge as a commercial material. However, the basic ingredients were still the natural resins and oils traditionally used by artists. These materials had a certain value as protective coatings, but were limited in effectiveness.

At the turn of the twentieth century, industrial development began to mean heavier demands on coatings, and scientists began to investigate the traditional art of paint making with an eye toward improvement. During the last 30 years, steadily increasing effort has virtually revolutionized the manufacture of coatings. Today, coatings are available that resist attack by nearly all chemicals and corrosive conditions. These protective coatings, which are usually distinguished from paints that are primarily decorative, are valuable engineering materials. They often make it feasible to use steel structures or other materials in environments where they would not otherwise have an economic life.

## 7.4.2 Wet Coatings

Paint can be defined as any fluid material that will spread over a solid surface and dry or harden to an adherent and coherent colored obscuring film. It usually consists of a powdered solid (the *pigment*) suspended in a liquid (the *vehicle*). The pigment provides the coloring and obscuring properties. The vehicle is the film-forming component that holds the pigment particles together and attaches them to the surface over which they are spread.

### Coverage of Coatings

If every drop in a gallon (U.S.) of liquid protective coating can be applied to a surface without any loss due to the application equipment, the material will cover 1604 ft$^2$ at a thickness of 1 mil (0.001 in.). In most instances, the liquid contains a volatile solvent that, upon evaporating, reduces the thickness of our film. For instance, if the wet coating contains 50% by volume of solvent, the film after drying will be 0.5 mil thick rather than 1 mil thick. The former is called the *wet thickness,* while the latter is the *dry thickness.*

Knowledge of the volume percent solvent—or more usually, the volume of non-volatiles, which we call the *percent volume solids*—enables us to calculate the dry theoretical coverage for a gallon of paint as follows:

$$\text{Theoretical coverage per U.S. gallon} = (1604 \times \% \text{ volume solids}) \div (100 \times \text{dry film thickness in mils})$$

Obviously, no applicator can get every drop of material out of the container, nor can one avoid leaving some of it in or on the application equipment. More important, there will be considerable losses, particularly in spray painting, due to air movement as well as missing the target. The magnitude of these losses will vary depending on what is being coated, the application equipment, and air movement. With the same coating, Painter A, spraying a flat wall indoors with an apparatus in good condition, will get considerably more mileage than Painter B, spraying 2-in. channel steel in a 20-mile wind with a badly maintained spray gun that must be unplugged every 10 min. Some contractors figure on a 20% loss during spraying, while others use 30% in estimating material requirements.

## Components of Wet Coatings

Most protective coatings contain a volatile component, the solvent. Its purpose is to keep the coating in a liquid condition suitable for uniform application and bonding to the target or substrate. The other major components of a coating are shown in Figure 7.28. They include the resin or binder, and pigments.

## Pigment Functions

Color—aesthetic effect, hides substrate.

Protection of resin binder—absorb and reflect solar radiation, which can cause breakdown of binder (chalking).

Corrosion inhibition—chromate salts and red lead in primers as passivators. Metallic zinc, when in high enough concentration, gives cathodic (sacrificial) protection.

Film reinforcement—finely divided fibrous and platey particles that increase hardness or tensile strength of film.

Nonskid properties—particles of silica or pumice that roughen film surface and increase abrasion resistance.

Sag control—so-called thixotropic agents that prevent sagging of the wet film and also reduce the tendency of other pigments to settle in the container during storage.

Hide and gloss control—increasing color pigment concentration improves hide, while an increase in either color or other pigmentation decreases gloss.

Increased coverage—properly selected filler pigments, sometimes called inerts, can increase the volume solids (or coverage) of a coating without reducing its chemical resistance. There is a limit to how much filler pigment can be used with a given resin composition. This constraint is termed the

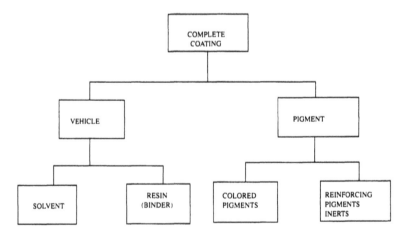

**FIGURE 7.28** The major components of a wet coating.

*critical pigment volume concentration* and indicates the volume of pigment that can be bound by the resin without leaving voids in the film.

The pigments, both those that contribute color and those that have other functions, must be uniformly dispersed and completely wetted in the resinous binder to function properly.

## Resinous Binder Functions

The binder, in addition to an ability to "glue" the pigments together in a homogeneous film, must be capable of wetting and adhering to the substrate, preventing penetration of aggressive chemicals, and maintaining its integrity in the corrosive environment. Since there is a wide variety of environments, it should not be surprising that there is a wide variety of binders, each with its special niche in the coatings industry.

## The Vehicle

To be capable of being applied in an even, essentially void-free film that will wet and adhere to the substrate, the binder-pigment mixture must be in a highly fluid condition. The use of low-molecular-weight liquids (solvents) to dissolve or to increase the fluidity of the binder is usually necessary. The combination of such solvents with the binder is termed the *vehicle*.

Each type of binder has specific solvent combinations that are most efficient in producing the desired application and film-forming properties. There is no universal solvent for protective coatings. The best solvent for one type of coating is water, while for another, a combination of expensive toxic organic compounds may be required. Use of the wrong solvent, or as it is usually called, "thinner," will cause precipitation of the binder. The result may range from instant gelation of the coating to films with substandard properties in application or performance. Some of the common solvents used in the formulation and in the thinning of protective coatings are listed below.

  *Mineral spirits*—often called the painter's naptha; high-boiling-point petroleum product used for oil and alkyd vehicles.
  *Aromatics*—compounds of the benzene family, including toluene, xylene, and higher-boiling homologs; major solvent for chlorinated rubber, coal tar, and certain alkyds; used in combination with other solvents in vinyl, epoxy, and polyurethane vehicles.
  *Ketones*—compounds of the acetone family, including methyl ethyl ketone (MEK), methylisobutyl ketone (MIBK), and cyclohexanone; most effective solvent for vinyls but sometimes used in epoxy and other formulations.
  *Esters*—have distinct, usually pleasant odors; commonly used in epoxy and polyurethane vehicles.
  *Esthers*—high-boiling solvents sometimes used to achieve vehicles with high flash points.

*Alcohols*—good solvents for highly polar resins such as phenolics and acetals. Also used in solvent-based self-curing silicates.

*Water*—the only thinner for latex and emulsion coatings, as well as for certain inorganic zinc silicate primers.

## Binders

In order to produce a film that will perform satisfactorily in a given environment, the coating, after application, must convert to a dense, solid membrane. Some materials can accomplish this simply by releasing their solvents, while others must go through a series of complicated chemical reactions, sometimes requiring the application of heat after evaporation of the solvent.

The ability of a resin to form a dense, tight film is directly related to its molecular size and complexity. The polymers capable of forming such films simply by the evaporation of solvent are initially of very high molecular weight and are not capable of further chemical reaction. Because of their large size, these polymers must be kept in a dilute solution, and coatings based on them have a low volume of solids. Resins of low molecular weight, although requiring chemical conversion to attain polymer structures of suitable size, have the advantage of being able to produce combinations of higher solids.

Five general types of binders are used to formulate protective coatings for the chemical processing and marine industries. These include lacquers, drying oil types, coreacting compounds, condensation coatings, and inorganic vehicles.

1. *Lacquers* are based on high molecular weight; chemically resistant polymers that form films by evaporation of solvents; low-volume solids that are sometimes increased by the addition of lower molecular weight resins or plasticizers. Examples include:

   *Polyvinyl chloride copolymers*—ranging from low-film-build materials (1–2 mils per coat) used as tank linings to high-build (5–10 mils per coat) maintenance coatings.

   *Chlorinated rubbers*—formulated in a variety of combinations with modifying resins to attain higher solids; chemical and water resistance varies with type and amount of modifier used; usually limited to applications at 2–3 mils per coat, although some newer materials are capable of building at 4–6 dry mils.

   *Polycrylics*—have excellent color and gloss stability in outdoor weathering; water and chemical resistances are not as high as for vinyl copolymers or chlorinated rubbers; often used in mixture with polyvinyl chloride copolymer or as final coat.

2. *Drying oil types*—low-molecular-weight resins capable of converting to tough films through intermolecular reactions with oxygen catalyzed by the presence of metal soaps, such as cobalt octoate. Examples include:

   *Natural oils,* such as linseed, tung, and soya oils—which contain reactive sites that are activated by oxygen in their molecules; very slow in converting to cured films.

*Alkyds*—natural oils chemically modified by reactions with synthetic acids or alcohols to improve rate of cure, chemical resistance, or ultimate hardness; degree of modification is designated by terms *long oil* (relatively minor modification), *medium oil* (moderate modification), and *short oil* (considerable modification).

*Varnishes*—natural oils or alkyds containing dissolved or reacted resins such as chlorinated rubber or phenolic to increase hardness and chemical resistance.

*Epoxy esters*—a type of alkyd in which a high-molecular-weight epoxy resin is used in the chemical modification; usually have much higher degree of chemical resistance than normal alkyds but have poor exterior weathering properties.

*Uralkyds (Urethane oils)*—a type of alkyd modified by reaction with tolylene diisocyanate that imparts excellent abrasion resistance, but also detracts from exterior weathering properties.

*Synthetic polyesters*—contain no drying oil, but curing mechanism is similar to that of the latter; organic peroxide added to the resin at the time of application; major use is in fabricating glass-reinforced structures, such as piping, but are occasionally employed with chopped glass or glass mat as tank linings; have advantage of being 100% solids since the solvents, styrene and vinyl toluene, coreact with the polyester resin upon the addition of the peroxide.

*Note:* The drying oil types, except for the synthetic polyester, convert to their cured state by taking up atmospheric oxygen; consequently, the rate of cure is most rapid at the air surface and slowest at the substrate interface. Because of the difference in cure rate through the thickness of the film, care must be taken to avoid *wrinkling* resulting from excessive film thickness or too-rapid recoating.

3. *Coreacting types*—unlike the lacquers and the drying oils, these materials have two or more separately packaged components that are combined just before application. Reaction rates between the components are modified somewhat by the presence of solvents. However, the coreacting coatings, after mixing of the components, have limited pot lives that are reduced by increased application temperatures and usually extended by lower ones. Examples include epoxy and polyurethane coatings.

*Epoxy coatings*—which cure at ambient temperature, are based on low-molecular-weight resins containing the reactive *epoxy group* at each end of the molecule. This group has a particular affinity for ammonia derivatives called *amines*. The amine group may be strung along a short molecule, the *polyamine,* or may be more widely scattered on a long chain, the *polyamide.* In both cases, the amine-containing chain ties the epoxy resin molecules together in a dense cross-linked structure. The film obtained with the polyamine is usually more dense, having better solvent and chemical resistance but decreased flexibility, and it is of a more brittle character than that of the polyamide. Modification of these *curing agents* by prereacting them with a portion of the epoxy or other resins or chemicals is sometimes undertaken by coatings manufacturers to improve cure properties. These modified amines are called *amine adducts.*

The water resistance and, to a degree, the acid resistance of the various room-temperature epoxy coatings can be improved by the incorporation of high concentrations of selected coal tar resins into the formulas. These *epoxy coal tar* coatings are often used in areas where a high order of water resistance is required without the need of the solvent resistance or appearance of the base epoxy system.

Epoxy coatings are characterized by hard films of limited flexibility with a high order of chemical and, except for the coal tar modifications, solvent resistance. They are, however, much more susceptible to attack by oxidizing agents such as chlorine gas, peroxides, and nitric acid than are coatings based on vinyl copolymers or chlorinated rubber. A major weakness is their high rate of chalking in exterior exposures.

*Polyurethane coatings* are based on the reaction of a group of chemicals, the *diisocyanates,* with resins or chemicals containing alcohol or amine substituents in their structures. The diisocyanates have a high order of toxicity and must be chemically modified to permit their use in protective coatings. The most common practice is to make a *urethane prepolymer* by reacting two molecules of the diisocyanate with one molecule of a resin or chemical compound containing two alcohol groups in its structure. The urethane prepolymer is capable of further reaction with resins or chemical compounds containing amine groups (a very rapid reaction) or alcohol groups (much slower). A third possible reaction that is sometimes useful but often a nuisance is with water. If a free isocyanate group comes in contact with water, it will decompose to form an amine and simultaneously release carbon dioxide gas. Neighboring isocyanate groups will react immediately with the amine to initiate the promotion of a complex polymer structure. This reaction with moisture is the basis for the one-package moisture-curing polyurethane.

Another approach to polyurethane coatings is to mix the prepolymer with a resin containing alcohol groups at the time of application to form a two-package polyurethane. The choice of the second resin will determine to a large degree physical properties such as hardness and flexibility in the cured film.

When properly formulated, applied, and cured, urethane coatings have outstanding toughness and abrasion resistance with chemical properties similar to the epoxies. Unlike the epoxies, they do not have good adhesive properties to steel and concrete. Epoxy primers are commonly used as part of a polyurethane system.

4. *Condensation (heat condensing) coatings*—are based on resins that interact to form cross-linked polymers when subjected to temperatures

of 350–400°F. The use of such materials is limited to tank linings and to objects or structures that can be handled in an oven. Many of the so-called powder coatings as well as high-temperature silicones fall into this category (see Subchapter 7.2).

The oldest of the condensation coatings is the *pure phenolic*. Applied in several coats to a thickness of 6 mils and baked to a temperature of 375°F, the phenolic film becomes extremely hard and chemically resistant. It is, however, extremely brittle and is rapidly attacked by caustic solutions.

5. *Inorganic silicates* are binders that do not contain organic (carbon) structures in their composition. For this reason, they are noncombustible and are unaffected by sunlight dust in *zinc inorganic primers*. There are several approaches to silicate binders, but each appears to depend on the development of an extremely adhesive form of silica during its cure. Examples follow.

*Postcured inorganic silicates* are based on water solutions of *alkali silicates* pigmented with zinc dust or zinc dust and a metal oxide. The dust or powder is mixed into the liquid at the time of application. A 2 1/2- to 3-mil film spray dries very rapidly, usually within 1 hr. At this point it is very hard but remains water soluble. Application of an acid curing solution is necessary to achieve conversion of the silicate to the insoluble silica.

*Self-curing, water-based silicates* are mixtures of alkali silicates and colloidal silica pigmented with zinc dust or zinc and a metal oxide. As with the postcured materials, the separately packaged powder is dispersed in the vehicle at the time of application. The development of water insolubilization for this type of film is dependent on the absorption of carbon dioxide from the atmosphere during the curing process.

*Self-curing ammonium silicates* are water-based dispersions of colloidal silica containing additives that are capable of generating ammonia during the cure of the film. The pigmentation, packaging, and mixing procedures are identical to those of postcured inorganic silicates and self-curing, water-based silicates. The initial hardness of the film is considerably lower than that of the latter coatings. However, as moisture is absorbed from the atmosphere, the curing mechanism proceeds to produce a film that is ultimately harder than the other types of inorganic films. Since ammonia is generated during the curing process, ammonium silicate films should not be topcoated until they have weathered to their ultimate hardness.

*Self-curing, solvent-based silicates* are organic esters that gradually hydrolyze upon exposure to moisture to produce a binder that appears to be essentially identical to those of the water-based types. A major advantage of this type of film is its almost instant resistance to rain or flowing water. Although most of the commercially available inorganic coatings in this category are sold as two separately packaged components, one-package products are proving to be practical.

The inorganic zinc primers are recognized as having many of the benefits of galvanizing while providing a much more suitable surface for topcoating than the latter. They should not be confused with the so-called zinc-rich organic primers.

The use of metallic zinc as the pigmentation in a variety of vehicles, most notably in epoxy–polymide combinations, does result in primers having outstanding anticorrosive properties. However, field experience has demonstrated that coating systems employing these zinc-rich organic primers do not have the longevity of those based on the inorganic zinc silicates.

## Primer Coats

A primer should meet most, if not all, of the following requirements:

1. Good adhesion to the surface to be protected when the latter has been cleaned or prepared according to specification
2. A satisfactory bonding surface for the next coat
3. The ability to stifle or retard the spread of corrosion from discontinuities such as pinholes, holidays, or breaks in the coating film
4. Enough chemical and weather resistance by itself to protect the surface for a time period in excess of that anticipated before application of the next coating in the system
5. Under certain conditions, notably tank linings, chemical resistance equivalent to the remainder of the system

Four types of primers commonly used over steel may be compared in their abilities to meet these requirements. (See Table 7.2.)

Intermediate coats may be required in a system to provide one or more of the following properties:

1. Adequate film thickness of the system (body coat)
2. A uniform bond between the primer and the topcoat (tie coat)
3. A superior barrier with respect to aggressive chemicals in the environment (may be too deficient with respect to appearance or physical properties to be a satisfactory finish coat)

Finish coats are the initial barriers to the environment, but they are also the surfaces seen by your management, the public, and ultimately the consumer. There are situations, however, where the barrier to the environment will primarily be a function of the body or primer coat, while the finish coat serves to provide a pleasing appearance, a nonskid surface, a matrix for antifouling agents, or other specialized purposes. Obviously, the chemical resistance of the finish coat in one of these situations must be sufficient to ensure its remaining intact in the environment.

TABLE 7.2  Major Types of Primer Coatings

| Premier type/ requirement | Alkyd or oil | Mixed resin | Resin identical to topcoats | Inorganic zinc |
|---|---|---|---|---|
| Bonding to surface | Usually have the ability to wet and bond to most surfaces and are somewhat tolerant of substandard surface preparation. | Adhesive properties are major consideration of formulation. Not quite as tolerant of substandard surface preparation as oil types. | Adequate for proposed use when surface properly prepared. | Outstanding adhesion to properly cleaned and roughened surfaces. |
| Adhesion of topcoats | Satisfactory for oil types. Usually unsatisfactory for vinyls, epoxies, and other synthetic polymers. Are softened and lose integrity by attack from solvent of these topcoats. | Formulated for a specific range of topcoats. | Usually part of specific generic system. Maximum permitted dry time before appplication of second coat must be observed. | Fits into wide range of systems. The coat may be required. Specific recomendation should be obtained for immersion systems. |
| Corrosion suppression | Limited alkali produced at cathode and corrosion battery attacks film (saponification) and cause disbonding. Results in spread of underfilm corrosion. | Usually formulated with good resistance to alkali undercut and contain chromate pigment for a degree of corrosion inhibition. | Often contain chromate pigments for degree of corrosion inhibition. Resistance to alkali undercut is variable. | Outstanding ability to resist disbonding and underfilm corrosion. Anodic property of metallic zinc protects minor film discontinuties. |
| Protection as single coat | Limited by severity of exposure. | Limited by severity of exposure. | Limited by severity of exposure. | With very few exceptions, will protect without topcoat. |
| Chemical | Typical of alkyds. | Usually of lower order of resistance than that of topcoat. | Typical of system. | Not resistant to strong acids and alkalis. Has outstanding solvent resistance. |

## Surface Preparation for Wet Coatings

*Why Surface Preparation?*
Industrial, commercial, and weekend painters have one thing in common: they resent the time or expense required to prepare surfaces properly for the application of protective or appearance coatings. However, if they ignore this most important step, the whole operation is likely to be a waste of time and money. A protective barrier, unless it is thick enough to be a self-supporting structure, must be uniformly bonded to the substrate for the following reasons:

1. Sufficient attachment is necessary to prevent dislodgement under the gravitational and mechanical forces to which it is exposed.
2. Under immersion or condensation conditions, water vapor will penetrate a barrier and condense on any unbonded surfaces. Progressive disbonding is almost sure to occur.

## Cleanliness

Primers spread on the surfaces they contact. If these are dirt, dust, scale, rust, oil, or moisture, the bond of the coating system to the structure can only be as good as the bond of the contaminant to the real surface. Furthermore, primers are formulated to stick to metals, concrete, wood, and masonry rather than to surface contaminants. The need for scrupulous surface cleaning prior to any coating has become more important since spraying has replaced brushing as the most common method of wet material application. Understanding of the corrosion factor is discussed in greater detail in Subchapter 7.1, and cleaning is discussed in Subchapter 7.2.

## Improving the Bonding Surface (Profile)

Mechanical roughening of a surface by the use of abrasives or acids can provide improvements in the degree of bond that will be developed by the initial coating material—whether this is a primer or some other coating. These processes expose fresh, chemically clean surfaces that are easy to wet. Furthermore, roughening increases the actual contact area. Imagine two surfaces, one smooth steel and the other sandblasted, with exactly the same dimensional area. Although both have exactly the same measured area, the sandblasted specimen has a much greater surface area for bonding with the coating. If the primer bond to the smooth surface has a strength of 2 lb, its bond to the sandblasted surface may be twice the actual number of square inches of surface, with a strength of 4 lb. If the sandblasting quadrupled the surface area, the bond strength should be 8 lb. Acid etching of steel, although less effective than sandblasting, also increases the bondable surface area. The latter can be further improved by the application of a phosphate conversion coating. In such a process, iron, zinc, or manganese dihydrogen phosphate is applied by dipping or spraying to clean steel. When this is done properly, a film of tightly adherent iron phosphate crystals forms at the surface to promote a more receptive substrate for the coating. The use of phosphate conversion coatings is very common to the product-finishing industry and other assembly-line types of operations such as coil coating.

## Application of Wet Coatings

The most important elements in producing good results and avoiding problems in application of wet coatings are described in the following paragraphs.

## Environmental Conditions

The ideal time for painting is when the weather is warm and dry with little wind. Obviously, many coating projects cannot be delayed until these ideal conditions prevail. Also, many shops either do not have an environmentally controlled spray booth, or the product cannot be brought inside, where conditions can be better controlled.

Under conditions of *high humidity*, condensation of moisture is likely to occur on surfaces. Condensation on the substrate interferes with bonding of the coating. Condensation on the surface of a freshly applied coating may alter its curing process.

*Very low humidities* can be a problem with water-based products. Rapid flash-off of the water may result in film cracking. It can also cause poor curing rates with certain types of inorganic coatings.

At *low temperatures*, the film thickness of high-build or *thixotropic* coatings becomes more difficult to achieve. Curing reactions slow down or stop for many materials. Water-based products may freeze. Solvents evaporate more slowly. Furthermore, when the relative humidity is high, condensation is likely to develop.

Although heat has many beneficial effects in the application of coatings, *high temperatures* often increase overspray (dry fallout), trapped air or solvent bubbles, and in the case of zinc inorganics, the incidence of film cracking. High temperatures also reduce the pot life of catalyzed materials.

*Wind* is a nuisance, particularly in spray painting. The material can be deflected from the target as it leaves the spray gun. Solvent tends to flash off, creating excessive dry spray at edges of the spray pattern. Lap marks become more evident. Dirt and other debris may become embedded in the wet film. The velocity of air inside a spray booth can create the same problems, as will be discussed later in this subchapter.

*Condensation* becomes a problem when humidities are high and surface temperatures are low. Unfortunately, on large-scale painting projects, which must be done outside, primers are often applied late in the workday, and sometimes at night. Abrasive blasting is a slow process, while applying a primer by spray goes very rapidly. Because of this wide difference in work rates, the contractor may take 6 hours of an 8-hour day to prepare the surface for 1 to 1 1/2 hours of primer application. Table 7.3 illustrates the relationship of air temperature, metal temperature, and percent relative humidity to condensation. The best procedure is to paint only when the surface temperature is at least 5°F above the temperature where condensation will form (the dew point).

Many of the application, drying, and curing problems created by weather conditions, either outside or inside the factory, can be reduced by lowering the viscosity

TABLE 7.3  Percent Relative Humidity Above Which Moisture Will Condense on Uninsulated Metal Surfaces

| Metal Surface Temp. | SURROUNDING AIR TEMPERATURE °F | | | | | | | | | | | | | | | | |
|---|---|---|---|---|---|---|---|---|---|---|---|---|---|---|---|---|---|
| | 40 | 45 | 50 | 55 | 60 | 65 | 70 | 75 | 80 | 85 | 90 | 95 | 100 | 105 | 110 | 115 | 120 |
| 35°F | 60 | 33 | 11 | | | | | | | | | | | | | | |
| 40 | | 69 | 39 | 20 | 8 | | | | | | | | | | | | |
| 45 | | | 69 | 45 | 27 | 14 | | | | | | | | | | | |
| 50 | | | | 71 | 49 | 32 | 20 | 11 | | | | | | | | | |
| 55 | | | | | 73 | 53 | 38 | 26 | 17 | 9 | | | | | | | |
| 60 | | | | | | 75 | 56 | 41 | 30 | 21 | 14 | 9 | | | | | |
| 65 | | | | | | | 78 | 59 | 45 | 34 | 25 | 18 | 13 | | | | |
| 70 | | | | | | | | 79 | 61 | 48 | 37 | 29 | 22 | 16 | 13 | | |
| 75 | | | | | | | | | 80 | 64 | 50 | 40 | 32 | 25 | 20 | 15 | |
| 80 | | | | | | | | | | 81 | 66 | 53 | 43 | 35 | 29 | 22 | 16 |
| 85 | | | | | | | | | | | 81 | 68 | 55 | 46 | 37 | 30 | 25 |
| 90 | | | | | | | | | | | | 82 | 69 | 58 | 49 | 40 | 32 |
| 95 | | | | | | | | | | | | | 83 | 70 | 58 | 50 | 40 |
| 100 | | | | | | | | | | | | | | 84 | 70 | 61 | 50 |
| 105 | | | | | | | | | | | | | | | 85 | 71 | 61 |
| 110 | | | | | | | | | | | | | | | | 85 | 72 |
| 115 | | | | | | | | | | | | | | | | | 86 |

of the material by the addition of the proper thinner. However, the limits shown in the product's application instructions should not be exceeded without checking with the paint manufacturer. A *thinner* is simply a mixture of solvents that are compatible with the resins in the coating. Thinning can provide these benefits:

1. Improved flow and uniformity in application of the material
2. Reduced overspray, lap marks, bubble entrapment, and film "mud crack-ing" caused by rapid solvent flash-off

With some materials, heating or warming has an effect similar to thinning. Several types of heating devices are available for use with spray equipment. The use of very thick lacquer, sprayed hot, is common in the furniture industries. This preheating permits a high-solids material to be sprayed easily in a single coat.

Since thinning reduces the volume solids of a coating, film build may be difficult to obtain. In that situation, reducing the thickness per coat and increasing the number of coats will result in a better job. This is often true in both cold and extremely hot weather. Thinner films permit easier escape of solvent under both conditions. Bubbles and pinholes in hot weather and extremely slow hardening rates of thick film in cold weather are the result of the solvent's difficulty in escaping at its ideal rate.

## Methods of Application—Spray Painting

The object in spray painting is to create a mist of atomized (finely dispersed) coating particles that will cling to the target in a uniform pattern and then flow into a continuous, even film. The three most common types of spray painting are the following:

1. Conventional (air atomization)
2. Airless (hydraulic pressure)
3. Electrostatic

## Air Atomization

The spray gun, which is the primary component in a spray system, brings the air and paint together. This is accomplished in such a way that the fluid is broken up into a spray that can then be directed at the surface to be coated. There are two adjustments in most spray guns: one that regulates the amount of fluid that passes through the gun when the trigger is pulled back, and a second that controls the amount of air passing through the gun, thus determining the width of the fan. In the external-mix spray gun, which is the most widely used, the air breaks up the fluid stream outside the gun after being directed through a specially designed air cap. The number, position, and size of the holes in the air cap determine the manner in which the air stream is broken up. This in turn governs the breakup or atomization of the fluid stream. See Figure 7.29, Figure 7.30, and Figure 7.31 for effects of fan and operator on the spray pattern.

The fluid leaves the gun through a small hole in the fluid tip. A needle operated by the gun's trigger controls the flow of material through this tip. As with air caps, fluid tips are manufactured in different sizes to accommodate various materials, the diameter of the orifice in the tip being the differentiating factor. Various coatings require different types of air caps and fluid tips in order to be properly atomized. Therefore, it is usually wise to follow the coating manufacturer's recommendations as to the proper spray fan, air cap, and fluid tip. As an example, vinyl, epoxy, and chlorinated rubber–based coatings can be successfully sprayed with a DeVilbiss MBC or JGA-type spray gun equipped with a 78 or a 765 air cap and an E fluid tip and needle. (The letter E denotes the diameter of the orifice.) Binks and others make similar equipment.

There are two methods for bringing fluid to the gun: suction or pressure feed. For *suction feed,* the gun is usually fitted with a 1-qt cup holding the fluid to be sprayed. When the trigger is pulled, suction is developed at the tip of the gun, drawing fluid out of the cup and up to the nozzle, where it is sprayed. This type of setup has severe limitations: the gun must be operated only pointing horizontally, it will

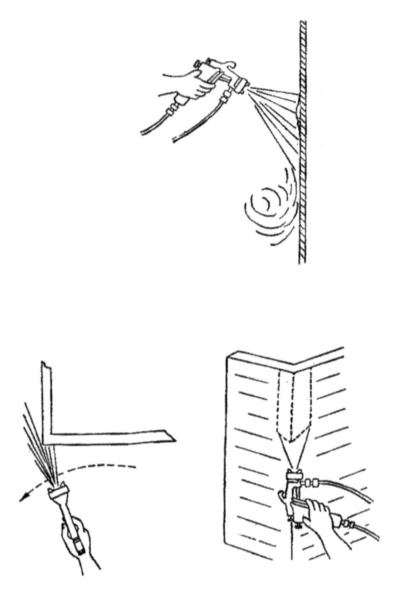

**FIGURE 7.29**  Paint spray gun techniques.

not spray viscous material, spraying is slow, and the cup must be filled frequently. (See Figure 7.32.)

With *pressure feed*, the fluid to be sprayed is forced to the gun under pressure. The pressure type has several advantages: material of higher viscosity can be sprayed, a heavier coat can be applied, and spraying is much faster. Although a 1-qt pressure cup is sometimes attached to the gun for small applications, the

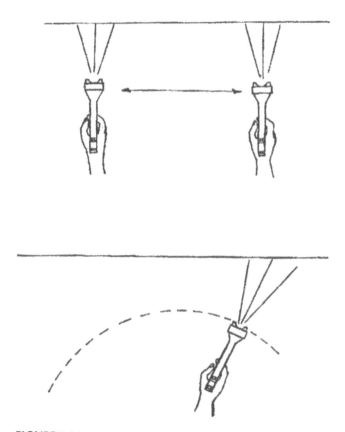

**FIGURE 7.30** Paint spray gun motions.

majority of pressure-fed spray operations are conducted with a separate pressure pot. This system ensures better pressure control and permits faster spraying. (See Figure 7.33.)

The *pressure pot* is a closed chamber of usually 2–10 gal capacity that contains the fluid to be sprayed. A large air hose, preferably 1-in. I.D., connects the pressure pot to the air source. A fluid hose and an air hose connect the pot to the spray gun. Air is directed into the pot through a regulator that maintains the proper pressure on the fluid. The fluid is forced under pressure out through the fluid hose to the gun. The adjustment of the pressure regulator on the pot determines the amount of fluid available to the spray gun. Air is also bypassed from the main source into the air hose leading to the gun.

It is extremely important that the correct size of fluid and air hoses be used. For best results, the gun should be equipped with a 1/2-in.-I.D. fluid hose and a 5/16-in. air hose. Smaller air hoses should be avoided, since they cause excessive pressure losses. For example, if an MBC gun with a 765 air cap is connected by 50 ft

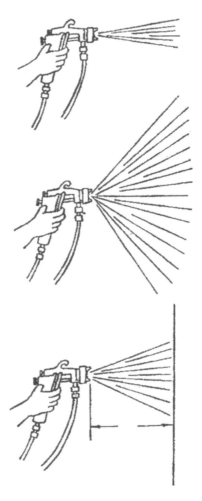

**FIGURE 7.31** Spray gun fan widths.

of 1/4-in.-I.D. hose to an air supply of 90 psi air pressure, the actual pressure at the gun will drop to 50 psi when the trigger is opened. This is not enough pressure for most coatings. However, when 50 ft of the recommended 5/16-in. air hose is used, under the same conditions, the pressure will drop only to 75 psi, which in most cases is satisfactory.

## Hydraulic (Airless) Spray

With hydraulic (airless) equipment, no air is used for atomization. The spray pattern is formed simply by forcing the material under high pressure through a very small orifice in the spray gun. As the material leaves the orifice, it expands and is broken up into fine droplets.

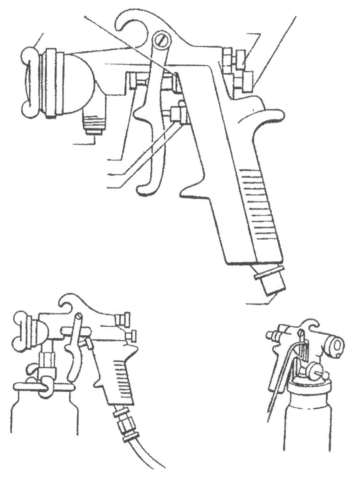

**FIGURE 7.32**  Typical spray gun showing suction cup and pressure cup.

The outstanding advantage of this type of equipment is the absence of overspray. Therefore, smoother applications can be made, especially to corners and crevices, and material loss to wind or other air flow is negligible.

Adaptability to various types of material is obtained by providing a series of interchangeable fluid tips for the gun, each of which has a fixed orifice and fan. Since the size and shape of the orifice determine the breakup of the material, the width of the fan, and the delivery rate, it is important that the proper tip be selected for spraying a particular coating. The only other adjustment is the pressure applied on the fluid. An air-driven pump with a ratio of approximately 28:1 or higher is commonly used to supply this pressure. Thus, if 100-psi air pressure is supplied to the pump, the resulting pressure on the material at the gun will be 2800 psi. As with the fluid tip, the pump pressure must be regulated to meet the application characteristics of various materials.

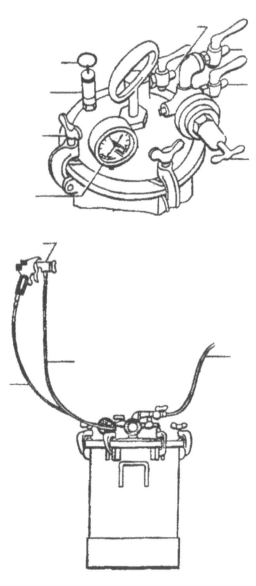

**FIGURE 7.33** Typical pressure pot setup.

## Ventilation of Work Areas

A few simple rules must be observed in the handling of coatings in enclosed areas where flammable solvents are a factor. Adequate ventilation is the most important safety rule. This is true whether materials are aliphatic hydrocarbons with low toxicity or aromatic hydrocarbons with more dangerous physiological properties. Ordinarily, no fire hazard will exist if the solvent vapor concentration is kept below 1% by volume.

Recommended ventilation rates for enclosed spaces are shown in Table 7.4, and general ventilation procedures for using suction blowers at the lower part of the enclosed area are shown in Figure 7.34.

## 7.4.3   Powder Coating

For industrial products, powder coating is rapidly becoming the finish of choice for many applications. While the principle is straightforward and easily understandable, success or failure lies in the detailed process control. The principle involved is to utilize the electrostatic phenomenon of attraction of oppositely charged particles. While the part to be coated could be charged and the powder particles sprayed onto it, the more popular system is to ground the product and charge the powder particles as they are sprayed (dry) on it. The product is then heated, which fuses the dry paint particles to make an adherent and coherent film.

The processes to be controlled in order to make this system viable—from a performance and cost point of view—start with product design. While the particles flow and are attracted to the sides and back of the product as well as the front facing the gun, there is a difference in the corona or force field. Sharp points and deep pockets are of concern in the powder coating process. The material, or substrate, also has its influence. It must be clean and dry, as in any other coating system. It usually needs a chemical coating after cleaning to improve adhesion and add corrosion protection. The size of the particles of powder and the type of powder affect the performance of the process and the finish coating. Perhaps the most important factors are the time and temperature control required for each step in the process. Although parts can be hand cleaned and hand sprayed in small shops or short production runs, the powder coating process lends itself to a conveyor of sorts, and precise controls on speed, cleanliness, and temperature at each step. The process we will discuss has the following steps, although all of them may not be necessary for some applications:

|               |            |
|---------------|------------|
| 1. Cleaning     | 5. Sealing   |
| 2. Rinsing      | 6. Drying    |
| 3. Phosphatizing | 7. Coating   |
| 4. Rinsing      | 8. Curing    |

### Cleaning

*What Is Soil?*
*Soil* is simply matter that is out of place. A *soil audit* should be performed to determine the cleaning required prior to powder coating. It should include the following:

What soils are incoming?
In-house applied?
Substrate types.
Substrate flow.
Process control.

TABLE 7.4 Recommended Ventilation Rates for Safe Application of Coatings in Enclosed Spaces

| | | | | | | | | |
|---|---|---|---|---|---|---|---|---|
| (1) Size of enclosed space, cu ft | 600 | 1200 | 2000 | 5000 | 10,000 | 25,000 | 50,000 | 100,000 |
| (2) 1 percent of enclosed space, cu ft | 6 | 12 | 20 | 50 | 100 | 250 | 500 | 1000 |
| (3) Typical coating application rate, gal per hr | 5 | 5 | 10 | 10 | 10 | 20 | 20 | 50 |
| (4) Volume of solvent vapor produced, cu ft per hr, @ 25 cu ft vapor per gal of coating material. (Line 3 × 25) | 125 | 125 | 250 | 250 | 250 | 500 | 500 | 1250 |
| (5) Air charges per hour to keep vapor concentration below 1 percent by volume* (Line 4 ÷ Line 2) | 20 | 10 | 12 | 5 | 2.5 | 2.0 | 1.0 | 1.25 |
| (6) Rate of air change, cfm, to keep vapor concentration below 1 percent by volume (Line 5 × Line 1 ÷ 60 min/hrs) | 200 | 200 | 400 | 41 | 41 | 83 | 83 | 28 |
| (7) Recommended air movement through suction fan, cim, to keep concentration for below 1 percent vapor by volume | 1000 | 2000 | 2000 | 3000 | 5000 | 10,000 | 15,000 | 20,000 |
| (8) Air changes per hr of recommended air movement (Line 7 × 60 min/hr ÷ Line 1) | 100 | 100 | 60 | 36 | 30 | 24 | 18 | 12 |

*1 percent concentration of vapor in air is below five lower explosive limit of most solvent mixtures used in point and coatings (except turpentine which has LEL of D.S percent)

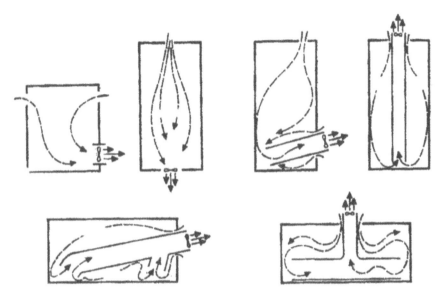

**FIGURE 7.34** Recommended ventilation procedure draws clean air from opening at top by exhausting air from lower part of enclosure.

Today you can demand a Material Safety Data Sheet from your metal supplier. What oil or preservative do they apply? Is it heat sensitive? Does it contain waxes? When you buy steel from many vendors, do you have control over the quality and consistency? When you determine that you can clean the incoming soils, make sure your vendors do not change soils without notifying you.

With in-house metal fabricating, what soils do you apply? Do you use rust inhibitors, forming oils, coolants, or lubricants? If so, pretest all these soils in your pretreatment system. Once you determine you can successfully clean these soils in a fresh or aged state, do not make changes unless you pretest again.

How many different types of metal substrates make up your product or products? Combinations of ferrous and nonferrous metals require different chemistries to clean effectively without metal attack. The use of zinc-bearing metals such as die-cast or galvanized metals may require posteffluent treatment.

Effective pretreatment and process control cannot be accomplished unless there is control over incoming soils, in-house applied soils, and substrates in use.

## Is the Part Clean?

If the part is clean, the powder will adhere. In most cases, if the chemical vendor can produce a clean part, a phosphate treatment will be sufficient prior to powder coating. Definitions and tests of cleanliness follow.

*Clean surface:* one that is free of oil and other unwanted contaminants.
*Organic soils:* oily, waxy films such as mill oils, rust inhibitors, coolants, lubricants, and drawing compounds. Alkaline cleaning solutions are most effective on organic soils: *alkalines clean organics.*

*Water break-free surface:* all organic soils have been removed. The parts exiting the last pretreatment stage prior to drying will show a uniform sheeting of the rinse water, indicating an organically clean surface. If the part exhibits a surface that resembles a freshly waxed car surface, there will be beads of water, indicating that the part is not organically clean.

*Inorganic soils:* rust, smut, heat, scale, and other inorganic particulates that reduce adhesion and gloss. They can most commonly be found after allowing the part to pass through the dry-off oven. Check for smut and other loosely adherent inorganic solids by using a clean white towel and wiping it over the dry surface. Smut is the black gritty substance found on weldments and hot-rolled pickled and oiled steel. Sometimes it is impossible to remove all soot, carbon, and smut without some form of mechanical or abrasive cleaning. Poor cleaning is most often found on or near weldments, or in areas that receive poor spray impingement to the part. Acidic cleaning solutions are most effective on inorganic soils: *acids clean inorganics.*

## Mechanical Cleaning

Mechanical or abrasive cleaning is suitable where steel surfaces have been subject to abuse such as severe corrosion and oxidation, or where steel surfaces exhibit large amounts of heat scale or controlled oxidation. This is especially true for the loosely adherent accumulations found in improperly stored steel, or hot-rolled steel of poor quality. These contaminants make it very difficult to achieve any form of quality adhesion. Three types of mechanical surface preparation have somewhat wide acceptance.

1. *Air/media blast. Sandblasting,* the most common type of air/media blasting, is a combination of compressed air and sand or other media. Sandblasting can be accomplished automatically or manually. Problems associated with employee safety, particularly silicosis, have decreased widespread use in open-air environments. A specially designed cabinet or enclosed area along with an air-induced breathing apparatus has ensured continuing growth of this method of mechanical surface preparation.
2. *Water/media blast.* This method is gaining popularity because of the reduction of silicosis-associated problems. Wet spot blasting of weldments has become accepted because of these reasons:
   During the welding process, oily soils are carbonized, creating an impossible cleaning condition using three- or five-stage washers.
   The surfaces of weldments are basically inert to the development of a conversion coating because of the scale and glassing developed. Wet spot blasting units are relatively inexpensive, low in labor requirements, and prove to provide the best possible substrate conditions prior to phosphatizing.
3. *Centrifugal wheel (airless) process.* This airless process is quite popular for larger, heavier-bodied parts where rust and scale must be removed.

Centrifugal wheel cleaning is most often done in enclosed cabinetry. The medium is normally steel shot of varying sizes, depending on the substrate profile required. For optimum results, centrifugal wheel cleaning should employ additional chemical pretreatment to ensure quality long-term finish life.

## Chemical Cleaning and Phosphatizing

Fremont Industries, Inc., of Shakopee, Minnesota, is one of the respected firms specializing in pretreatment systems to clean and phosphatize parts prior to powder coating. Its 5-stage cleaning and phosphatizing system is shown in Figure 7.35. A six-stage system most often includes a final deionized water mist rinse.

*Stage 1: Cleaning.* Typically, alkaline cleaning produces a metal surface free of organic and inorganic reactive soils. These cleaning products incorporate detergents and surfactants to wet the soil; alkaline builders to dergrade, emulsify, and saponify organics; and water conditioners to soften and control contaminants.

*Stage 2: Fresh water rinsing.* The purpose of stage 2 is to flush all remaining organic soil from the part, neutralize alkalinity, and prevent pH contamination to stage 3.

*Stage 3: Phosphatizing.* Iron phosphatizing is the most common form of conversion coating in general industry for powder coating. The clean and rinsed part enters the phosphate stage and receives a uniform acidic attack. Chemical reactions occur at the substrate solution interface. Most five-stage iron phosphates deliver 40–70 mg/ft$^2$ of coating.

*Stage 4: Fresh water rinsing.* The purpose of stage 4 is to flush any remaining phosphate solution and prevent the subsequent stage from being chlorinated.

*Stage 5: Seal rinsing.* The purpose of final seal rinsing is to remove any unreacted phosphate and other contaminants, to cover bare spots in the coating,

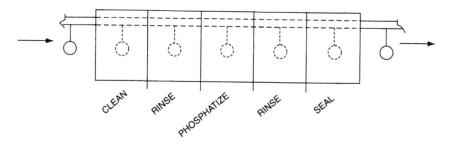

**FIGURE 7.35** Typical five-stage cleaning and phosphatizing precoating system for powder coating. (Courtesy of Fremont Industries, Inc. With permission.)

to prevent the surface from flash corrosion, and to extend the salt spray performance. The selection of the type of seal rinse, whether a deionized water rinse, a chromic acid rinse, or a reactive rinse, depends on the type of substrate as well as the type of powder coating system planned.

### Nonferrous Metals

Nonferrous materials require slightly different chemical treatment, although the principles are the same. One approach includes cleaning of organic soils without attack to the nonferrous metal. This is followed by a rinse and the application of an acidic cleaner to slightly etch or remove the oxide layer of the metal. Careful consideration must be given to the choice of both the alkaline cleaner and the acidic material. The limiting factor is the alloy and the amount of etch that can be done without overattack and the subsequent development of smut.

The second approach for pretreatment of nonferrous metals includes a third step of conversion coating. This approach is common when the metal finisher runs a combination of ferrous and nonferrous metals through the same system. The etch portion and conversion coating portion are usually accomplished in the same stage by incorporating fluoride accelerators into the iron phosphate bath. The ultimate deposited coating is a combination of surface etch and a combination conversion coating of the dissolved alloys bound in the particular nonferrous substrate.

The third approach to nonferrous metal pretreatment for powder coating is the chromate process, associated with the highest-quality underfilm corrosion protection. Chromate conversion coatings serve as effective pretreatments for powder coatings. These pretreatments are used extensively on aluminum and also find application with zinc and magnesium. Two types of chromate coatings are in use: chrome oxide (amorphous chrome) and chrome phosphate. Coatings formed with chrome oxide are based primarily on hexavalent chrome, which is extremely corrosion resistant. Chrome phosphate coatings contain primarily trivalent chrome, which is less corrosion resistant than hexavalent, but generally more mechanically sound and stress durable than chrome oxide types.

The parts should be dried at this point, usually in an oven. One approach is to duct the hot air exiting the curing oven and control the dry-off oven temperature by varying the fresh air intake. The moisture-laden air is vented outside. Another approach is to provide the dry-off oven with its own gas burner, separate from the curing oven. This allows the option to preheat the parts prior to application of powder. In some cases this can solve some of the problems related to Faraday cage and outgassing. Use of air knives to remove moisture can reduce the time and temperature for drying the parts, unless they are too small to withstand the air velocity.

Additional preparation prior to powder coating can include masking of areas not needing the coating. Also, finished bolt holes, threads, and so forth may require plugs or caps to keep the powder out. There is a wide assortment of products to aid in this process step, and usually they are rated as to the temperature they will withstand. Tapes may be made of paper, polyester, or glass cloth. In general, the cost of a 400°F tape is twice the cost of one that will withstand 300°F, and half the cost of one that will withstand 500°F. The selection must be based on the correct temperature for the product powder cure temperature, ease of application and removal, and cost of the masking material. Some masking materials will withstand chemical or abrasive cleaning operations, and others will withstand temperatures from 150°F to 600°F.

One of the more important remaining steps is the racking arrangement, including hooks. This is necessary to present the part to the powder coating operation correctly, and to provide a continuous ground to the part. The measured resistance between the part to be coated and a ground wire connected to the booth ground should not exceed 1 MΩ. Since the hooks as well as the part will become coated, the cured coatings must be removed from the hooks and racks in order to maintain this ground path. This is often accomplished in a burn-off oven, followed by brushing or some other cleaning step, prior to reuse. In many cases it is more economical to replace the hooks rather than clean them.

## Powder Coating Process

The powder coating spray booth may be a simple one, using a manual powder spray gun. Binks makes such a gun, using low-voltage cable and an integral-cascade generator that maximizes tip voltage. The gun's two-position trigger enables the operator to easily control voltages to combat Faraday cages. Pattern-shaping air can also be controlled at the gun. Other suppliers have somewhat similar products available.

Most finishers utilize an automatic system with an overhead conveyor that carries the racked parts through the pretreatment phase, the drying phase (and preheating if required), and on into the powder application booth. This booth contains several corona-charging spray guns, which apply the powder the same way on each part.

Each spray booth has reclaim modules to collect the overspray powder, using filters similar to an industrial vacuum cleaner. The booth, and particularly the reclaim filters, must be thoroughly cleaned prior to changing color in the system. The reclaimed powder is then processed for reuse, usually with about 50% new powder added each time.

## Curing Ovens

There are as many solutions to curing ovens as there are in the other sections of the powder finishing system. Gas convection ovens are probably the most common today. They generally operate at 350 to 425°F and can cure parts in 15 to 30 min, depending on the coating and the mass of the part. In production operations, the conveyor carries the parts through the oven, where the length of the oven, the conveyor speed, the temperature required to fuse the coating, and the mass of the parts are influencing factors.

With new product and process technology under continuous development, production equipment is required to give precise control over product quality while operating efficiently with minimum downtime. Infrared radiation, with wavelengths expressed in micrometers (microns), can be accurately measured, controlled, and applied to the product. Figure 7.36a illustrates the electromagnetic energy spectrum

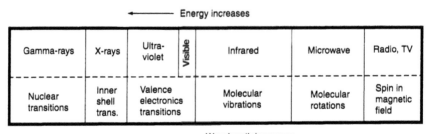

**(a)**

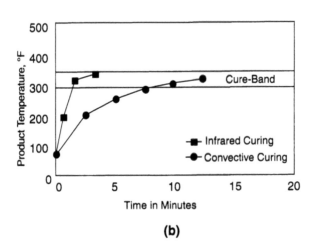

**(b)**

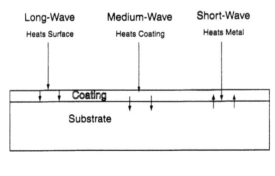

**(c)**

**FIGURE 7.36** (a) Electromagnetic scale showing the position of infrared radiation relative to other types. (b) Product temperature versus time for curing a powder in a convection oven compared with an infrared oven. (c) Selection of proper infrared wavelength. (Courtesy Casso-Solar Corp. With permission.)

in which infrared is centered. Figure 7.36b shows the infrared theory of quickly bringing the powder coating up to curing temperature, and Figure 7.36c shows the proper infrared wavelength for the most efficient process.

Every organic material has a unique energy spectral absorption curve. There will be peak wavelengths where the material absorbs very well, and valley wavelengths where it is almost transparent to the applied energy. Most materials have absorption peaks falling between the wavelengths of 2.0 and 5.0 μm. Peak efficiencies are achieved on most products by matching the emission wavelength of the heater to the absorption wavelength of the product.

Casso-Solar Infrared (Pomona, New York) uses the proprietary selective wavelength method (SWM). For example, with a water-base coating on a polyethylene film, the goal is to transfer maximum energy to the water in the coating, with minimal energy pickup by the film substrate. Water, with peak absorption wavelengths between 2.5 and 2.6 μm, will receive energy from an infrared heater operating at an emitter temperature of 1600°F with efficiencies of greater than 80%. The substrate film, with a peak absorption wavelength at 3.4 μm, will be almost transparent to the emission energy, preventing heating of the film and subsequent distortion.

On thick materials, the SWM theory can be applied in reverse. Penetration within a material can be realized by selecting an off-peak emission wavelength, allowing the radiant energy to pass through the outer layers of the material. With the trend toward new high-solids coatings, 100% solids coatings, and powder coatings, the infrared energy can be absorbed in the coatings directly, without substantial heat absorption by the substrate, saving power and cooling time. For sensitive solvent coatings, short-wavelength infrared is often utilized to penetrate the coating, heating from the inside out and eliminating blistering and other surface defects.

## Troubleshooting Basics for Powder Coatings

As mentioned in the introduction to this subsection on powder coatings, the basic principles of powder coating are simple and quite understandable. However, the success or failure of a system may lie in close attention to the details actually involved in the system. The following material is adapted from an article in *Powder Coating Magazine*. Matt Matheny, technical services manager of O'Brien Powder Products (Houston, Texas), is the author, and was most helpful in the preparation of this subsection.

The application of any chemical coating, especially in a relatively young and evolving technology, tends to be fairly problematic. Coaters simply have so many variables to control that chances are anyone who sprays powder will sooner or later have some problems. Arming yourself with some basic tools and guidelines to overcome these problems will help you run a smooth powder coating day to day.

### Basic Troubleshooting Tools

The test kit outlined in Table 7.5 will help trace and prevent many of the problems you will encounter. Although it is true that you could spend the entire family inheritance on sophisticated, state-of-the-art troubleshooting equipment, you will find that you can accomplish a lot with just the basic tools in this kit.

TABLE 7.5  Basic Troubleshooting Test
Kit for Powder Coatings. (From Powder
Coating Magazine.)

Film thickness gauge

Small bottle of MEK

Cotton swabs with wooden stems

Razor knife

Tape with strong adhesive backing

Jeweler's magnifying lens (10X loupe)

Clean white cloth, tissues, or gloves

Commercial test panels

Sharpie marker

Razor blades

Ohmmeter

## Film Thickness Problems

Both thin and thick films can cause problems, as shown in Table 7.6. Aside from these problems, excess film build simply costs money—the larger the job, the more impact there is on cost. For example, if you spent $3000 on powder to coat a job at 3 mils thick when 2 mils would have met your requirements, you just gave away $1000!

Even veteran powder coaters and other experts can be fooled by a coating's visual appearance and perceived thickness. To remove doubt, use a film thickness gauge. Most gauges cost from $175 to $2900, depending on style and options, but it is more difficult to prevent and diagnose coating problems without one.

## Film Thickness Troubleshooting

As a first step in troubleshooting, you should get into the habit of checking film thickness first. In many cases, this will seem unnecessary (for instance, when film color is incorrect). Many times, however, film thickness will turn out to be the unexpected cause of a problem or at least a contributing factor. The problem with incorrect film color, for example, could be caused by the substrate showing through a thin coating, which affects the perceived color.

Considering that film thickness measurement is probably the easiest and fastest test you can do with powder coatings, it makes little sense not to check it. Even if a coating problem requires the assistance of your suppliers, you should have all film thickness data on hand before calling them, especially when calling your powder coatings supplier. Some of the most common factors in film thickness control are listed in Table 7.7.

TABLE 7.6 Thin and Thick Films Can Cause These Problems. (From Powder Coating Magazine.)

| Thin films | Thick films |
| --- | --- |
| Reduced corrosion resistance | Reduced impact resistance |
| Reduced chemical resistance | Reduced flexibility |
| Pinhole rusting | Reduced chip resistance |
| Reduced electrical insulation | Inconsistent or incorrect |
| Inconsistent or incorrect | appearance, texture, and gloss |
| appearance, texture, and gloss | Orange peel |
| Orange peel | |
| Coating seeds | |
| Reduced edge coverage | |
| Inconsistent or incorrect color | |

TABLE 7.7 Common Factors in Film Thickness Control. (From Powder Coating Magazine.)

Line speed

Electrical grounding of parts

Part presentation when coated

Gun charging problems

Gun kilovolt settings

Operator training

Powder particle size

Powder charging characteristics (formulation)

## Curing Problems

During the powder coating baking process, a chemical reaction called *cross-linking* occurs. When complete, this chemical reaction provides a fully cured thermoset powder coating film with the physical attributes that were designed into it. Incomplete cross-linking, or cure, provides a final product with reduced physical properties, depending on the actual degree of undercure. In short, a fully cured powder coating prevents a host of postapplication problems.

As with film thickness measurement, cure is a cornerstone of powder performance, yet easy enough to check at the start of most problem investigations.
A strong solvent, such as methyl ethyl ketone (MEK), provides a fast and easy
way to measure the chemical resistance of a powder coating. Because chemical
resistance generally develops in relation to the degree of cross-linking that has
occurred, it provides a workable assessment of cure. See Table 7.8 for instructions
on how to use MEK to test cure. Because of exceptions, you must compare your
test results against a fully cured powder coating to determine if the results are
normal or not. A dulling of gloss at the test point is normal. Rapid softening of the
coating usually indicates undercure. The quicker and more pronounced the softening, the less the cure. Heavy discoloration of the cotton swab will be evident. In
cases of severely undercured films, the coating can be completely removed from
the metal substrate.

When the coating is close to fully cured, it can be difficult or even impossible
to determine the exact degree of cure. Fortunately, powder coatings that are close
to fully cured seldom create significant problems. Remember, failure to compare
all test results with a fully cured coating of the same powder can lead to false
conclusions.

---

TABLE 7.8  Solvent Cure Test for Powder Coatings. (From Powder Coating
Magazine.)

**Equipment required**

1.  A few milliters (or ounces) of clean MEK, available at most paint and hardware stores

    Note: A few ounces of MEK will be enough to do hundreds of tests

2.  Cotton-tipped swabs with wooden stems

    Note: Swabs with plastic stems dissolve in strong solvents

**Instructions**

1.  Wet the swab generously in the MEK bottle.

2.  With the thoroughly soaked swab, rub a small, 1-inch-long area of the coated surface with 50 double
    rubs. A double rub is once up the coated surface and once down the coated surface, as if erasing a
    pencil mark. Try to use a force similar to erasing with a pencil.

3.  After 50 double rubs, most cured coatings will permit very little or no removal of the film on the
    cotton swab.

    Note: Urethanes, textures, and other select powder coatings may not resist strong solvents well, even
    when fully cured. Therefore, it's important to establish how a fully cured coating will react to MEK
    before conducting this test. To establish such a baseline, get a fully cured standard panel from your
    powder coatings supplier or bake a coated part two or even three times to ensure that you have a cured
    coating to use for comparison.

    (Test kits are avaliable from O'Brien Powder Products)

## Oven Temperature

The most frequent problem regarding oven temperature is usually not a lack of knowledge regarding the correct temperature range to use for a powder, but rather a blind faith in the oven temperature readout. Powders usually have a cure curve developed by the coating supplier. Figure 7.37 shows typical temperature/time curves for an epoxy, and Figure 7.38 for a polyester. The lower the temperature, the longer cure time required. More often than not, oven readouts tend to be at least a little off from the actual average oven temperature. This is due to normal drifting of the calibration and to the fact that most oven temperature probes are located in only one spot in the oven, which may or may not reflect the overall average temperature (usually it doesn't).

Most ovens have hot and cold spots that are not registered on a single probe located at the other end of the oven. Although it is convenient if the oven read-out and the actual oven temperature are the same, it is not mandatory in curing powder-coated products successfully. Just follow two simple rules for curing success: (1) if the powder is turning yellow or brown, reduce the oven temperature setting (ignore the readout) or decrease the oven bake time; and (2) if the powder is not cured (fails the MEK testing), increase the oven temperature (ignore the readout) or increase the bake time.

## Metal Mass

Most powder manufacturers report cure schedules for their products at metal temperature. For example, if a powder is supposed to be cured for 10 min at 400°F and you are coating a 16-gauge steel part, you will probably need to add another 5–10 min to

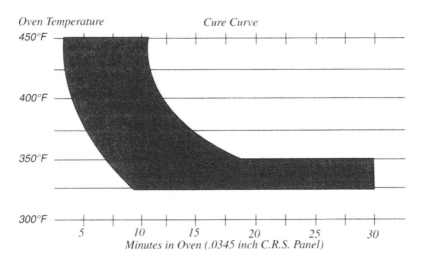

**FIGURE 7.37** Typical epoxy powder temperature versus time cure curve. (Courtesy Morton Powder Coatings. With permission.)

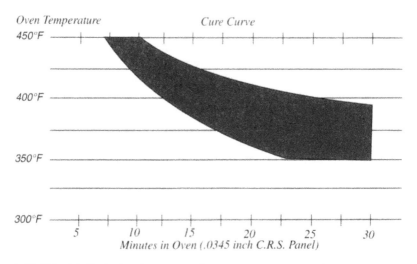

**FIGURE 7.38** Typical polyester powder temperature versus time cure curve. (Courtesy Morton Powder Coatings. With permission.)

the bake time to allow the metal to come up to temperature and to achieve full cure of the powder coating.

## Metal Thickness

As you might expect, the energy required to heat two drastically different metal cross sections is also different. If your production setup permits it, the best solution for this problem is to preheat the part first in a range between 250 and 350°F, apply the powder, and then bake at the time and temperature required for the thinnest metal section. The next best option is to reduce your oven temperature as low as permissible for the powder you are using and then extend the bake time as long as required to cure the thickest metal section. You may still be unsuccessful with this approach, however. Some yellowing of the thin section could still occur. In that case, you may need to contact your powder supplier for other options, such as changing to a powder with increased heat stability.

## Adhesion Problems

If a coating is to meet performance expectations, it must adhere to the substrate. Because the substrate composition and its preparation have a profound effect on adhesion, a method to evaluate adhesion is useful for troubleshooting and in-process quality control. Some measure of adhesion can be done by cutting through the coating in a latticed pattern and then trying to pull the film away with tape. Table 7.9 describes this test procedure.

TABLE 7.9  Powder Coating Adhesion Test Procedure. (From Powder
Coating Magazine.)

1.  Use a razor knife to cut through the coating to the substrate in a latticed pattern. Make six or more
    cuts in each direction. Each parallel scribed line should be 2 millimeters apart for coatings that are at
    least 2 mils thick. For coatings thinner than 2 mils space the lines 1 millimeter apart. A steel
    straightedge is useful for making these cuts.

2.  Brush off loose bits of coating.

3.  Apply adhesive tape firmly over the lattice. Use a pencil eraser to rub the back of the tape. This
    ensures good contact with the coating. Then, quickly pull back the tape at about a 180 degree angle.

If failure occurs during the test, look at the back of the removed coating for evidence that might provide clues. Red-brown rust from steel parts or white oxidation from aluminum and galvanized parts is frequently found on the back of the coating. If you are testing freshly coated parts, these oxides were present before the parts were coated. Continue your investigation, examining parts about to be coated and working backwards through the process as needed. If you are testing parts that were coated some time ago, you will have to conduct a more complex investigation. Of course, it is still possible that the parts had oxidation present when coated but just were not caught at that time. If the coating has blisters, is still intact with visible pinhole rust, or both, then this is a strong possibility. If pinhole rust is visible, the powder coating may just be too thin or have had contaminants present when the coating was applied.

For more demanding application requirements or for more test details, consult the procedures in American Society for Testing and Materials Test D-3359. This procedure also provides a useful pictorial table for classification of test results. For a copy, call ASTM at (215) 299-5400.

## Contamination Problems

Many clues to coating problems are not easily visible to the naked eye. A jeweler's lens (10× loupe) is especially useful in identifying foreign matter on an intact film or on the back of a film that has peeled off a substrate. The root of a crater (fisheye) problem may also become apparent by closely examining the bottom of a film depression for a contaminant.

Powder-coated panels that have not been cured can be hung in strategic places around the coating area, collected later, and then cured and checked for contaminants. Panels should be marked for later identification of the test sites. Airstreams from outdoors and other plant operations are especially suspect. The most common contaminant types and sources in a powder coating operation are discussed next.

### Cross-Contamination (from within the Powder Coating System)

Inadequate cleanup between different powders is a major culprit. A quick test for determining the source of colored specks from another powder is to thoroughly clean a manual spray gun and then spray a sample of virgin powder right from the box. Next, spray a powder sample from the hopper. If the virgin powder looks good and the hopper powder looks contaminated, reclean the powder system and recharge it with virgin powder.

## Dirt and Powder Falling from the Conveyor

When dirt and powder fall from your conveyor, you should notice most of the contaminants at the top of the parts. If this is the case, clean the conveyor and install shields to protect the parts.

## Surface Preparation Problems

Regardless of the mechanical and chemical processes you have selected, it is important to monitor them closely to get the best and most consistent results. If you need help determining the best approach for your application, contact your powder supplier, your pretreatment supplier, or both.

The adhesion test described previously can be used to alert you to serious surface preparation problems. A part with more subtle surface preparation problems may pass initial adhesion tests, such as the cross-hatched method, but will eventually cause premature failure of the coating while in use. Detection of these subtle problems requires special testing, such as salt spray (ASTM B-117) or hot-water immersion (ASTM D-870). If performed at temperatures of 180 to 200°F, the hot-water test has the advantage of being fairly quick to perform—usually 2–6 hr—and can be done with inexpensive equipment right in your plant.

The most common problems associated with chemical pretreatment systems are poor adhesion and premature corrosion failure. Frequently these problems are caused by the following.

*Residual soils* may be caused by (1) conveyor line speed that exceeds the design limits of the cleaning system, causing low dwell time, (2) inappropriate cleaner for the soils present (bear in mind that the soils on your metal may have changed), and (3) incorrect temperatures—120–130°F is best for good cleaning unless you are using a low-temperature cleaner. In that case, high temperatures can be detrimental.

To determine the temperature that emulsifies the soils present on parts, immerse an uncleaned part into a container of water and begin heating it. Use a thermometer to watch the temperature rise. Keep an eye on the point where the water line touches the part. At some point, the water will become hot enough to visibly loosen the soils, causing globules to float to the surface.

*Flash rust* can be caused by (1) excessive line speeds that prevent adequate exposure to the sealer in the final rinse, (2) line stops that overexpose parts to chemicals or allow them to dry off between stages, and (3) lack of sealer in the final rinse. When using a solvent-type cleaning system or an iron phosphate conversion process, wiping with a clean, white cloth is an ideal way to check parts cleanliness before powder coating.

*Aluminum oxide:* When coating aluminum, remember that a natural oxide is present on the surface that will interfere with adhesion if it is not removed. If using a combination iron phosphate and cleaner, be sure it is designed for steel and aluminum.

*Inadequate rinsing* is one of the great sins of metal cleaning. It is caused by increased line speeds that reduce rinse-stage dwell time and inadequate rinse-water overflow (excessive dissolved solids in the final rinse). Simple tests can include slowing down the production line or hand rinsing parts in deionized water.

If you suspect that your surface preparation system is causing a problem, clean test parts with clean rags dipped in a solvent, such as MEK, instead of running the parts through your normal cleaning process. If this fixes the problem, you should focus your investigation on the surface preparation system.

## Powder Fluidization Problems

Consistent application of powder starts with consistent powder flow through the spray guns. Consistent powder flow requires good powder fluidization. If the powder bed has dead spots or blow holes (geysers) or just does not fluidize well, a smooth flow through the hoses and guns is jeopardized. The most common causes of poor fluidization are as follows.

*Moisture from the compressed-air supply.* If your powder is initially free of soft lumps but starts forming lumps while fluidizing, it probably means that your air supply is contaminated with moisture, oil, or both. Compressed-air dryers and filters are recommended for all powder coating operations. A white cloth can be used to detect oil or water in the coating operation's compressed-air source. After blowing the compressed air into a cloth for a couple of minutes, you will see that oil- or water-laden airstreams leave wet or discolored deposits.

*Lumps in the powder.* Aside from moisture in the air supply, lumps can be caused by exposure of the powder to high temperatures (above 75°F) during storage or transportation. Usually, sieving the powder before use breaks up or removes the lumps. A common window screen will often suffice. The best way to check for lumps is to fluidize the powder well and run your hand through it. While you have your hand in there, check the bottom and the corners of the bed for dead spots.

*Plugged porous plate in the fluidizer.* Anything less than clean, dry air will eventually plug the porous plate in the bottom of the fluidizer. You will notice the air pressure rise to maintain good fluidization. If possible, examine the bottom of the porous plate for plugging, oil residue, and so on. If the powder does not fluidize well, try using another fluid bed to see if the problem goes away.

## Grounding Problems

Because electrostatic equipment is used to apply most powder coatings, adequate electrical ground is required. The measured resistance between the part to be coated and a ground wire connected to the booth ground should not exceed 1 mΩ. Take measurements inside the booth where parts are actually coated. Following are the most common problems associated with poor ground:

*Poor powder transfer efficiency.* This can be characterized by slow or low film build.

*Electrical arcing from the gun to the part or hanger.* This indicates a
grounding or gun problem and is a safety hazard. Under specific conditions, electrical arcing can start fires or cause explosions. This situation
should be corrected immediately.

*Premature electrostatic rejection of the powder from the part.* In premature
electrostatic rejection, powder is repelled from the part, usually leaving pits
in the powder surface. With some powder formulations, this problem can
occur with good grounding. In those cases, reduce the gun's kilovolt setting.
In general, electrostatic rejection occurs with all powders if spraying time is
excessive. Poor grounding aggravates the condition, accelerating it.

*Bare areas on the part that are difficult to coat.* This occurs especially at
edges, around holes, and near hooks. The most common cause for these
grounding problems is dirty hangers and hooks. Cleaning or replacing these
fixtures should be part of your regular maintenance program.

### 7.4.4 Metallizing by Flame Spraying

Corrosion is constantly taking a heavy toll on costly equipment and structures.
*Metallizing,* a cold process of coating with flame-sprayed metal, presents a flexible, practical, and very effective means of protection against corrosive attack. The
great advantage of the metallizing process is that it employs portable equipment. See
Figure 7.39 for equipment layout for flame spraying using metal wire. This makes
it adaptable for applying protective metallic coatings to large, complex assemblies
and structures. It is not economically feasible to attempt hot-dip galvanizing or electroplating of assembled structures, especially in the field, but pure metal protective
coats of aluminum or zinc can be applied to such structures by metallizing.

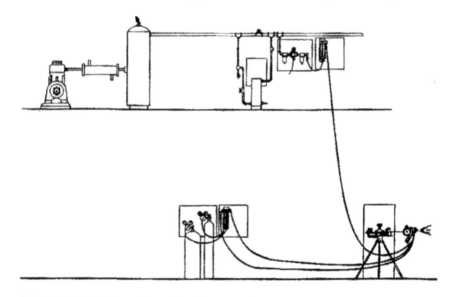

**FIGURE 7.39** Equipment layout for flame spraying.

## The Process

Metallized coatings have unique structures and desirable properties. They are formed by layers of thin flakes as the atomized globules of the sprayed metal strike the target area with considerable speed and impact. Initially, each particle is a tiny casting that has solidified very rapidly and acquired an oxide film during its short journey from the gun nozzle to the metal surface. On impact, the particles are instantly deformed into thin platelets or flakes. Their impact energy permits them to conform to the surface contour and bond to the prepared surface by mechanical interlocking with surface irregularities. Where the oxide film is disrupted on impact, it is reestablished by cold welding and metal-to-metal contact is regained.

The bond is almost entirely mechanical in nature. The base metal presents a degree of roughness depending on the method used for surface preparation. Most of the sprayed metal particles that strike the surface are sufficiently plastic to conform to and interlock with the surface irregularities. Although only a small amount of fusion may take place between the particles of sprayed metal and the base metal, the overall effect of such fusion is very significant. This unique structure results in a change of the physical characteristics of the metal. The ductility, elongation, and tensile strength of the sprayed metal are greatly reduced when compared with the same metal in cast form.

## Materials

All metals that are available in wire form can be sprayed, and several of them are quite effective in controlling corrosion. These include stainless steel, aluminum, zinc, cadmium, lead, tin, nichrome, nickel, nonel, tantalum, molybdenum, silver, and gold. These metals are applied for decoration, wear resistance, or corrosion control. The corrosion control that can be obtained from properly applied sprayed metals can be rated as good to excellent, even for a harsh environment. (See Table 7.10.) Tests have demonstrated that metallized aluminum sealed with vinyl is an excellent corrosion-control method for salt-water exposure. These coatings provide protection against corrosion not only by covering the surface with a corrosion-resistant metal, but also by cathodic action since both aluminum and zinc are anodic to iron and steel. Twelve-year tests by the American Welding Society of sprayed aluminum and zinc coatings on steels showed good results on all panels. However, the sprayed metal that was sealed with a wash primer and vinyl top coat was in the best condition. In salt-water immersion tests, the staying power of zinc-sprayed metal ran a very poor second to aluminum. The zinc oxide coating does not appreciably retard the oxidation rate; the zinc continues to sacrifice or oxidize. The aluminum-sprayed coating oxidizes to form an inert film ($Al_2O_3$), which greatly retards further attack. The film is stable unless broken or ruptured.

Metallizing has the advantage that much thicker coatings of the protective metals can be applied more rapidly than other conventional metal coatings applied by hot dipping or electroplating. This is significant because corrosion protection is often in direct relation to the protective coating thickness. The life of a flame-sprayed zinc coating is almost directly proportional to the thickness of the coat. Aluminum and

TABLE 7.10 Examples of Environments Where Flame Sprayed Metals Are
Used. (I is Goldstone, CA; IV is Bermuda.)

| Environment | | Physical Limits | |
| --- | --- | --- | --- |
| Class | Type | | |
| I | Dry (Minimum corrosion rate) | Rel. Humidity—1.5% or less<br>Temperature Range below 0°F to 130°F | 95% of time |
| II | Normal | Rel. Humidity—1.5% or 50%<br>Temperature Range below 0°F to 120°F | 95% of time |
| III | Humid | Rel. Humidity—41% or 100%<br>Temperature Range below 32°F to 120°F | 95% of time |
| IV | Harsh (Minimum corrosion rate) | Rel. Humidity—41% or 100%<br>Temperature Range below 35°F to 120°F<br>One or more saline aerations/24 hrs. | 95% of time |

cadmium, the other two metals commonly sprayed for corrosion control, also show
a fairly direct relationship between the life of the coating and its thickness. Because
of the porosity of metallized coatings, it is customary to apply a greater thickness of
protective metal than for dipped or plated coatings.

The best corrosion-control system to protect steel may be the aluminum-sprayed
coating, sealed with a wash primer, and vinyl paint. Wash primer coatings are useful
because they reduce the danger of unfavorable reaction from a doubtful paint combination. They have the added advantage of displacing moisture to some extent and should
always be used if there is any question of moisture in the sprayed metal. Unfortunately,
some wash primers are too reactive for use on sprayed metal. Formulations suitable
for galvanized sheet may be too acidic for use with sprayed coatings of zinc. No wash
primer should be used until it has been positively determined that it is suitable for
use on sprayed metal. In no instance should a wash primer containing more than 4%
$H_3PO_4$ be used. In the case of aluminum, since it is slightly porous, the sealing materials fill up the pores of the coating to establish a permanent barrier.

### Surface Preparation

In metallizing, as in painting, the preparation of the substrate is critical. In preparing
a surface for flame-spraying aluminum, there are no options regarding the grade of
surface preparation; it must be white-blasted with washed, salt-free angular silica
sand or crushed garnet that will pass through mesh sizes 20 to 40. In addition,
the white-blasted surface must be perfectly dry before the flame-sprayed metal is
applied. The drying process can be accomplished by the metallizing gun merely
by releasing the metal-feeding trigger and drying the desired section with the gas
flame. Whenever there is any doubt whether a surface is dry, it should be warmed
before flame spraying. The ambient temperature may be well above the dew-point
temperature, but the metal itself, because of shadows, the mass involved, and so

forth, may be below the dew point, creating the danger of an invisible moisture film forming on the metal. Heating a surface to 100°F will prevent atmospheric moisture. Some highly experienced flame-spray craftspersons prefer to heat the surface from 175 to 200°F before starting to metallize; this results in an improved bond. In addition to atmospheric moisture condensation, the higher preheat prevents water vapor originating in the combustion of the gas-oxygen flame from condensing on the base metal or the preceding pass of sprayed metal.

## 7.4.5 Porcelain Enameling

Enamel is a vitreous glaze of inorganic composition (chiefly oxides) fused on a metallic surface. Glass is particularly resistant to corrosion by atmospheric influences and chemicals, and has a smooth and very strong surface. But glass is fragile. When the good properties of glass are combined with the strength of steel or cast iron, the objects made from these materials (kitchen utensils, bathtubs, pipes, basins, laundry equipment, etc.) have excellent service properties. The name "vitreous enameling" or "porcelain enameling" is applied to such materials.

In particular cases the two materials supplement each other so well that entirely new material properties are obtained. Certain parts of jet engines and marine propulsion engines are enameled in order to make the surfaces resistant to high temperatures. Some of these applications are now considered to be in the field of ceramics; however, enameling still has wide usage as a finish for various products.

Enamel has been known since ancient times, when it was used (as it still is) for ornamental purposes on precious and nonferrous metals. In the last few hundred years, however, it has been used chiefly for improving the surface properties of steel and cast-iron objects and protecting them against corrosion.

### The Material

An enamel consists of glass-forming oxides and oxides that produce adhesion or give the enamel its color. A normal enamel may consist, for example, of 34 (23) parts borax, 28 (52) parts felspar, 5 (5) parts fluorspar, 20 (5) parts quartz, 6 (5) parts soda, 5 (2.5) parts sodium nitrate, 0.5–1.5 parts cobalt, manganese, and nickel oxide, and (6.5) parts of cryolite. The figures not in parentheses relate to a ground-coat enamel, while those in parentheses relate to a cover enamel, to which 6–10% of an opacifier (a substance that makes the enamel coating opaque, e.g., tin oxide, titanium silicate, antimony trioxide) and a color oxide are added. This mixture is ground to a very fine powder and melted. The hot melt is quenched by pouring it into water, and the glasslike frit that is thus produced is ground fine again. During grinding, water (35–40%), clay, and quartz powder are added. Opacifiers and pigments may also be added. The enamel slip (thick slurry) obtained in this way must be left to stand for a few days before use.

### The Process

The metal objects to be enameled are heated thoroughly, pickled in acid, neutralized in an alkaline bath, and rinsed. Next, the ground-coat enamel slip is applied to them

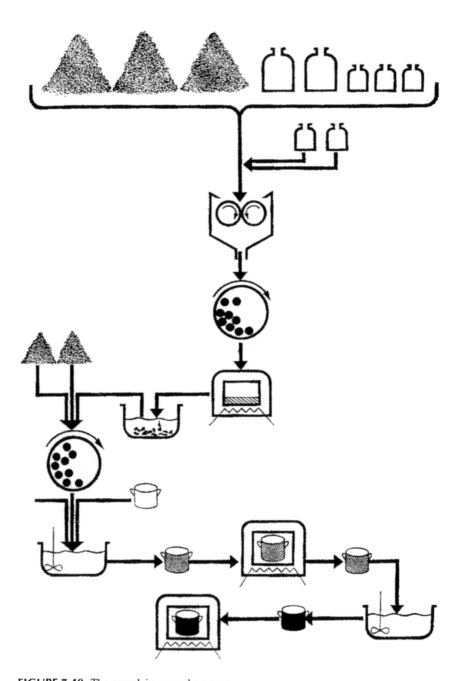

**FIGURE 7.40** The porcelain enamel process.

by dipping or spraying and the material is fired at 850–900°C, so that it fuses to form a glass coating. The ground-coated objects are then provided with one or more coats of cover enamel, each coat being fired at 800–850°C in a muffle furnace. (See Figure 7.40.)

Because an enamel coat is always more brittle than the underlying metal, the enamel will crack or spall if the object is deformed or roughly knocked.

From the chemical point of view, enamel is a melted mixture of silicates, borates, and fluorides of the metals sodium, potassium, lead, and aluminum. Color effects are produced, for example, by the admixture of various oxides to the melt (oxides of iron, chromium, cadmium, cobalt, nickel, gold, uranium, and antimony).

# Index

Milton Keynes UK
Ingram Content Group UK Ltd.
UKHW020320111024
449327UK00040B/1462